Study Guide

to accompany

Elementary Algebra

Second Edition

Larson/Hostetler

Jay R. Wiestling
Palomar College

D. C. Heath and Company
Lexington, Massachusetts   Toronto

Address editorial correspondence to:
D. C. Heath and Company
125 Spring Street
Lexington, MA 02173

Dedication

To my parents, Ron and Nova, without whom this would not have been possible.

Published simultaneously in Canada.

Printed in the United States of America.

International Standard Book Number: 0–669–41636–3

10 9 8 7 6 5 4 3 2 1

PREFACE

This study guide has been prepared to supplement the sections of each chapter of *Elementary Algebra*, Second Edition, by Roland E. Larson and Robert P. Hostetler.

Features

Each section starts with *Section Highlights,* which review definitions, facts, formulas, and properties that students should understand in order to be proficient in the section.

The workbook contains many examples with detailed step-by-step solutions. These are followed by *Starter Exercises*, in which a partial solution has been provided and students are expected to complete the solution. Detailed solutions to the Starter Exercises are provided at the end of each section.

Each section has an *exercise set*. Answers to all exercises have been included at the end of each chapter.

Each chapter has a *Cumulative Practice Test*. These tests contain exercises on the topics covered in that chapter and all chapters back through Chapter 1. Answers to all of these exercises have also been provided at the end of each chapter.

At the end of the Study Guide you will find a set of ten Warm-Up exercises for each section of the text starting with Chapter 1. These allow you to practice the skills you learned previously that are necessary to master the "new skills" presented in the section. All Warm-Up exercises are answered in the section following the Warm-Ups in the back of the text.

How to Use This Workbook

It is assumed that the corresponding sections in the text have been read.

Each section of the workbook corresponds to a text section. After reading a section in the text, read the examples in this workbook and work the accompanying Starter Exercises. Then work the exercises in the text. If more exercises (with answers) are desired, then work the exercises in this workbook.

The Cumulative Practice Test exercises can be used as a study aid for exams. These exercises will help you maintain proficiency with all the material covered in the course.

How to Study Mathematics

A lot of students find success in working with a study partner. Partners can help by answering questions and pushing each other to become better. Study partners seem to work particularly well at exam time. When reviewing for an exam, questions may arise (when an instructor is not available) and someone in the study group may know the answer.

Here are four steps for section-by-section studying.

1. Read the section in your textbook before you go to a lecture on that section. You may not completely understand the material at this point, but you will be able to pinpoint your problems and ask precise questions in class.
2. Go to class regularly and take good notes. Your notes should include the important ideas and examples of these ideas. Try recopying your notes after class. This gives you a complete, understandable set of notes to review later. Also, you may be surprised at how much more information you retain.
3. Reread the section in your textbook before starting the exercises. This allows you to see how ideas relate to each other and enhances your understanding of the material. In addition, you may want to read the examples and try the Starter Exercises in this study guide as they correspond to the text material.
4. Now it is time to work the exercises. Some students try to work the exercises before having adequately prepared and find it to be a frustrating experience. The exercises are there to improve your understanding of the subject; if you are overly frustrated, you will not

benefit from your work. You may also want to work the exercises in this study guide. The combination should give you the maximum benefit.

Finally, a few remarks on *studying for exams*. You should start studying well in advance of the exam (at least a week). The first day or two, only study for about two hours. Gradually increase the study time each day. Be completely prepared for the exam two days in advance. Spend the final day just building confidence so you can be relaxed during the exam.

The first things to look at when studying for an exam are all definitions, properties, and formulas. After these are known, then work as many exercises as you can. Working exercises is very important in the learning process as well as studying for an exam. Make sure you work a lot of the exercises that have given you trouble in the past.

When taking the exam itself, go through the exam and answer only the questions that you know immediately. Do not struggle with any one question on the first time through. If you do, you may not get to all the questions that you know by the end of the exam. Also, you may begin to panic if you struggle and not be able to recall the things you know.

After answering all the questions that you know immediately, then go back and work on the questions you skipped. After doing this, go back and check your work.

Good luck.

Jay R. Wiestling
Palomar College
Department of Mathematics
San Marcos, CA 92069

CONTENTS

C H A P T E R P
Prerequisites: Arithmetic Review

| P.1 | Real Numbers: Order and Absolute Value |

Section Highlights

1. The set of *natural numbers* is {1, 2, 3, ...}.
2. The set of *integers* is {..., −3, −2, −1, 0, 1, 2, 3...}.
3. The set of *rational numbers* is the set of all numbers that can be written as the ratio of two integers.
4. The set of *irrational numbers* is the set of numbers that cannot be written as the ratio of two integers.
5. The set of *real numbers* is the set of all rational numbers together with all the irrational numbers.
6. The *real number line*: each real number corresponds to one and only one point on the real number line and each point on the real number line corresponds to one and only one real number.
7. The *absolute value* of a number is the distance from that number to the origin on the real number line.

EXAMPLE 1 ■ Classifying Real Numbers

Determine which of the following are (a) natural numbers, (b) integers, (c) rational numbers, and (d) real numbers.

$$\left\{-2, \ -\tfrac{1}{2}, \ 6, \ 579, \ -14, \ \tfrac{11}{12}\right\}$$

Solution

(a) Natural numbers: $\{6, 579\}$

(b) Integers: $\{-2, \ 6, \ 579, \ -14\}$

(c) Rational numbers: $\left\{-2, \ -\tfrac{1}{2}, \ 6, \ 579, \ -14, \ \tfrac{11}{12}\right\}$

[Note that $-2 = \tfrac{-2}{1}$, $6 = \tfrac{6}{1}$, $579 = \tfrac{579}{1}$, and $-14 = \tfrac{-14}{1}$. This makes these numbers rational numbers.]

(d) $\left\{-2, \ -\tfrac{1}{2}, \ 6, \ 579, \ -14, \ \tfrac{11}{12}\right\}$

All elements of this set are real numbers.

Starter Exercise 1 *Classifying Real Numbers*

Determine which of the following are (a) natural numbers, (b) integers, (c) rational numbers, and (d) real numbers.

$$\left\{-\tfrac{4}{11}, \ 13, \ \tfrac{1}{12}, \ -\tfrac{3}{1}, \ -16, \ 3\right\}$$

D $-\frac{4}{11}, 13, \frac{1}{12}, \frac{3}{1}, -16, 3)$

(a) $13, 3$

(B) $13, -16, 3$

(c) $13, \frac{1}{12}, -\frac{3}{1}, -16, 3$

1

EXAMPLE 2 ■ **Ordering Real Numbers**

Place the correct inequality symbol ($<$ or $>$) between the two numbers.

(a) $2 \boxed{<} 5$

(b) $-3 \boxed{<} 2$

(c) $-\frac{1}{2} \boxed{>} -6$

(d) $\frac{1}{12} \boxed{>} -\frac{1}{13}$

(e) $-2.7 \boxed{<} 2.6$

Solution

(a) $2 < 5$

(b) $-3 < 2$

(c) $-\frac{1}{2} > -6$

(d) $\frac{1}{12} > -\frac{1}{13}$

(e) $-2.7 < 2.6$

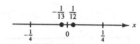

Starter Exercise 2 | *Fill in the blanks.*

Place the correct inequality symbol ($<$ or $>$) between the two numbers.

(a) $-2 \boxed{<} 1$

(b) $\frac{1}{2} \boxed{>} \frac{1}{3}$

(c) $-1.1 \boxed{<} -1$

EXAMPLE 3 ■ **Evaluating Absolute Values**

Evaluate the following expressions.

(a) $|2|$

(b) $|-3|$

(c) $-|-5|$

Solution

(a) $|2| = 2$ because the distance between 2 and 0 is 2.

(b) $|-3| = 3$ because the distance between -3 and 0 is 3.

(c) $-|-5| = -(5) = -5$

Starter Exercise 3 | *Fill in the blanks.*

Correctly place a positive sign ($+$) or a negative sign ($-$) in the blank box.

(a) $|6| = \boxed{+} 6$

(b) $|-1| = \boxed{+} 1$

(c) $-|-3| = \boxed{-} 3$

In Exercises 27–30, determine whether the numbers are relatively prime.

27. 70, 121 *yes*

28. 429, 350 *N*

29. 69, 75 *N*

30. 63, 72, 95 *yes*

$$\frac{\begin{array}{r}1\\16\\36\end{array}}{42}$$

| **P.3** | Adding and Subtracting Integers |

Section Highlights

1. To add two real numbers with like signs, add their absolute values and attach the common sign of the two terms to this sum.
2. To add two real numbers with unlike signs, find the absolute value of each term, subtract the smaller absolute value from the larger absolute value, and attach the sign of the term with the larger absolute value to the result.
3. To subtract one number (the subtrahend) from another number (the minuend), add the opposite of the subtrahend to the minuend.

EXAMPLE 1 ■ Adding Integers

Find the following sums.

(a) $-12 + (-6)$ *−18*

(b) $13 + (-16)$ *−3*

(c) $-6 + 9$ *3*

Solution

(a) $-12 + (-6) = -18$

(b) $13 + (-16) = -3$

(c) $-6 + 9 = 3$

Starter Exercise 1   *Fill in the blanks.*

Find the following sums.

(a) $-3 + 11 =$ *8*

(b) $-2 + (-5) =$ *−7*

(c) $4 + (-9) =$ *−5*

EXAMPLE 2 ■ Subtracting Integers

Find the following differences.

(a) $-2 - (-4)$

(b) $16 - (-36)$ *16+36 = 52*

(c) $-10 - 9$ *−10 + 9 = −1*

Solution

(a) $-2 - (-4) = -2 + 4 = 2$

(b) $16 - (-36) = 16 + 36 = 52$

(c) $-10 - 9 = -10 + (-9) = -19$

■ **Starter Exercise 2**   *Fill in the blanks.*

Find the following differences.

(a) $-17 - 5 = -17 + (-5) = \boxed{-22}$ (b) $13 - (-2) = \boxed{13} + \boxed{2} = 15$

(c) $-6 - (-8) = -6 + \boxed{8} = \boxed{2}$

■   **Solutions to Starter Exercises**   ■

1. (a) $-3 + 11 = \boxed{8}$

(b) $-2 + (-5) = \boxed{-7}$

(c) $4 + (-9) = \boxed{-5}$

2. (a) $-17 - 5 = -17 + (-5) = \boxed{-22}$

(b) $13 - (-2) = \boxed{13} + \boxed{2} = 15$

(c) $-6 - (-8) = -6 + \boxed{8} = \boxed{2}$

P.3   EXERCISES

In Exercises 1–9, find the sum.

1. $-2 + (-3)$ $-2 + -3 = -5$

2. $24 + (-16)$ 8

3. $-42 + 9 = -33$

4. $-37 + 17 = -20$

5. $21 + (-12)$ 9

6. $|6| + |-3| = 9$
$6 + 3$

7. $|-4| + |-15|$
$4 + 15 = 19$

8. $7 + (-|-13|) = -6$

9. $(-|5|) + (-7)$
$-5 + -7 = -12$

In Exercises 10–18, find the difference.

10. $-7 - 12$
$-7 + -12 = -19$

11. $-8 - 6$ $-8 + -6 = -14$

12. $-19 - (-6)$
$-19 + 6 = -13$

13. $47 - (-19)$
$47 + 19 = 66$

14. $119 - (-5)$ 124
$119 + 5$

15. $|10| - |-10|$
$10 + 10 = 0$

16. $|-3| - |-9|$
$3 + 9 = -6$

17. $17 - (-|-16|)$
$17 + 16 = 33$

18. $-26 - (-|-4|)$
$-26 + 4 = -4$

$-26 + = -22$

P.4   Multiplying and Dividing Integers

Section Highlights

1. To multiply two numbers with like signs, find the product of their absolute values.
2. To multiply two numbers with unlike signs, find the product of their absolute values, and attach a negative sign to this product.
3. To divide two numbers, multiply the dividend by the reciprocal of the divisor.

EXAMPLE 1 ■ Multiplying Integers

Find the following products.

(a) $(-3)(-6) = 18$

(b) $-4 \cdot 7 = -28$

(c) $24 \cdot (-36)$

-864

Solution

(a) $(-3)(-6) = 18$

(b) $-4 \cdot 7 = -28$

(c) The vertical algorithm is the best method for finding the product of the two absolute values.

$$
\begin{array}{r}
24 \\
\times \quad 36 \\
\hline
144 \quad (6 \times 24) \\
72 \quad (3 \times 24) \\
\hline
864 \quad \text{(sum of columns)}
\end{array}
$$

$24 \cdot (-36) = -864$

| Starter Exercise 1 | *Fill in the blanks.* |

Find the following products.

(a) $-3 \cdot 7 = \boxed{-21}$

(b) $(-5) \cdot (-4) = \boxed{20}$

EXAMPLE 2 ■ Division of Integers

Find the following quotients.

(a) $-12 \div 4$ 3

(b) $-32 \div (-2)$ 16

(c) $\dfrac{480}{-32} = 15$

(d) $0 \div (-6)$ 0

(e) $-15 \div 0$ 0

Solution

(a) $-12 \div 4 = -3$ because $-12 = 4 \cdot (-3)$.

(b) $-32 \div (-2) = 16$ because $-32 = (-2) \cdot (16)$.

(c) We will use the long division algorithm.

$$
\begin{array}{r}
15 \\
32 \overline{)\, 480} \\
32 \\
\hline
160 \\
160 \\
\hline
0
\end{array}
$$

$\dfrac{480}{-32} = -15$

(d) $0 \div (-6) = 0$ because $0 = -6 \cdot 0$.

(e) $-15 \div 0$ is undefined because division by zero is undefined.

Starter Exercise 2   *Fill in the blanks.*

Find the following quotients.

(a) $\dfrac{-15}{-3} = \boxed{5}$ (b) $14 \div (-7) = \boxed{-2}$ (c) $-312 \div 13 = \boxed{-24}$

                                                                        Use long division algorithm.

■   Solutions to Starter Exercises   ■

1. (a) $-3 \cdot 7 = \boxed{-21}$ **2.** (a) $\dfrac{-15}{-3} = \boxed{5}$

   (b) $(-5)(-4) = \boxed{20}$ (b) $14 \div (-7) = \boxed{-2}$

                                               (c) $-312 \div 13 = \boxed{-24}$

$$
\begin{array}{r}
24 \\
13\overline{)312} \\
\underline{26} \\
52 \\
\underline{52} \\
0
\end{array}
$$

P.4   EXERCISES

In Exercises 1–9, find the product.

1. $(-9) \cdot (-4) = 36$ **2.** $(-2) \cdot (-7) = 14$ **3.** $(-5) \cdot (10)\ -50$

4. $4 \cdot (-3) = -12$ **5.** $|-19| \times 2 = 38$ **6.** $4 \times (-|8|) = -32$

7. $|-7| \cdot |-1|$ **8.** $-22 \times 19 = -418$ **9.** $91 \cdot (-21) =$

In Exercises 10–18, find the quotient.

10. $-\frac{8}{4} = -2$ **11.** $12 \div (-4) = -3$ **12.** $(-30) \div (-3) = +10$

13. $\frac{-24}{-6} = 4$ **14.** $-|20| \div (-4) = 5$ **15.** $|-32| \div (-4) = 8$

16. $27 \div (-|-3|)$ **17.** $1235 \div (-19)$ **18.** $-744 \div (-31)$

19. *Area of a Rectangle*   The area of a rectangle is the product of its length and its width. Joe has a rectangular patio that has a length of 17 feet and a width of 16 feet. Find the area of Joe's patio.

20. *Area of a Rectangle*   A room is 30 feet long by 24 feet wide. How much carpet is needed to cover the entire floor?

Practice Test for Chapter P

In Exercises 1–4, label each with *all* letters that apply from the list below.

(a) Natural number   (b) Integer   (c) Rational number   (d) Real number

1. What kind of number is $\frac{3}{4}$?

2. What kind of number is 2?

3. What kind of number is -6?

4. What kind of number is 2.49?

In Exercises 5–16, evaluate the expression.

5. $(-2)(-5)$ $= 10$

6. $2 + ^-17$ $= ^-15$

7. $|-11 \cdot 2|$ $= 22$

8. $-6 + 17$ $= 11$

9. $25 \div (-5)$ $= ^-5$

10. $|-6 \cdot 6|$ $= 32$

11. $|-2 + (-3)|$ $= 5$

12. $-14 \cdot 10$ $= ^-140$

13. $-1692 \div (-36)$

14. $-172 - (-43)$ $= ^-129$

15. $(-13) \cdot (-24)$ $= 352$

16. $96 - 126$

In Exercises 17–20, place the correct symbol ($<$, $>$, or $=$) between the two numbers.

17. $\left|-\frac{1}{2}\right|$ $\boxed{<}$ $\frac{3}{4}$

18. -11 $\boxed{>}$ -14

19. $|-3|$ $\boxed{=}$ $|3|$

20. $-|-6|$ $\boxed{<}$ -6

In Exercises 21–23, write the prime factorization.

21. 18
$2 \cdot 3 \cdot 3$

22. 56

23. 136,125

In Exercises 24–26, find the greatest common factor.

24. 126, 147

25. 108, 288

26. 60, 105, 135

In Exercises 27–29, find the least common multiple.

27. 18, 27

28. 28, 35

29. 12, 20, 24

30. *Price per Foot*   Bob put 481 feet of electrical wire in his house. He paid $134.68 for the wire. What is the price per foot of the wire?

| P.1 | **Answers to Exercises** |

1. 2

2. $\frac{1}{2}$

3. 3

4. 16

5. −4

6. −7

7. −14

8. −6

9. 2 < 4

10. 5 > 1

11. $\frac{1}{2}$ < 3.5

12. $0.96 > \frac{3}{4}$

13. −6 < −2

14. −4 < −1

15. |3| = |−3|

16. 2 = |−2|

17. $\left|-\frac{1}{2}\right| < |−4|$

18. |−1| < |−5|

19. $-\left|-\frac{3}{5}\right| = -\frac{3}{5}$

20. −|0.6| = −|−0.6|

21. $-\frac{5}{2} > -|3|$

22. −|3| < |3|

23. 0 = |0|

24. |0| = −|0|

25. $|0.75| = \left|-\frac{3}{4}\right|$

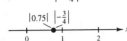

26. $-|0.5| < \frac{1}{2}$

27. (d)   **28.** (c)   **29.** (e)   **30.** (b)   **31.** (a)

| P.2 | **Answers to Exercises** |

1. composite   **2.** prime   **3.** composite

4. composite   **5.** composite   **6.** prime

7. $2 \cdot 3 \cdot 13$ **8.** $2 \cdot 13 \cdot 13$ **9.** $2 \cdot 2 \cdot 2 \cdot 7 \cdot 7$

10. $2 \cdot 2 \cdot 2 \cdot 2 \cdot 3 \cdot 7 \cdot 7$ **11.** $2 \cdot 491$ **12.** $3 \cdot 5 \cdot 163$

13. 19   **14.** 12   **15.** 21

16. 96   **17.** 5   **18.** 28

19. 8, 16, 24, 32   **20.** 23, 46, 69, 92   **21.** 30

22. 60   **23.** 180   **24.** 210

25. 24   **26.** 336   **27.** yes

28. yes   **29.** no   **30.** yes

| P.3 | **Answers to Exercises** |

1. -5 **2.** 8   **3.** -33

4. -20 **5.** 9   **6.** 9

7. 19   **8.** -6 **9.** -12

10. -19 **11.** -14 **12.** -13

13. 66   **14.** 124   **15.** 0

16. -6 **17.** 33   **18.** -22

P.4 | Answers to Exercises

1. 36
2. 14
3. −50
4. −12

5. 38
6. −32
7. 7
8. −418

9. −1911
10. −2
11. −3
12. 10

13. 4
14. −5
15. −8
16. −9

17. −65
18. 24
19. 272 sq ft
20. 720 sq ft

P | Answers to Practice Test

1. c, d
2. a, b, c, d
3. b, c, d
4. c, d

5. 10
6. −15
7. 22
8. 11

9. −5
10. 36
11. 5
12. −140

13. 47
14. −129
15. 312
16. −30

17. <
18. >
19. =
20. =

21. 2 · 3 · 3
22. 2 · 2 · 2 · 7
23. 3 · 3 · 5 · 5 · 5 · 11 · 11

24. 21
25. 36
26. 15

27. 54
28. 140
29. 120

30. $0.28

CHAPTER ONE
The Real Number System

1.1 Adding and Subtracting Fractions

Section Highlights

1. A fraction is in reduced form if the numerator and denominator have no common factors.
2. To add (or subtract) two fractions with the same denominators, add (or subtract) their numerators and write the sum (or difference) over the common denominator.
3. To add (or subtract) two fractions with different denominators, rewrite both fractions equivalently so that they have the same denominators. Then use the rules for adding (or subtracting) fractions with the same denominators.

EXAMPLE 1 ■ Adding Fractions

Find the sums.

(a) $\dfrac{3}{11} + \dfrac{2}{11}$ $\dfrac{5}{11}$

(b) $\dfrac{1}{4} + \dfrac{1}{6}$ $\dfrac{1}{4} \cdot \dfrac{3}{3} + \dfrac{1}{6}$

(c) $-3 + \dfrac{1}{5}$

(d) $2\dfrac{1}{3} + \dfrac{1}{12}$

Solution

(a) Since the denominators are the same, we just add the numerators.

$$\frac{3}{11} + \frac{2}{11} = \frac{3+2}{11} = \frac{5}{11}$$

(b) We need a common denominator and 12 is the smallest.

$$\frac{1}{4} + \frac{1}{6} = \frac{1}{4} \cdot \frac{3}{3} + \frac{1}{6} \cdot \frac{2}{2}$$

$$= \frac{1(3)}{4(3)} + \frac{1(2)}{6(2)}$$

$$= \frac{3}{12} + \frac{2}{12}$$

$$= \frac{3+2}{12}$$

$$= \frac{5}{12}$$

(c) $-3 + \dfrac{1}{5} = \dfrac{-3}{1} + \dfrac{1}{5}$ Convert –3 to a fraction.

$$= \frac{-3}{1} \cdot \frac{5}{5} + \frac{1}{5} \qquad \text{Find a common denominator.}$$

$$= \frac{-3(5)}{1(5)} + \frac{1}{5}$$

$$= \frac{-15}{5} + \frac{1}{5}$$

$$= \frac{-15+1}{5} \qquad \text{Add numerators.}$$

$$= \frac{-14}{5}$$

(d) $2\dfrac{1}{3} + \dfrac{1}{12} = 2 + \dfrac{1}{3} + \dfrac{1}{12}$ 　　Convert the mixed number into a sum of fractions.

$$= \dfrac{2}{1} + \dfrac{1}{3} + \dfrac{1}{12}$$

$$= \dfrac{2}{1} \cdot \dfrac{12}{12} + \dfrac{1}{3} \cdot \dfrac{4}{4} + \dfrac{1}{12}$$ 　　Find a common denominator.

$$= \dfrac{2(12)}{1(12)} + \dfrac{1(4)}{3(4)} + \dfrac{1}{12}$$

$$= \dfrac{24}{12} + \dfrac{4}{12} + \dfrac{1}{12}$$

$$= \dfrac{24 + 4 + 1}{12}$$ 　　Add numerators.

$$= \dfrac{29}{12}$$

Starter Exercise 1 　*Fill in the blanks.*

Find the following sums.

(a) $\dfrac{1}{23} + \dfrac{5}{23} = \dfrac{1 + \boxed{5}}{23} = \dfrac{\boxed{6}}{23}$

(b) $-\dfrac{2}{3} + \dfrac{1}{5} = -\dfrac{2}{3} \cdot \dfrac{5}{5} + \dfrac{1}{5} \cdot \dfrac{\boxed{3}}{\boxed{3}}$

$$= -\dfrac{2(5)}{3(\boxed{5})} + \dfrac{1(\boxed{3})}{5(\boxed{3})}$$

$$= -\dfrac{10}{15} + \dfrac{3}{15}$$

$$= \dfrac{\boxed{10} + \boxed{3}}{15}$$

$$= \dfrac{\boxed{7}}{15}$$

(c) $5\dfrac{1}{2} + \dfrac{3}{8} = 5 + \dfrac{1}{2} + \dfrac{3}{8}$

$$= \dfrac{5}{1} + \dfrac{1}{2} + \dfrac{3}{8}$$

$$= \dfrac{5}{1} \cdot \dfrac{\boxed{8}}{\boxed{8}} + \dfrac{1}{2} \cdot \dfrac{\boxed{4}}{\boxed{4}} + \dfrac{3}{8}$$

$$= \dfrac{5 \cdot \boxed{40}}{1 \cdot \boxed{8}} + \dfrac{1 \cdot \boxed{4}}{2 \cdot \boxed{8}} + \dfrac{3}{8}$$

$$= \dfrac{\boxed{40}}{8} + \dfrac{\boxed{4}}{8} + \dfrac{3}{8}$$

$$= \dfrac{\boxed{40} + \boxed{4} + \boxed{3}}{8}$$

$$= \boxed{87}$$

EXAMPLE 2 ■ **Subtracting Fractions**

Find the following differences.

(a) $\dfrac{17}{19} - \dfrac{2}{19}$
                  (b) $\dfrac{5}{12} - \dfrac{7}{16}$
                  (c) $-\dfrac{1}{6} - \left(-\dfrac{8}{15}\right)$

Solution

(a) $\dfrac{17}{19} - \dfrac{2}{19} = \dfrac{17}{19} + \dfrac{-2}{19} = \dfrac{17 + (-2)}{19} = \dfrac{15}{19}$

(b) $\dfrac{5}{12} - \dfrac{7}{16} = \dfrac{5}{12} + \dfrac{-7}{16}$

$\qquad\qquad\quad = \dfrac{5}{12} \cdot \dfrac{4}{4} + \dfrac{-7}{16} \cdot \dfrac{3}{3}$ Find a common denominator.

$\qquad\qquad\quad = \dfrac{5(4)}{12(4)} + \dfrac{-7(3)}{16(3)}$

$\qquad\qquad\quad = \dfrac{20}{48} + \dfrac{-21}{48}$

$\qquad\qquad\quad = \dfrac{20 + (-21)}{48}$ Subtract numerators.

$\qquad\qquad\quad = \dfrac{-1}{48}$

$\qquad\qquad\quad = -\dfrac{1}{48}$

(c) $-\dfrac{1}{6} + \left(-\dfrac{8}{15}\right) = -\dfrac{1}{6} + \dfrac{8}{15}$ Convert to addition.

$\qquad\qquad\qquad\quad = -\dfrac{1}{6} \cdot \dfrac{5}{5} + \dfrac{8}{15} \cdot \dfrac{2}{2}$ Find a common denominator.

$\qquad\qquad\qquad\quad = \dfrac{-1(5)}{6(5)} + \dfrac{8(2)}{15(2)}$

$\qquad\qquad\qquad\quad = \dfrac{-5}{30} + \dfrac{16}{30}$

$\qquad\qquad\qquad\quad = \dfrac{-5 + 16}{30}$

$\qquad\qquad\qquad\quad = \dfrac{11}{30}$

Starter Exercise 2 *Fill in the blanks.*

Find the following differences.

(a) $\dfrac{2}{9} - \dfrac{7}{15} = \dfrac{2}{9} + \dfrac{-7}{15}$

$\qquad = \dfrac{2}{9} \cdot \dfrac{5}{\boxed{5}} + \dfrac{-7}{15} \cdot \dfrac{\boxed{3}}{\boxed{3}}$

$\qquad = \dfrac{2(5)}{9(\boxed{45})} + \dfrac{-7(\boxed{21})}{15(\boxed{45})}$

$\qquad = \dfrac{\boxed{10}}{45} + \dfrac{\boxed{21}}{45}$

$\qquad = \dfrac{\boxed{10} + \boxed{21}}{45}$

$\qquad = \boxed{\dfrac{11}{45}}$

(b) $6 - \dfrac{2}{7} = 6 + \dfrac{-2}{7}$

$\qquad = \dfrac{6}{1} \cdot \dfrac{7}{7} + \dfrac{-2}{7}$

$\qquad = \dfrac{\boxed{42}}{7} + \dfrac{-2}{7}$

$\qquad = \dfrac{42}{7} + \dfrac{-2}{7}$

$\qquad = \dfrac{\boxed{42} + \boxed{-2}}{7}$

$\qquad = \boxed{\dfrac{40}{7}}$

■ **Solutions to Starter Exercises** ■

1. (a) $\dfrac{1}{23} + \dfrac{5}{23} = \dfrac{1 + \boxed{5}}{23} = \dfrac{\boxed{6}}{23}$

(b) $-\dfrac{2}{3} + \dfrac{1}{5} = -\dfrac{2}{3} \cdot \dfrac{5}{5} + \dfrac{1}{5} \cdot \dfrac{\boxed{3}}{\boxed{3}} = -\dfrac{2(5)}{3(\boxed{5})} + \dfrac{1(\boxed{3})}{5(\boxed{3})}$

$\qquad = -\dfrac{10}{15} + \dfrac{3}{15} = \dfrac{\boxed{-10} + \boxed{3}}{15} = \boxed{-\dfrac{7}{15}}$

(c) $5\dfrac{1}{2} + \dfrac{3}{8} = 5 + \dfrac{1}{2} + \dfrac{3}{8} = \dfrac{5}{1} + \dfrac{1}{2} + \dfrac{3}{8} = \dfrac{5}{1} \cdot \dfrac{\boxed{8}}{\boxed{8}} + \dfrac{1}{2} \cdot \dfrac{\boxed{4}}{\boxed{4}} + \dfrac{3}{8}$

$\qquad = \dfrac{5 \cdot \boxed{8}}{1 \cdot \boxed{8}} + \dfrac{1 \cdot \boxed{4}}{2 \cdot \boxed{4}} + \dfrac{3}{8} = \dfrac{\boxed{40}}{8} + \dfrac{\boxed{4}}{8} + \dfrac{3}{8}$

$\qquad = \dfrac{\boxed{40} + \boxed{4} + \boxed{3}}{8} = \boxed{\dfrac{47}{8}}$

(handwritten top margin: $\frac{24}{8} = 2$, $\frac{24}{8}$)

■ **Solutions to Starter Exercises** ■

2. (a) $\dfrac{2}{9} - \dfrac{7}{15} = \dfrac{2}{9} + \dfrac{-7}{15} = \dfrac{2}{9} \cdot \dfrac{\boxed{3}}{\boxed{5}} + \dfrac{-7}{15} \cdot \dfrac{\boxed{3}}{3} = \dfrac{2(5)}{9(\boxed{5})} + \dfrac{-7(\boxed{3})}{15(\boxed{3})}$

$= \dfrac{\boxed{10}}{45} + \dfrac{\boxed{-21}}{45} = \dfrac{\boxed{10} + \boxed{(-21)}}{45} = \boxed{\dfrac{-11}{45}}$

(b) $6 - \dfrac{2}{7} = \dfrac{6}{1} + \dfrac{-2}{7} = \dfrac{6}{1} \cdot \dfrac{7}{7} + \dfrac{-2}{7}$

$= \dfrac{\boxed{6(7)}}{1(7)} + \dfrac{-2}{7} = \dfrac{42}{7} + \dfrac{-2}{7} = \dfrac{\boxed{42} + \boxed{(-2)}}{7} = \boxed{\dfrac{40}{7}}$

1.1 EXERCISES

In Exercises 1–4, write the given fraction in reduced form.

1. $\frac{24}{48}$ *$\frac{12\cdot2}{12\cdot4} = \frac{2}{4} = \frac{2\cdot1}{2\cdot2}$*
2. $\frac{15}{27}$ *$\frac{3\cdot5}{3\cdot9} = \frac{5}{9}$*
3. $\frac{57}{95}$ *$= \frac{19\cdot3}{19\cdot5} = \frac{3}{5}$*
4. $\frac{106}{154}$ *$\frac{2\cdot53}{2\cdot77} = \frac{53}{77}$*

In Exercises 5–25, perform the indicated operations and write your answer in reduced form.

5. $\frac{1}{5} + \frac{3}{5}$ *$\frac{4}{5}$*
6. $\frac{2}{7} + \frac{3}{7}$ *$\frac{5}{7}$*
7. $\frac{5}{6} + \frac{1}{15}$
8. $2\frac{1}{3} - \frac{7}{8}$

9. $-\frac{1}{4} + \frac{11}{9}$
10. $-16\frac{3}{4} + \left(-1\frac{5}{8}\right)$
11. $\frac{10}{37} - \frac{14}{23}$
12. $4 - \frac{1}{2}$

13. $-\frac{3}{2} - \frac{3}{4}$
14. $2\frac{7}{16} - 12$ *$\frac{39 - 12}{}$*
15. $\left|-\frac{3}{5}\right| + \frac{1}{8}$
16. $7\frac{1}{4} - 10\frac{3}{8}$

17. $\left|5\frac{4}{5} - 7\frac{9}{10}\right|$ *$\frac{29}{5} - \frac{79}{10}$*
18. $9 - \frac{5}{8} + 2\frac{7}{8}$
19. $-\left|\frac{7}{9}\right| - \left|-\frac{11}{20}\right|$
20. $-3\frac{1}{11} + 9$ *$\frac{34}{11} + \frac{9}{1}$*

21. $\frac{63}{16} - \frac{7}{10} + \frac{3}{25}$
22. $6\frac{11}{20} - \left|-\frac{1}{2}\right|$
23. $5\frac{5}{6} + 11$ *$\frac{35 + 11}{6}$*
24. $-\frac{1}{2} - \frac{7}{8} - \frac{11}{16}$

25. $\frac{13}{20} - \frac{2}{3} + 8 - \frac{3}{10}$

In Exercises 26–28, write the mixed number as a fraction.

26. $5\frac{1}{12}$ *$= \frac{61}{12}$*
27. $9\frac{11}{15}$ *$= \frac{146}{15} =$*
28. $13\frac{3}{100}$ *$\frac{1303}{100}$*

In Exercises 29–31, fill in the box to make the two fractions equal.

29. $\frac{3}{4} = \frac{3 \cdot \boxed{2}}{4 \cdot 2}$ *$\frac{6}{8}$*
30. $\frac{5}{6} = \frac{\boxed{25}}{30}$ *$\frac{5\cdot5}{5\cdot6}$*
31. $\frac{11}{12} = \frac{121}{\boxed{132}}$ *$\frac{11\cdot11}{11\cdot12}$*

1.2 Multiplying and Dividing Fractions

Section Highlights

1. $\dfrac{a}{b} \cdot \dfrac{c}{d} = \dfrac{ac}{bd}$

2. $\dfrac{a}{b} \div \dfrac{c}{d} = \dfrac{a}{b} \cdot \dfrac{d}{c}$

EXAMPLE 1 ■   Multiplying Fractions

Find the following products and give your answers in reduced form.

(a) $\dfrac{7}{16} \cdot \dfrac{3}{5}$

(b) $-\dfrac{16}{25} \cdot \dfrac{5}{8}$

(c) $\left(-2\dfrac{1}{2}\right)\left(-\dfrac{6}{7}\right)$

Solution

(a) $\dfrac{7}{16} \cdot \dfrac{3}{5} = \dfrac{7(3)}{16(5)} = \dfrac{21}{80}$

(b) $-\dfrac{16}{25} \cdot \dfrac{5}{8} = -\dfrac{16(5)}{25(8)} = -\dfrac{2 \cdot \cancel{8} \cdot \cancel{5}}{5 \cdot \cancel{5} \cdot \cancel{8}} = -\dfrac{2}{5}$

Alternative Method: I have found that most students can cancel effectively in arithmetic, but few can cancel effectively in algebra. Hence, I will offer another method that may work better in the future.

$$-\dfrac{16}{25} \cdot \dfrac{5}{8} = -\dfrac{16 \cdot 5}{25 \cdot 8}$$

$$= -\dfrac{2 \cdot 8 \cdot 5}{5 \cdot 5 \cdot 8}$$
Find the common factor.

$$= -\dfrac{2}{5} \cdot \dfrac{8 \cdot 5}{5 \cdot 8}$$
Separate the fraction into the product of two fractions, where one is $\dfrac{\text{common factor}}{\text{common factor}}$.

$$= -\dfrac{2}{5} \cdot 1$$
$\dfrac{\text{common factor}}{\text{common factor}} = 1$

$$= -\dfrac{2}{5}$$

(c) $\left(-2\frac{1}{2}\right)\left(-\frac{6}{7}\right) = \left(-\frac{5}{2}\right)\left(-\frac{6}{7}\right) = \frac{5\cdot 6}{2\cdot 7} = \frac{5\cdot \cancel{2}\cdot 3}{\cancel{2}\cdot 7} = \frac{15}{7}$

Alternative Method:

$$\left(-2\frac{1}{2}\right)\left(-\frac{6}{7}\right) = \left(-\frac{5}{2}\right)\left(-\frac{6}{7}\right)$$

$$= \frac{5\cdot 6}{2\cdot 7}$$

$$= \frac{5\cdot 2\cdot 3}{2\cdot 7} \qquad\qquad \text{Find common factor.}$$

$$= \frac{5\cdot 3}{7}\cdot\frac{2}{2} \qquad\qquad \begin{array}{l}\text{Separate the fraction into the}\\ \text{product of two fractions, where}\\ \text{one is } \frac{\text{common factor}}{\text{common factor}}.\end{array}$$

$$= \frac{15}{7}\cdot 1 \qquad\qquad \frac{\text{common factor}}{\text{common factor}} = 1$$

$$= \frac{15}{7}$$

Starter Exercise 1   *Fill in the blanks.*

Find the following products and give your answers in reduced form.

(a) $\dfrac{5}{6}\cdot\dfrac{7}{11} = \dfrac{5\cdot 7}{\boxed{66}} = \dfrac{35}{\boxed{66}}$

(b) $-\dfrac{5}{24}\cdot\dfrac{8}{25} = -\dfrac{5\cdot 8}{24\cdot 25} = -\dfrac{\overset{40}{5\cdot 8}}{3\cdot 8\cdot 5\boxed{}} = \boxed{}$

(handwritten:) $\dfrac{40}{600} = \dfrac{1}{15}$

EXAMPLE 2 ■   Dividing Fractions

Find the following quotients and give your answers in reduced form.

(a) $\dfrac{2}{7}\div\dfrac{4}{21}$ \qquad\qquad (b) $-\dfrac{12}{23}\div\dfrac{6}{7}$ \qquad\qquad (c) $\dfrac{8}{35}\div 2$

Solution

(a) $\dfrac{2}{7}\div\dfrac{4}{21} = \dfrac{2}{7}\cdot\dfrac{21}{4}$ \qquad\qquad $\begin{array}{l}\text{Multiply by the reciprocal}\\ \text{of the divisor.}\end{array}$

$$= \frac{2\cdot 21}{7\cdot 4}$$

$$= \frac{\cancel{2}\cdot 7\cdot 3}{7\cdot \cancel{2}\cdot 2} \qquad\qquad \text{Reduce.}$$

$$= \frac{3}{2}$$

Alternative Method:

$$\frac{2}{7}\div\frac{4}{21} = \frac{2}{7}\cdot\frac{21}{4} = \frac{2\cdot 21}{7\cdot 4} = \frac{2\cdot 7\cdot 3}{7\cdot 2\cdot 2} = \frac{2\cdot 7}{7\cdot 2}\cdot\frac{3}{2} = 1\cdot\frac{3}{2} = \frac{3}{2}$$

(b) $-\dfrac{12}{23} \div \dfrac{6}{7} = -\dfrac{12}{23} \cdot \dfrac{7}{6}$ Multiply by the reciprocal of the divisor.

$= -\dfrac{12 \cdot 7}{23 \cdot 6}$

$= -\dfrac{2 \cdot \cancel{6} \cdot 7}{23 \cdot \cancel{6}}$ Reduce.

$= -\dfrac{14}{23}$

Alternative Method:

$-\dfrac{12}{23} \div \dfrac{6}{7} = -\dfrac{12}{23} \cdot \dfrac{7}{6}$ Multiply by the reciprocal of the divisor.

$= -\dfrac{12 \cdot 7}{23 \cdot 6}$

$= -\dfrac{6 \cdot 2 \cdot 7}{6 \cdot 23}$

$= -\dfrac{6}{6} \cdot \dfrac{2 \cdot 7}{23}$ $\dfrac{6}{6}$ is $\dfrac{\text{common factor}}{\text{common factor}}$.

$= -1 \cdot \dfrac{14}{23}$

$= -\dfrac{14}{23}$

(c) $\dfrac{8}{35} \div 2 = \dfrac{8}{35} \cdot \dfrac{1}{2}$ Multiply by the reciprocal of the divisor.

$= \dfrac{8 \cdot 1}{35 \cdot 2}$

$= \dfrac{4 \cdot \cancel{2} \cdot 1}{35 \cdot \cancel{2}}$ Reduce.

$= \dfrac{4}{35}$

Alternative Method:

$\dfrac{8}{35} \div 2 = \dfrac{8}{35} \cdot \dfrac{1}{2}$

$= \dfrac{8 \cdot 1}{35 \cdot 2}$

$= \dfrac{4 \cdot 2 \cdot 1}{35 \cdot 2}$

$= \dfrac{4 \cdot 1}{35} \cdot \dfrac{2}{2}$ $\dfrac{2}{2}$ is $\dfrac{\text{common factor}}{\text{common factor}}$.

$= \dfrac{4}{35} \cdot 1$

$= \dfrac{4}{35}$

Starter Exercise 2   *Fill in the blanks.*

Find the following quotients and give your answers in reduced form.

(a) $-\dfrac{8}{3} \div \dfrac{12}{11} = -\dfrac{8}{3} \cdot \dfrac{11}{12} = -\dfrac{8 \cdot \boxed{11}}{\boxed{3} \cdot 12} = \dfrac{2 \cdot 4 \cdot \boxed{11}}{\boxed{3} \cdot 4 \cdot 3} = \boxed{}$

(b) $\left(-1\dfrac{3}{4}\right) \div \left(-\dfrac{15}{16}\right) = \left(-\dfrac{7}{4}\right) \div \left(-\dfrac{15}{16}\right)$

$= \left(-\dfrac{7}{4}\right) \cdot \boxed{\dfrac{16}{15}} = \dfrac{7 \cdot \boxed{16}}{4 \cdot \boxed{15}} = \boxed{} = \boxed{}$

EXAMPLE 3 ■   Operations with Decimal Fractions

Perform the indicated operation.

(a) $2.7931 + 0.637$ (b) $(-16.21)(0.301)$ (c) $23.564 \div 2.4$

Solution

(a)
```
  2.7931
+ 0.637      Align the decimal points.
  3.4301     Align the decimal point in sum also.
```

(b)
```
   -16.21
    0.301
    1621
    0000
   4863
  -4.87921    Five decimal places and negative
```

(c) Recall that the divisor must be an integer. Hence, move the decimal point one place to the right in both the divisor and the dividend.

```
        9 .8 1 8 3 3    Align decimal point.
  24) 2 3 5 .6 4 0 0 0   Move decimal point.
      2 1 6
        1 9 6
        1 9 2
          4 4
          2 4
          2 0 0
          1 9 2
            8 0
            7 2
            8 0
            7 2
             8
```

Note that the 3 in the quotient is repeating. Therefore, we write $23.564 \div 2.4 = 9.818\overline{3}$. Also note that if we round off to two decimal places $23.564 \div 2.4 \approx 9.82$.

■ **Starter Exercise 3**   *Fill in the blanks.*

Perform the indicated operations.

(a)   2.36
     + 16.1
     ⬚

(b)   39.67
     ×   2.3
        11901
        7934
        ⬚

(c) 65 ⟌ 317.85 with box above

■   **Solutions to Starter Exercises**   ■

1. (a) $\dfrac{5}{6} \cdot \dfrac{7}{11} = \dfrac{5 \cdot 7}{\boxed{6 \cdot 11}} = \dfrac{35}{\boxed{66}}$

(b) $-\dfrac{5}{24} \cdot \dfrac{8}{25} = -\dfrac{5 \cdot 8}{24 \cdot 25}$

$= -\dfrac{\not{5} \cdot \not{8}}{3 \cdot \not{8} \cdot \not{5} \cdot \boxed{5}}$

$= \boxed{-\dfrac{1}{15}}$

2. (a) $-\dfrac{8}{3} \div \dfrac{12}{11} = -\dfrac{8}{3} \cdot \dfrac{11}{12} = -\dfrac{8 \cdot \boxed{11}}{\boxed{3} \cdot 12} = -\dfrac{2 \cdot 4 \cdot \boxed{11}}{\boxed{3} \cdot 4 \cdot 3} = \boxed{-\dfrac{22}{9}}$

(b) $\left(-1\dfrac{3}{4}\right) \div \left(-\dfrac{15}{16}\right) = \left(-\dfrac{7}{4}\right) \div \left(-\dfrac{15}{16}\right) = \left(-\dfrac{7}{4}\right) \cdot \boxed{\left(-\dfrac{16}{15}\right)}$

$= \dfrac{7 \cdot \boxed{16}}{4 \cdot \boxed{15}} = \boxed{\dfrac{7 \cdot 4 \cdot 4}{4 \cdot 15}} = \boxed{\dfrac{28}{15}}$

3. (a)   2.36
          +   16.1
          ⎯⎯⎯⎯⎯
          $\boxed{18.46}$

(b)   39.67
     ×   2.3
     ⎯⎯⎯⎯⎯
        11901
        7934
     ⎯⎯⎯⎯⎯
        $\boxed{91.241}$

(c) $\boxed{4.89}$
     65 ⟌ 317.85
        260
        ⎯⎯⎯
        57 8
        52 0
        ⎯⎯⎯
        5 85
        5 85
        ⎯⎯⎯
          0

1.2   **EXERCISES**

In Exercises 1–10, evaluate the expression and write your answer in reduced form.

1. $\left(\dfrac{1}{5}\right)\left(\dfrac{7}{8}\right)$ **2.** $\left(-\dfrac{3}{10}\right)\left(\dfrac{5}{9}\right)$ **3.** $\left(\dfrac{13}{6}\right)\left(-\dfrac{4}{39}\right)$ **4.** $\left(-\dfrac{14}{19}\right)\left(-\dfrac{5}{21}\right)$

5. $(-2)\left(\frac{3}{4}\right)$ **6.** $\left(\frac{9}{5}\right)\left(-\frac{25}{36}\right)$ **7.** $\left(\frac{1}{3}\right)\left(\frac{6}{5}\right)\left(-\frac{5}{12}\right)$ **8.** $\left(2\frac{1}{6}\right)\left(3\frac{4}{5}\right)$

9. $-\left(-3\frac{4}{9}\right)\left|-\frac{6}{28}\right|$ **10.** $\left(3\frac{1}{5}\right)\left(-2\frac{7}{8}\right)\left(1\frac{16}{25}\right)$

In Exercises 11–14, write the reciprocal of the number.

11. 6 **12.** $\frac{1}{2}$ **13.** $\frac{7}{8}$ **14.** $2\frac{1}{5}$

In Exercises 15–20, evaluate the expression. Write your answer in reduced form.

15. $\frac{6}{7} \div \frac{9}{14} = \frac{6}{7} \cdot \frac{14}{9} = \frac{84}{63} = \frac{12}{7}$ **16.** $\frac{-\frac{5}{8}}{\frac{25}{48}} = -\frac{5}{8} \cdot \frac{48}{25} = \frac{240}{200} = \frac{6}{5}$ **17.** $\frac{11}{13} \div \frac{1}{2} = \frac{11}{12} \cdot \frac{2}{1} = \frac{22}{12} = \frac{11}{6}$

18. $-\frac{12}{25} \div \frac{24}{5}$ **19.** $-\left|-\frac{5}{9}\right| \div \left(-\frac{4}{27}\right)$ **20.** $\frac{2}{6/7}$

In Exercises 21–24, write the rational number in decimal form.

21. $\frac{1}{2} = .5$ **22.** $\frac{5}{4}$ 8 **23.** $\frac{1}{3}$ .3 **24.** $\frac{4}{15}$ .26

In Exercises 25–34, perform the indicated operations and round your answer to two decimal places.

25. $2.36 + 1.57$ **26.** $1.3 - 2.56$ **27.** $-172.1 + (16.75)$

28. $-|+11.03| - |-14.6|$ **29.** $(2.11)(1.07)$ **30.** $-|9.37|(-4.91)$

31. $-7.1397 \div 0.24$ **32.** $-16.35 \div (-9.2)$ **33.** $-\frac{76.551}{21}$

34. $\frac{-|-22.76|}{|-1.9|}$

35. *Purchase Price*   If you buy 1.37 pounds of fish that cost \$3.97 per pound, what is the total cost of the fish?

36. *Cost per Foot*   A 50-foot spool of loudspeaker cable costs \$16.97. How much is the cable per foot?

In Exercises 37–40, determine whether the statement is true or false.

37. The quotient of any two integers is an integer.

38. The quotient of any two rational numbers is a rational number.

39. Decimal numbers are rational numbers.

40. The reciprocal of zero is zero.

1.3 | Exponents and Order of Operations

Section Highlights

1. In the expression 4^6, 4 is called the **base** and 6 is called the **exponent**.
2. **Order of Operations**
    i) Perform operations inside *symbols of groupings*, starting with the innermost symbol.
    ii) Evaluate all *exponential* expressions.
    iii) Perform all *multiplication* and *division* as they appear from left to right.
    iv) Perform all *addition* and *subtraction* as they appear from left to right.

EXAMPLE 1 ■   Evaluating Exponential Expressions

Evaluate each of the following expressions.

(a) 2^3 (b) 3^2 (c) $(-2)^4$ (d) -2^4 (e) 1^5 (f) 0^6 (g) $\left(\frac{3}{5}\right)^3$

Solution

(a) $2^3 = 2 \cdot 2 \cdot 2 = 8$ (b) $3^2 = 3 \cdot 3 = 9$

(c) $(-2)^4 = (-2) \cdot (-2) \cdot (-2) \cdot (-2) = 16$ (d) $-2^4 = -(2 \cdot 2 \cdot 2 \cdot 2) = -(16) = -16$

(e) $1^5 = 1 \cdot 1 \cdot 1 \cdot 1 \cdot 1 = 1$ (f) $0^6 = 0 \cdot 0 \cdot 0 \cdot 0 \cdot 0 \cdot 0 = 0$

(g) $\left(\frac{3}{5}\right)^3 = \frac{3}{5} \cdot \frac{3}{5} \cdot \frac{3}{5} = \frac{3 \cdot 3 \cdot 3}{5 \cdot 5 \cdot 5} = \frac{27}{125}$

Starter Exercise 1   *Fill in the blanks.*

Evaluate the following expressions.

(a) $(-2)^6 = (-2) \cdot (-2) \cdot (-2) \cdot (-2) \cdot (-2) \cdot (-2) = \boxed{64}$

(b) $-2^6 = -(2 \cdot 2 \cdot 2 \cdot 2 \cdot 2 \cdot 2) = \boxed{64}$

(c) $-1^3 = -(1 \cdot 1 \cdot 1) = \boxed{-1}$

(d) $0^4 = 0 \cdot 0 \cdot \boxed{} \cdot \boxed{} = \boxed{0}$

(e) $\left(\frac{3}{4}\right)^4 = \frac{3}{4} \cdot \frac{3}{4} \cdot \frac{3}{4} \cdot \frac{3}{4} = \frac{3 \cdot 3 \cdot 3 \cdot 3}{4 \cdot 4 \cdot \boxed{} \cdot \boxed{}} = \boxed{\frac{81}{256}}$

EXAMPLE 2 ■   Order of Operations

Evaluate the following expressions.

(a) $(36 \div 6)^2 + [(-2)(4) + 1]$ (b) $2 + \left[11 \div \frac{1}{2} - 3^3\right]$

Solution

(a) $(36 \div 6)^2 + [(-2)(4) + 1] = 6^2 + [(-2)(4) + 1]$ Divide inside the parentheses.

$= 36 + [(-2)(4) + 1]$ Evaluate the exponential expression.

$= 36 + [-8 + 1]$ Multiply inside the parentheses.

$= 36 + [-7]$ Add inside the parentheses.

$= 29$ Add.

(b) $2 + \left[11 \div \frac{1}{2} - 3^3 \right] = 2 + \left[11 \div \frac{1}{2} - 27 \right]$ Evaluate the exponential expression.

$= 2 + [22 - 27]$ Divide.

$= 2 + [-5]$ Subtract.

$= -3$ Add.

(handwritten: $\frac{11 \cdot 2 - 22 \cdot 27}{1 \quad 1 \quad 1}$ $2 + -5$)

Starter Exercise 2   *Fill in the blanks.*

Evaluate the following expressions.

(a) $1 + 2(2 \cdot 3^2 + 1) = 1 + 2(2 \cdot 9 + 1)$

$= 1 + 2(18 + 1)$

$= 1 + 2\boxed{19}$

$= 1 + \boxed{38}$

$= \boxed{39}$

(b) $2^2(2 + 3) \div 10 = 2^2(5) \div 10$

(handwritten: $4 \cdot 5$ $20 - 10 = 2$)

$= \boxed{4}(5) \div 10$

$= \boxed{20} \div 10$

$= \boxed{2}$

EXAMPLE 3 ■ Order of Operations

Evaluate the following expressions.

(a) $[4(2 - 1) + 3]^2 \div 7$

(b) $12 - \dfrac{3 - 6^2}{2^3 + 3} - 4^2$

(handwritten near (b): $\frac{3 - 36}{16 + 3}$ $12 \cdot \frac{-9}{9} - 16$ $12 - 1 - 16$)

Solution

(handwritten: 6)

(a) $[4(2 - 1) + 3]^2 \div 7 = [4(1) + 3]^2 \div 7$ Subtract inside the parentheses.

$= [4 + 3]^2 \div 7$ Multiply inside the parentheses.

$= 7^2 \div 7$ Add inside the parentheses.

$= 49 \div 7$ Evaluate the exponential expression.

$= 7$ Divide.

(b) $12 - \dfrac{3 - 6^2}{2^3 + 3} - 4^2 = 12 - \dfrac{3 - 36}{2^3 + 3} - 4^2$ Evaluate the exponential. expression in the numerator.

$$= 12 - \dfrac{-33}{2^3 + 3} - 4^2$$ Subtract in the numerator.

$$= 12 - \dfrac{-33}{8 + 3} - 4^2$$ Evaluate the exponential. expression in the denominator.

$$= 12 - \dfrac{-33}{11} - 4^2$$ Add in the denominator.

$$= 12 - \dfrac{-33}{11} - 16$$ Evaluate the exponential expression.

$$= 12 - (-3) - 16$$ Divide.

$$= 15 - 16$$ Subtract.

$$= -1$$ Subtract.

Starter Exercise 3   *Fill in the blanks.*

Evaluate the following expression.

$$7 + \dfrac{26}{2^4 - 3} \div 2 = 7 + \dfrac{26}{16 - 3} \div 2$$

$$= 7 + \dfrac{26}{13} \div 2$$

$$= 7 + \boxed{2} \div 2$$

$$= 7 + \boxed{1}$$

$$= \boxed{8}$$

EXAMPLE 4 ■   Evaluating Expressions with a Calculator

(a) To evaluate the expression $2^3 + 4 \cdot 5$ use the following keystrokes.

Keystrokes   *Display*

2 $\boxed{y^x}$ 3 $\boxed{=}$ $\boxed{+}$ 4 $\boxed{\times}$ 5 $\boxed{=}$ 28   Scientific

2 $\boxed{\wedge}$ 3 $\boxed{+}$ 4 $\boxed{\times}$ 5 $\boxed{\text{ENTER}}$ 28   Graphing

(b) To evaluate the expression $6^2(3 - 4^2) \div (-2)$ use the following keystrokes.

Keystrokes   *Display*

6 $\boxed{x^2}$ $\boxed{\times}$ $\boxed{(}$ 3 $\boxed{-}$ 4 $\boxed{x^2}$ $\boxed{)}$ $\boxed{\div}$ 2 $\boxed{+/-}$ $\boxed{=}$ 234   Scientific

6 $\boxed{x^2}$ $\boxed{\times}$ $\boxed{(}$ 3 $\boxed{-}$ 4 $\boxed{x^2}$ $\boxed{)}$ $\boxed{\div}$ $\boxed{(-)}$ 2 $\boxed{\text{ENTER}}$ 234   Graphing

---

■ Solutions to Starter Exercises ■

1. (a) $(-2)^6 = (-2) \cdot (-2) \cdot (-2) \cdot (-2) \cdot (-2) \cdot (-2) = \boxed{64}$

(b) $-2^6 = -(2 \cdot 2 \cdot 2 \cdot 2 \cdot 2 \cdot 2) = \boxed{-64}$

(c) $-1^3 = -(1 \cdot 1 \cdot 1) = \boxed{-1}$

(d) $0^4 = 0 \cdot 0 \cdot \boxed{0} \cdot \boxed{0} = \boxed{0}$

(e) $\left(\frac{3}{4}\right)^4 = \frac{3}{4} \cdot \frac{3}{4} \cdot \frac{3}{4} \cdot \frac{3}{4} = \frac{3 \cdot 3 \cdot 3 \cdot 3}{4 \cdot 4 \cdot \boxed{4} \cdot \boxed{4}} = \boxed{\frac{81}{256}}$

2. (a) $1 + 2(2 \cdot 3^2 + 1) = 1 + 2(2 \cdot 9 + 1) = 1 + 2(18 + 1)$

$= 1 + 2 \boxed{19} = 1 + \boxed{38} = \boxed{39}$

(b) $2^2(2 + 3) \div 10 = 2^2(5) \div 10 = \boxed{4}(5) \div 10 = \boxed{20} \div 10 = \boxed{2}$

3. $7 + \dfrac{26}{2^4 - 3} \div 2 = 7 + \dfrac{26}{16 - 3} \div 2 = 7 + \dfrac{26}{13} \div 2$

$= 7 + \boxed{2} \div 2 = 7 + \boxed{1} = \boxed{8}$

1.3   EXERCISES

In Exercises 1–6, evaluate the given exponential expression.

1. 2^5

2. $(-3)^4$

3. -3^4

4. $\left(\frac{1}{4}\right)^3$

5. $(1.2)^2$

6. $\frac{1}{3^4}$

In Exercises 7–24, evaluate the given expression.

7. $(1 + 2) \cdot 4$

8. $5 - (2 \cdot 7)$

9. $4 \cdot 3 - 1$

10. $4 \div 2 - 9 \cdot 2$

11. $2 \cdot (1 - 10 \div 2) + 1$

12. $-3 + 5 \cdot 2^2$

13. $\left(\frac{1}{2}\right)^2 + (2 \cdot 3)^2$

14. $\left(\frac{4}{3} + \frac{5}{3}\right) \cdot 3$

15. $2 \cdot 4^2 - 5$

16. $[29 - (14 + 8)] \cdot 2$

17. $2^2 - 4[32 \div (13 - 5)]$

18. $21 - 3^2[16 - (2^5 - 12)]$

19. $\frac{1}{2} \cdot \frac{1}{5} - \frac{7}{9}$

20. $\left(-\frac{4}{5}\right)^2 + \left[5 + \frac{2}{4}\left(\frac{1}{4} \cdot 2\right)\right]$

21. $2 + \dfrac{5^2 - 11}{3 + 4} \cdot 3^3$

22. $15 - 2 \cdot \dfrac{21}{10^2 - 93} + 6^2$

23. $97 + 4 \cdot \dfrac{14 - 3^2}{2 + 3^2}$

24. $-\dfrac{5}{6} + \dfrac{23 - 5^2}{14 + (3)(-4)} + \dfrac{5}{12}$

 In Exercises 25 and 26, use a calculator to evaluate the given expression.

25. $5.6 - 2.1(-3.7)^3$

26. $\dfrac{75.1 - 34.3}{15} - 9.2^2$

In Exercises 27–29, show why the two quantities are not equal.

27. $-2^4 \neq (-2)^4$

28. $3 \cdot (4 + 5) \neq 3 \cdot 4 + 5$

29. $17 - 3 \cdot 4 \neq 14 \cdot 4$

In Exercises 30–33, place the correct inequality symbol (> or <) between the quantities.

30. $-2^4 \ \square \ (-2)^4$

31. $\left(\dfrac{2}{3}\right)^5 \ \square \ \left(\dfrac{2}{3^5}\right)$

32. $\dfrac{5}{6} \ \square \ \left(\dfrac{5}{6}\right)^2$

33. $(-3)^4 \ \square \ 2^4$

In Exercises 34 and 35, find the total area of the figure.

34.

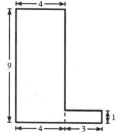

35.

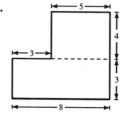

36. *Perimeter of a Rectangle*   The perimeter of a rectangle with a length of 6 inches and a width of 4 inches is given by $P = 2 \cdot 4 + 2 \cdot 6$. Calculate this value.

37. *Balance in an Account*   If $100 is invested at an annual interest rate of 12% compounded yearly, the balance in the account after 6 years is $A = 100(1.12)^6$. Calculate the amount.

1.4 Algebra and Problem Solving

Section Highlights

1. *Basic Concept:* A lot of algebra is arithmetic done with symbols that represent numbers.
2. *Properties of the Real Numbers:* Let a, b, and c be real numbers.

Commutative Property of Addition: $a + b = b + a$

Commutative Property of Multiplication: $a \cdot b = b \cdot a$

Associative Property of Addition: $(a + b) + c = a + (b + c)$

Associative Property of Multiplication: $(a \cdot b) \cdot c = a \cdot (b \cdot c)$

Identity Property of Addition: Zero is the additive identity since
$$a + 0 = 0 + a = a.$$
Identity Property of Multiplication: One is the multiplicative identity
since $a \cdot 1 = 1 \cdot a = a.$

Inverse Property of Addition: $-a$ is the additive inverse of a since
$$a + (-a) = -a + a = 0.$$
Inverse Property of Multiplication: $1/a$ is the multiplicative inverse of a
$(a \neq 0)$ since $a \cdot (1/a) = (1/a) \cdot a = 1.$

Distributive Property: $a \cdot (b + c) = a \cdot b + a \cdot c$
$$(a + b) \cdot c = a \cdot c + b \cdot c$$

EXAMPLE 1 ▪ Application of the Distributive Property

The following are uses of the Distributive Property.

(a) $2x + 5x = (2 + 5)x$ Distributive Property

       $= 7x$ Addition

(b) $(x + 2)x = x \cdot x + 2x$ Distributive Property

       $= x^2 + 2x$ $x \cdot x = x^2$

EXAMPLE 2 ▪ Identifying Properties of Real Numbers

Name the property of real numbers that justifies the given statement.

(a) $2x + (3x + 4) = (2x + 3x) + 4$ (b) $x + 7x = 1x + 7x$

(c) $(5a) \cdot (-4) = -4(5a)$ (d) $-6x + 14x = (-6 + 14)x$

Solution

(a) Associative Property of Addition

(b) Multiplicative Identity Property

(c) Commutative Property of Multiplication

(d) Distributive Property

EXAMPLE 3 ■   Using the Properties of Real Numbers

Complete the following statements using the specified property of real numbers.

(a) Associative Property of Multiplication: $-5(16a) = \boxed{}$

(b) Distributive Property: $-9x + 12x = \boxed{}$

(c) Additive Inverse Property: $0 = 6 + \boxed{}$

(d) Multiplicative Inverse Property: $\left(\frac{1}{3} \cdot 3\right)x = \boxed{}$

Solution

(a) $-5(16a) = (-5 \cdot 16)a$

(b) $-9x + 12x = (-9 + 12)x$

(c) $0 = 6 + (-6)$

(d) $\left(\frac{1}{3} \cdot 3\right)x = 1x$

■

Starter Exercise 1   *Fill in the blanks.*

Complete the following statements by using the specified property of real numbers.

(a) Associative Property of Addition: $-21y + (4y + 6) = \left(\boxed{} + \boxed{}\right) + 6$

(b) Distributive Property: $-3a + 5a = \left(\boxed{} + \boxed{}\right)a$

(c) Commutative Property of Multiplication: $(15z) \cdot 7 = 7 \cdot \boxed{}$

(d) Distributive Property: $6(2x + 8) = \boxed{} \cdot (2x) + \boxed{} \cdot \boxed{}$

■ **Solutions to Starter Exercises** ■

1. (a) $= \left(\boxed{-21y} + \boxed{4y}\right) + 6$ (b) $= \left(\boxed{-3} + \boxed{5}\right)a$

(c) $= 7 \cdot \boxed{(15z)}$ (d) $= \boxed{6} \cdot (2x) + \boxed{6} \cdot \boxed{(8)}$

1.4   EXERCISES

In Exercises 1–12, state the property of real numbers that justifies the given statement.

1. $6 + 3x = 3x + 6$

2. $1x = x$

3. $-4a + 9a = (-4 + 9)a$

4. $14y + 3 + (-3) = 14y + 0$

5. $\left(\frac{1}{5} \cdot 5\right)c = 1c$

6. $x + 3x = 1x + 3x$

7. $\frac{1}{6}(6a) = \left(\frac{1}{6} \cdot 6\right)a$

8. $5(2z + 7) = 5(2z) + 5(7)$

9. $(17m)9 = 9(17m)$

10. $-16k + \left(8k + \frac{1}{5}\right) = (-16k + 8k) + \frac{1}{5}$

11. $5x + 0 = 5x$

12. $\frac{1}{2} + (3n + 7) = (3n + 7) + \frac{1}{2}$

In Exercises 13–22, complete the statement by using the specified property of real numbers.

13. Multiplicative Identity Property:

$x = \boxed{} x$

14. Associative Property of Addition:

$-5x + (2x + 1) = (-5x + 2x) + \boxed{}$

15. Distributive Property:

$6a + (-2a) = (6 + (-2))\boxed{}$

16. Additive Inverse Property:

$9y + 4 + (-4) = 9y + \boxed{}$

17. Commutative Property of Multiplication: $(6a) \cdot 14 = 14 \cdot \boxed{}$

18. Distributive Property: $8(2x + 12) = 8 \cdot \boxed{} + 8 \cdot \boxed{}$

19. Additive Identity Property:

$6x + 0 = \boxed{}$

20. Commutative Property of Addition:

$-13z + 9 + 26z = \boxed{} + \boxed{} + 9$

21. Associative Property of Multiplication: $\frac{1}{3}(3a) = (\boxed{} \cdot \boxed{})a$

22. Multiplicative Inverse Property: $\left(\frac{1}{3} \cdot 3\right)a = \boxed{} \cdot a$

In Exercises 23–28, rewrite the following statements using the Distributive Property.

23. $2x + 4x$

24. $-6a + \frac{2}{3}a$

25. $2(x + 2)$

26. $13(z + 1)$

27. $\frac{3}{7}(5x + 19)$

28. $2x + 2y$

In Exercises 29–32, identify the property of real numbers that justifies each step of the rewriting of the expression.

29. $4x + 7x = (4 + 7)x$
$ = 11x$

30. $-3x + 2 + 5x = -3x + 5x + 2$
$ = (-3 + 5)x + 2$
$ = 2x + 2$

31. $(a + 3) + (6a + 1)$
$ = a + (3 + 6a) + 1$
$ = a + (6a + 3) + 1$
$ = (a + 6a) + (3 + 1)$
$ = (1a + 6a) + (3 + 1)$
$ = (1 + 6)a + (3 + 1)$
$ = 7a + 4$

32. $2(4z + 3) + 8z$
$ = 2(4z) + 2(3) + 8z$
$ = (2 \cdot 4)z + 2(3) + 8z$
$ = 8z + 6 + 8z$
$ = 8z + 8z + 6$
$ = (8 + 8)z + 6$
$ = 16z + 6$

Cumulative Practice Test for Chapters P–1

In Exercises 1–4, label each with *all* letters that apply from the list below.

    (a) Natural number   (b) Integer   (c) Rational number   (d) Real number

1. What kind of number is $\frac{3}{4}$?

2. What kind of number is 2?

3. What kind of number is -6?

4. What kind of number is 2.49?

In Exercises 5–16, evaluate the expression.

5. $(-2)(-5)$

6. $2 - 17$

7. $|-11 \cdot 2|$

8. $4^2 \div (-8)$

9. $\frac{2}{3} + \frac{4}{5}$

10. $\frac{9}{13} \cdot \frac{26}{27}$

11. $2(6 - 14)^2 + 3$

12. $4 - |5 - 7| + 3^3$

13. $\frac{1}{5} \cdot \left(-\left|-\frac{1}{2}\right|^2\right)$

14. $(-2.6) \cdot (1.3)$

15. $2 - 4(3^2 - 5) \div 8$

16. $\dfrac{3}{8} \div \dfrac{2^3 - 1}{4 + 2^2} - 7$

In Exercises 17–20, place the correct symbol ($<$, $>$, or $=$) between the two numbers.

17. $\left|-\frac{1}{12}\right|$ ☐ $\frac{3}{4}$

18. -11 ☐ -14

19. -2^2 ☐ $(-2)^2$

20. $|-3^4|$ ☐ 3^4

In Exercises 21–24, indicate which property of real numbers justifies the statement.

21. $-5x + 12x = (-5x + 12)x$

22. $x + 16x = 1x + 16x$

23. $-2(4z + 7) = -2(4z) + (-2)(7)$

24. $\frac{1}{3}(3x) = \left(\frac{1}{3} \cdot 3\right)x$

In Exercises 25–28, complete each statement by using the indicated property of real numbers.

25. Commutative Property of Addition:

    $16x + (-4 + 9x) = $ ☐ $+ 16x$

26. Multiplicative Inverse Property:

    $\left(\frac{1}{3} \cdot 3\right)x = $ ☐ x

27. Distributive Property:

    $16a + 5a = (16 + 5)$ ☐

28. Associative Property of Addition:

    $12n + (3n + 1) = $ ☐

29. *Balance in an Account*   If $2000 is deposited in an account at an annual interest rate of 13.25% compounded annually, the balance in the account after 4 years is $2000(1.1325)^4$. Calculate this amount.

30. *Price per Foot*   Bob put 481 feet of electrical wire in his house. He paid $134.68 for the wire. What is the price per foot of the wire?

| 1.1 | **Answers to Exercises** |

1. $\frac{1}{2}$ **2.** $\frac{5}{9}$ **3.** $\frac{3}{5}$ **4.** $\frac{58}{77}$

5. $\frac{4}{5}$ **6.** $\frac{5}{7}$ **7.** $\frac{9}{10}$ **8.** $\frac{35}{24}$

9. $\frac{35}{36}$ **10.** $-18\frac{3}{8}$ **11.** $-\frac{288}{851}$ **12.** $3\frac{1}{2}$

13. $-\frac{9}{4}$ **14.** $-9\frac{9}{16}$ **15.** $\frac{29}{40}$ **16.** $-3\frac{1}{8}$

17. $2\frac{1}{10}$ **18.** $11\frac{1}{4}$ **19.** $-\frac{239}{180}$ **20.** $5\frac{10}{11}$

21. $\frac{1343}{400}$ **22.** $6\frac{1}{20}$ **23.** $16\frac{5}{6}$ **24.** $-\frac{33}{16}$

25. $7\frac{41}{60}$ **26.** $\frac{61}{12}$ **27.** $\frac{146}{15}$ **28.** $\frac{1303}{100}$

29. 2   **30.** 25   **31.** 132

| 1.2 | **Answers to Exercises** |

1. $\frac{7}{40}$ **2.** $-\frac{1}{6}$ **3.** $-\frac{2}{9}$ **4.** $\frac{10}{57}$

5. $-\frac{3}{2}$ **6.** $-\frac{5}{4}$ **7.** $-\frac{1}{6}$ **8.** $8\frac{7}{30}$

9. $\frac{31}{42}$ **10.** $-15\frac{11}{125}$ **11.** $\frac{1}{6}$ **12.** 2

13. $\frac{8}{7}$ **14.** $\frac{5}{11}$ **15.** $\frac{4}{3}$ **16.** $-\frac{6}{5}$

17. $\frac{22}{13}$ **18.** $-\frac{1}{10}$ **19.** $\frac{15}{4}$ **20.** $\frac{7}{3}$

21. 0.5   **22.** 1.25   **23.** $0.\overline{3}$ **24.** $0.2\overline{6}$

25. 3.93   **26.** -1.26 **27.** -155.35 **28.** -25.63

29. 2.26   **30.** 46.01   **31.** -29.75 **32.** 1.78

33. -3.65 **34.** -11.98 **35.** \$5.44   **36.** \$0.34

37. False   **38.** True   **39.** True   **40.** False

| 1.3 | **Answers to Exercises** |

1. 32

2. 81

3. -81

4. $\frac{1}{64}$

5. 1.44

6. $\frac{1}{81}$

7. 12

8. -9

9. 11

10. -16

11. -7

12. 17

13. $36\frac{1}{4}$

14. 9

15. 27

16. 14

17. -12

18. 57

19. $-\frac{61}{90}$

20. $-\frac{243}{50}$

21. 56

22. 45

23. $\frac{1,087}{11}$

24. $-\frac{17}{12}$

25. 111.9713

26. -81.92

27. $-2^4 = -16$ and $(-2)^4 = 16$

28. $3(4+5) = 27$ and $3 \cdot 4 + 5 = 17$

29. $17 - 3 \cdot 4 = 5$ and $14 \cdot 4 = 56$

30. $<$

31. $>$

32. $>$

33. $>$

34. 39

35. 44

36. 20

37. $197.38

| 1.4 | **Answers to Exercises** |

1. Commutative Property of Addition

2. Multiplicative Identity Property

3. Distributive Property

4. Additive Inverse Property

5. Multiplicative Inverse Property

6. Multiplicative Identity Property

7. Associative Property of Multiplication

8. Distributive Property

9. Commutative Property of Multiplication

10. Associative Property of Addition

11. Additive Identity Property

12. Commutative Property of Addition

13. $1x$

14. $(-5x + 2x) + 1$

15. $(6 + (-2))a$

16. $9y + 0$

17. $14 \cdot (6a)$

18. $8 \cdot (2x) + 8 \cdot (12)$

19. $6x$

20. $-13z + 26z + 9$

21. $\left(\frac{1}{3} \cdot 3\right)a$

22. $1a$

23. $(2+4)x$

24. $\left(-6 + \frac{2}{3}\right)a$

25. $2x + 2 \cdot 2$

26. $13z + 13 \cdot 1$

27. $\frac{3}{7}(5x) + \frac{3}{7}(19)$

28. $2(x + y)$

29. Distributive Property
Closure Property of Addition

30. Commutative Property of Addition
Distributive Property
Closure Property of Addition

31. Associative Property of Addition
Commutative Property of Addition
Associative Property of Addition
Multiplicative Identity Property
Distributive Property
Closure Property of Addition

32. Distributive Property
Associative Property of Multiplication
Closure Property of Multiplication
Commutative Property of Addition
Distributive Property
Closure Property of Addition

Answers to Practice Test

1. c, d

2. a, b, c, d

3. b, c, d

4. c, d

5. 10

6. −15

7. 22

8. −2

9. $\frac{22}{15}$

10. $\frac{2}{3}$

11. 131

12. 29

13. $-\frac{1}{20}$

14. −3.38

15. 0

16. $-\frac{46}{7}$

17. <

18. >

19. <

20. =

21. Distributive Property

22. Multiplicative Identity Property

23. Distributive Property

24. Associative Property of Multiplication

25. $(-4 + 9x) + 16x$

26. $1x$

27. $(16 + 5)a$

28. $(12n + 3n) + 1$

29. $3,289.90

30. $0.28

CHAPTER TWO
Fundamentals of Algebra

| 2.1 | Algebraic Expressions and Exponents

Section Highlights

1. Algebra uses symbols to represent quantities whose numerical value is unknown.
2. Let m and n be positive integers, and let a and b be real numbers, or algebraic expressions. Then:

   i) $a^m \cdot a^n = a^{m+n}$

   ii) $(a^m)^n = a^{m \cdot n}$

   iii) $(ab)^m = a^m b^m$

EXAMPLE 1 ■   **Writing Products in Exponential Form**

Write the following products in exponential form.

(a) $2 \cdot y \cdot y \cdot y$ (b) $5x \cdot 5x \cdot 5x$ (c) $xxyzzzy$

Solution

(a) $2 \cdot \underbrace{y \cdot y \cdot y}_{\text{3 factors}} = 2y^3$ (b) $\underbrace{5x \cdot 5x \cdot 5x}_{\text{3 factors}} = (5x)^3$

(c) $xxyzzzy = xxyyzzz = x^2 y^2 z^3$

EXAMPLE 2 ■   **Writing Exponential Forms as Products**

Write each expression as a product.

(a) $5x^2$ $5 \cdot x \cdot x$ (b) $x^3 y^2$ $x \cdot x \cdot x \cdot y \cdot y$ (c) $(-y)^3 z^2$

Solution

(a) $5x^2 = 5 \cdot x \cdot x$ (b) $x^3 y^2 = x \cdot x \cdot x \cdot y \cdot y$

(c) $(-y)^3 z^2 = (-y)(-y)(-y)z \cdot z = -y \cdot y \cdot y \cdot z \cdot z$

EXAMPLE 3 ■   **Simplifying Products Involving Exponential Forms**

Simplify the following expressions.

(a) $x^2 \cdot x^3$ $x5$ (b) $(x^2)^5$ $x^{15 = 13}$

(c) $(2a^3)(3a^7)$ 6^{10} (d) $4xy^5 (2x^3 y^4)^2$ $8xy^{13}$

Solution

(a) $x^2 \cdot x^3 = x^{2+3} = x^5$ (b) $(x^2)^5 = x^{2 \cdot 5} = x^{10}$

x

(c) $(2a^3)(3a^7) = (2 \cdot 3)(a^3 \cdot a^7)$

$\qquad = 6a^{3+7}$

$\qquad = 6a^{10}$

(d) $4xy^5(2x^3y^4)^2 = 4xy^5 \cdot 2^2(x^3)^2(y^4)^2$

$\qquad = 4xy^5 \cdot 4x^{3 \cdot 2}y^{4 \cdot 2}$

$\qquad = 4 \cdot xy^5 \cdot 4x^6y^8$

$\qquad = 4 \cdot 4xx^6y^5y^8$

$\qquad = 16x^{1+6}y^{5+8}$

$\qquad = 16x^7y^{13}$

Starter Exercise 1 | *Fill in the blanks.*

Simplify the following expressions.

(a) $(2b^2)^3 = 2^3(b^2)^3 = 2^3b^{2 \cdot 3} = 8\boxed{6^6}$

(b) $(xy^3)(x^6y^4) = (xx^6)(y^3y^4) = x^{1+6}\boxed{x^7} = \boxed{}$

(c) $(3x^5)(4x^8)^2 = (3x^5)4^2(x^8)^2 = (3x^5)16\boxed{x} = (3 \cdot 16) \cdot \boxed{}$

$3 \cdot 3 = 9 \cdot 3 = 27x \qquad 3 \cdot 3 \cdot 3 = 27 \qquad 4 \cdot 4 = 16 \text{ C}$

$\qquad\qquad\qquad\qquad 8 \cdot 8$

EXAMPLE 4 ■ Simplifying Expressions with More Than One Term

Simplify the following expressions.

(a) $16(x^4)^5 - (4y^5)^3$

(b) $(-a^2b^5)^5 + 6(ab^3)^4$

Solution

(a) $16(x^4)^5 - (4y^5)^3 = 16(x^4)^5 - 4^3(y^5)^3$

$\qquad\qquad = 16x^{4 \cdot 5} - 4^3y^{5 \cdot 3}$

$\qquad\qquad = 16x^{20} - 64y^{15}$

(b) $(-a^2b^5)^5 + 6(ab^3)^4 = (-a^2)^5(b^5)^5 + 6a^4(b^3)^4$

$\qquad\qquad = (-1)^5(a^2)^5(b^5)^5 + 6a^4(b^3)^4$

$\qquad\qquad = -1a^{2 \cdot 5}b^{5 \cdot 5} + 6a^4b^{3 \cdot 4}$

$\qquad\qquad = -a^{10}b^{25} + 6a^4b^{12}$

Starter Exercise 2 | *Fill in the blanks.*

Simplify the following expressions.

(a) $(x^2y)^4 + 7x^3(x^2y^9) = (x^2)^4y^4 + 7(x^3x^2)y^9$

$x^8y^4 + 7x^3 \text{C}$

$\qquad\qquad = \boxed{x^8}y^4 + 7\boxed{x^6}y^9$

$\qquad\qquad = \boxed{xy}$

(b) $(4u^5v^2)^2 - 5a^5(6u^4v) = 4^2(u^5)^2(v^2)^2 - 5 \cdot 6u^5u^4v$

$\qquad\qquad = 4^2\boxed{16} \cdot \boxed{} - 5 \cdot 6\boxed{}v$

$\qquad\qquad = 16\boxed{}\boxed{} - 30\boxed{}\boxed{}$

■   **Solutions to Starter Exercises**   ■

1. (a) $8\boxed{b^6}$

(b) $x^{1+6}\boxed{y^{3+4}} = \boxed{x^7y^7}$

(c) $(3x^5)16\boxed{x^{8\cdot2}} = (3\cdot16)\cdot\boxed{x^5x^{16}} = 48x^{5+16} = 48x^{21}$

2. (a) $\boxed{x^{2\cdot4}}y^4 + 7\boxed{x^{3+2}}y^9 = \boxed{x^8y^4 + 7x^5y^9}$

(b) $4^2\boxed{u^{5\cdot2}}\cdot\boxed{v^{2\cdot2}} - 5\cdot6\boxed{u^{5+4}}v = 16\boxed{u^{10}}\boxed{v^4} - 30\boxed{u^7}\boxed{v}$

2.1 | EXERCISES

In Exercises 1–6, rewrite the given product in exponential form.

1. $-2x\cdot x\cdot y\cdot y\cdot y$ $-2x^2y^3$

2. $x\cdot y\cdot x\cdot y\cdot x$ $x^3\cdot y^2$

3. $6\cdot u\cdot u\cdot 6\cdot v\cdot 6$ 6^3u^2v

4. $a\cdot a\cdot a\cdot a\cdot a\cdot a\cdot b\cdot b$ a^6b^2

5. $(x-y)(x-y)(x-y)$ $(x-y)^3$

6. $2\cdot(2x+1)\cdot(2x+1)\cdot2$

In Exercises 7–12, rewrite the given exponential form as a product.

7. 2^3u^2 $2\cdot2\cdot2\cdot u\cdot u$

8. $(xy)^4$ $xy\cdot xy\cdot xy\cdot xy$

9. $3ab^2$

10. $7^2x^3y^2$ $7\cdot7\cdot x\cdot x\cdot x\cdot y\cdot y$

11. $(x+3)^2$ $x^{3\cdot2}$

12. $6(2a-1)^3$

In Exercises 13–27, simplify the following expressions.

13. $x^2\cdot x^5$ x^{10}

14. $(a^2)^3$ a^6

15. $-17x^2(x^6)$

16. $u(uv)^3$ u^4v^3

17. $-5a^2(4a^3b)^2$

18. $(3s^2t^3)(5s)^2$

19. $(3x^2y^6)^3(4x^2)(2xy)^5$

20. $(u^7v^6)^5(-2uv^3)^4$

21. $(-4a^2b^3)^2(-2ab^7)^3a^2b$

22. $(x-y)(x-y)$

23. $(2x+3)^2(2x+3)^5$

24. $2(a+b)^5(a+b)$

25. $x^2(xy^3) - (uv)^2$

26. $(-2x^5y^6)^3 + 2x(x^2y)^4$

27. $3a(a^2)^3b^3 + (ab^3)^6(2a)^3$

In Exercises 28 and 29, perform the indicated operations.

28. $4\cdot10^3 + 2\cdot10^2 + 9\cdot10^1$

29. $7\cdot10^5 + 6\cdot10^2 + 5\cdot10^1$

30. What power of 10 is 1,000?

31. What power of 10 is 100,000?

32. *Balance in an Account*   The balance in an account that has an annual interest rate r, compounded yearly for four years, is given by $P(1 + r)(1 + r)(1 + r)(1 + r)$. Write the expression in exponential form.

33. *Volume of a Box*   Find an exponential expression that represents the volume of a box that has length x, width x, and height y.

2.2 Basic Rules of Algebra

Section Highlights

1. The properties of real numbers that were discussed in Section 1.5 can be used to rewrite algebraic expressions.

2. In a variable expression, two terms are said to be *like terms* if they have identical variable parts (two constant terms are considered to have the same variable part).

EXAMPLE 1 ■ Identifying the Basic Rules of Algebra

Identify the rule of algebra illustrated in each of the following.

(a) $\frac{1}{2}(2x^2) = \left(\frac{1}{2} \cdot 2\right) x^2$

(b) $-3x^2 + 2 + 5x^2 = -3x^2 + 5x^2 + 2$

(c) $13y + (6y + 12) = (13y + 6y) + 12$

(d) $(x - 1)(x + 5) = (x - 1)x + (x - 1)5$

Solution

(a) $\frac{1}{2}(2x^2) = \left(\frac{1}{2} \cdot 2\right) x^2$ is an illustration of the Associative Property of Multiplication.

(b) $-3x^2 + 2 + 5x^2 = -3x^2 + 5x^2 + 2$ is an illustration of the Commutative Property of Addition.

(c) $13y + (6y + 12) = (13y + 6y) + 12$ is an illustration of the Associative Property of Addition. Note that the order of the terms did not change, just the grouping symbols.

(d) $(x - 1)(x + 5) = (x - 1)x + (x - 1)5$ is an illustration of the Distributive Property, with $a = x - 1, b = x$, and $c = 5$ in $a \cdot (b + c) = a \cdot b + a \cdot c$.

EXAMPLE 2 ■ Applying the Basic Rules of Algebra

Use the given rule to complete the following statement.

(a) Distributive Property: $5x^2 + 7x^2 = \left(\boxed{} + \boxed{}\right) x^2$

(b) Commutative Property of Multiplication: $(6x)3 = 3 \cdot \left(\boxed{}\right)$

(c) Multiplicative Identity Property: $x + 8x = \boxed{} x + 8x$

Solution

(a) $5x^2 + 7x^2 = (5 + 7)x^2$

(b) $(6x)3 = 3(6x)$

(c) $x + 8x = 1x + 8x$

EXAMPLE 3 ■ Using the Distributive Property

Each of the following is a use of the Distributive Property.

(a) $-3(2x + 5) = (-3)(2x) + (-3)5$

(b) $6y^2 + 15y^2 = (6 + 15)y^2$

(c) $(x + y)(4x + y) = (x + y)(4x) + (x + y)y$

(d) $2(a + 1) - a(a + 1) = (2 - a)(a + 1)$ ■

Starter Exercise 1 | *Fill in the blanks.*

Use the Distributive Property to complete the following.

(a) $(3x + 7)(2) = (3x)\boxed{} + 7\boxed{}$

(b) $-15y + 6y = (\boxed{} + \boxed{})y$

(c) $(4x + 3)(9x - 1) = 4x\boxed{} + 3\boxed{}$

(d) $a(a^2 + 1) + 9(a^2 + 1) = (\boxed{} + \boxed{})(a^2 + 1)$

EXAMPLE 4 ■ Using the Distributive Property to Add Like Terms

Simplify the following expressions.

(a) $-3x + 5x$ (b) $6x + 2x + 1$ (c) $17 - 4a^2 - 8a^2$

Solution

(a) $-3x + 5x = (-3 + 5)x$ Distributive Property

$\qquad\qquad = 2x$ Simplest form

(b) $6x + 2x + 1 = (6 + 2)x + 1$ Distributive Property

$\qquad\qquad\quad = 8x + 1$ Simplest form

(c) $17 - 4a^2 - 8a^2 = 17 + (-4a^2) + (-8a^2)$ Change subtraction to addition.

$\qquad\qquad\qquad = 17 + (-4 + (-8))a^2$ Distributive Property

$\qquad\qquad\qquad = 17 + (-12)a^2$

$\qquad\qquad\qquad = 17 - 12a^2$ Simplest form ■

Starter Exercise 2 | *Fill in the blanks.*

Simplify the following expressions.

(a) $11y^2 - 6y = 11y^2 + (-6y^2)$

$\qquad\qquad = \left(\boxed{} + \boxed{}\right)y^2$

$\qquad\qquad = \boxed{}\,y^2$

(b) $7x + 3x + 10 = (7 + 3)\boxed{} + 10$

$\qquad\qquad = \boxed{} + 10$

EXAMPLE 5 ■   **Using Rules of Algebra to Add Like Terms**

Simplify the following expressions.

(a) $4z + 17 + 6z$ (b) $-5x + y + 3x - 2y$ (c) $14y + x + 3y - 8y$

Solution

(a) $4z + 17 + 6z = 4z + 6z + 17$ Commutative Property

$= (4 + 6)z + 17$ Distributive Property

$= 10z + 17$ Simplest form

(b) $-5x + y + 3x - 2y = -5x + y + 3x + (-2y)$ Change to addition.

$= -5x + 3x + y + (-2y)$ Commutative Property

$= -5x + 3x + 1y + (-2y)$ Identity Property

$= (-5 + 3)x + (1 + (-2))y$ Distributive Property

$= -2x + (-1)y$

$= -2x - y$ Simplest form

(c) $14y + x + 3y - 8y = x + 14y + 3y - 8y$ Commutative Property

$= x + (14 + 3 - 8)y$ Distributive Property

$= x + 9y$ Simplest form

Starter Exercise 3   *Fill in the blanks.*

Simplify the following expressions.

(a) $2x + 4y + 5x + 7y = 2x + 5x + 4y + 7y$

$= (\boxed{2} + \boxed{5})x + (4 + 7)\boxed{y}$

$= \boxed{7}x + 11\boxed{x}$

(b) $(-6z + 4u) + (2u - 8z) = -6z + (4u + 2u) + (-8z)$

$= -6z + \boxed{8z} + (4u + 2u)$

$= (-6 + \boxed{8})z + (\boxed{4} + \boxed{2})u$

$= \boxed{14}z + \boxed{6}u$

■ **Solutions to Starter Exercises** ■

1. (a) $= (3x)\boxed{(2)} + 7\boxed{(2)}$ (b) $= (\boxed{-15} + \boxed{6})y$

(c) $= 4x\boxed{(9x-1)} + 3\boxed{(9x-1)}$ (d) $= (\boxed{a} + \boxed{9})(a^2 + 1)$

2. (a) $= (\boxed{11} + \boxed{(-6)})y^2 = \boxed{5}\,y^2$ (b) $= (7+3)\boxed{x} + 10 = \boxed{10x} + 10$

3. (a) $= (\boxed{2+5})x + (4+7)\boxed{y} = \boxed{7}\,x + 11\boxed{y}$

(b) $= -6z + \boxed{(-8z)} + (4u + 2u)$

$= (-6 + \boxed{(-8)})z + (\boxed{4} + \boxed{2})u = \boxed{-14}\,z + \boxed{6}\,u$

2.2 EXERCISES

In Exercises 1–4, identify the rule of algebra illustrated by the given equation.

1. $-3x + 1 + 2x = -3x + 2x + 1$

2. $(5y + 7y) = (5 + 7)y$

3. $-4(13a) = (-4 \cdot 13)a$

4. $b + 0 = b$

In Exercises 5–8, use the given property to complete the given statement.

5. Multiplicative Identity Property: $y^2 = \boxed{}\,y^2$

6. Multiplicative Inverse Property: $(\frac{1}{2} \cdot 2)x = \boxed{1}\,x$

7. Distributive Property: $3(x + y) = \boxed{}\,x + \boxed{}\,y$

8. Commutative Property of Multiplication: $(9x)(-3) = -3\boxed{}$

In Exercises 9–14, rewrite the given expression using the Distributive Property.

9. $5(x + y)$

10. $-3y + 8y$

11. $4x^2 + 9x^2$

12. $2(3x + y + z)$

13. $(6a + 2b)(-5)$

14. $(2x + 3)(x + 4)$

In Exercises 15–24, simplify the given expression by adding like terms.

15. $15x + 7x$ $= 22x$

16. $-3y^2 + 9y^2$ $= 6y^2$

17. $5a - 2a + 1$ $3a + 1$

18. $11z - 16z - 9$ $= -5z - 9$

19. $12x^2 + 2x^2 + 3y + 6y$

$14x^2 + 9y$

20. $7a - 13a + 8b + 4b$

$-6a + 12b$

21. $\frac{1}{2}z + 6a - \frac{3}{4}z + 7a$

22. $\frac{4}{5}b - \frac{4}{5}c + \frac{5}{4}c + 2b^3 + \frac{4}{5}b$

23. $4xy + 6z - 3xy - 4z + xy$

24. $7a + 3b - c - 2a + 5b + c$

In Exercises 25–30, use the Basic Rules of Algebra to determine whether the following statements are true or false.

25. $2(2x + 1) = 2(2x) + 1$

26. $x(yz) = (xy)(xz)$

27. $5x + (-6 + 6) = 5x$

28. $2x + (3y + 9) = (2x + 3y) + (2x + 9)$

29. $\left(\frac{1}{3} \cdot 3\right)x = x$

30. $8(2x + 7y) = 8(2x) + 8(7y)$

In Exercises 31–36, use the Distributive Property to perform the arithmetic **mentally**.

31. $20(1.5) = 2(1 + 0.5)$

32. $50(0.15) = 50(0.1 + 0.05)$

33. $12(14) = 12(12 + 2)$

34. $8(48) = 8(50 - 2)$

35. $4(14.95) = 4(15 - 0.05)$

36. $6(38.5) = 6(40 - 1 - 0.5)$

2.3 Rewriting and Evaluating Algebraic Expressions

Section Highlights

1. We use the multiplication and exponent properties, along with the Distributive Property, to remove symbols of grouping when simplifying algebraic expressions.

2. To evaluate a variable expression means to find its value when the variables are replaced by given real numbers.

EXAMPLE 1 ■ Removing Symbols of Grouping

Simplify the following expressions.

(a) $2(2x + 4)$

(b) $-5(y - 6) - (3y - 5)$

Solution

(a) $2(2x + 4) = 2(2x) + 2(4)$ Distributive Property

$= (2 \cdot 2)x + 2(4)$ Associative Property

$= 4x + 8$ Simplest form

(b) $-5(y-6) \ -(3y-5)$

$$= -5(y+(-6)) + (-1)(3y+(-5))$$ Change to addition and Identity Property.

$$= -5y + (-5)(-6) + (-1)(3y) + (-1)(-5)$$ Distributive Property

$$= -5y + (-5)(-6) + (-1 \cdot 3)y + (-1)(-5)$$ Associative Property

$$= -5y + 30 + (-3y) + 5$$

$$= -5y + (-3y) + 30 + 5$$ Commutative Property

$$= (-5 + (-3))y + 30 + 5$$ Distributive Property

$$= -8y + 35$$ Simplest form

Starter Exercise 1   *Fill in the blanks.*

Simplify the following expressions.

(a) $-4(z+6) = \boxed{} + \boxed{} 6 = \boxed{} + \boxed{} = \boxed{} - \boxed{}$

(b) $2(7x+1) - 3(x-2) = 2(7x+1) + (-3)(x+(-2))$

$$= \boxed{}(7x) + \boxed{}1 + (-3x) + \boxed{}(-2)$$

$$= (\boxed{} \cdot \boxed{})x + \boxed{}1 + (-3x) + \boxed{}(-2)$$

$$= \boxed{}x + 2 + (-3x) + \boxed{}$$

$$= \boxed{}x + (-3x) + 2 + \boxed{}$$

$$= (\boxed{} + (-3))\boxed{} + 2 + \boxed{}$$

$$= \boxed{}x + \boxed{}$$

EXAMPLE 2 ■   Removing Nested Symbols of Grouping

Simplify the following expressions.

(a) $4[x + 2(3x+6)]$

(b) $-8x - [2x + 4(x-1)]$

Solution

(a) $4[x + 2(3x+6)] = 4[x + 2(3x) + 2(6)]$ Distributive Property

$$= 4[x + (2 \cdot 3)x + 2(6)]$$ Associative Property

$$= 4[1x + 6x + 12]$$ Identity Property

$$= 4[(1+6)x + 12]$$ Distributive Property

$$= 4[7x + 12]$$

$$= 4(7x) + 4(12)$$ Distributive Property

$$= (4 \cdot 7)x + 4(12)$$ Associative Property

$$= 28x + 48$$

(b) $-8x - [2x + 4(x - 1)] = -8x + (-1)[2x + 4(x + (-1)]$ Change to addition and Identity Property.

$$= -8x + (-1)[2x + 4x + (-4)]$$ Distributive Property

$$= -8x + (-1)[(2x + 4)x + (-4)]$$ Distributive Property

$$= -8x + (-1)[6x + (-4)]$$

$$= -8x + (-1)(6x) + (-1)(-4)$$ Distributive Property

$$= -8x + (-1 \cdot 6)x + (-1)(-4)$$ Associative Property

$$= -8x + (-6)x + 4$$

$$= [-8 + (-6)]x + 4$$ Distributive Property

$$= -14x + 4$$

Starter Exercise 2   *Fill in the blanks.*

Simplify the following expressions.

(a) $3[x - 2(5x + 7)] = 3[x + (-2)(5x + 7)]$

$$= 3[x + (-2)\boxed{5x} + (-2)\boxed{7}]$$

$$= 3[x + (-2 \cdot \boxed{5})\boxed{x} + (-2)\boxed{7}]$$

$$= 3[1x + \boxed{-10}x + \boxed{-14}]$$

$$= 3[(1 + \boxed{-10})x + \boxed{14}]$$

$$= 3[\boxed{-9}x + \boxed{-14}]$$

$$= \boxed{3}(\boxed{-9}x) + \boxed{3} \cdot \boxed{-14}$$

$$= (\boxed{3} \cdot \boxed{9}) + \boxed{3} \cdot \boxed{-14}$$

$$= \boxed{} \cdot -27x - 42$$

(b) $-2[4(y + 6)] - 6y + 14 = -2[4(y + 6)] + (-6y) + 14$

$$= -2[4y + 4 \cdot \boxed{}] + (-6y) + 14$$

$$= -2[4y + \boxed{}] + (-6y) + 14$$

$$= -2(\boxed{}) + \boxed{} \cdot \boxed{} + (-6y) + 14$$

$$= (-2 \cdot \boxed{})y + \boxed{} \cdot \boxed{} + (-6y) + 14$$

$$= \boxed{}$$

$$\vdots$$

EXAMPLE 3 ■   Simplify Algebraic Expressions

Simplify the following expressions.

(a) $(-3z^2)z^4 + 17z^6$ (b) $2x(x + y) - 5(x^2 + y)$ (c) $\dfrac{3x}{5} + \dfrac{x}{3}$

Solution

(a) $(-3z^2)z^4 + 17z^6 = -3(z^2 \cdot z^4) + 17z^6$ Associative Property

$\qquad\qquad\qquad\quad = -3z^{2+4} + 17z^6$ Exponent Property

$\qquad\qquad\qquad\quad = -3z^6 + 17z^6$

$\qquad\qquad\qquad\quad = (-3 + 17)z^6$ Distributive Property

$\qquad\qquad\qquad\quad = 14z^6$

(b) $2x(x + y) - 5(x^2 + y) = 2x(x + y) + (-5)(x^2 + y)$

$\qquad\qquad\qquad\qquad\quad = 2x(x) + 2xy + (-5)x^2 + (-5)y$ Distributive Property

$\qquad\qquad\qquad\qquad\quad = 3x^{1+1} + 2xy + (-5x^2) + (-5y)$ Exponent Property

$\qquad\qquad\qquad\qquad\quad = 3x^2 + 2xy + (-5x^2) + (-5y)$

$\qquad\qquad\qquad\qquad\quad = 3x^2 + (-5x^2) + 2xy + (-5y)$ Commutative Property

$\qquad\qquad\qquad\qquad\quad = (3 + (-5))x^2 + 2xy + (-5y)$ Distributive Property

$\qquad\qquad\qquad\qquad\quad = -2x^2 + 2xy - 5y$ Simplest form

(c) $\dfrac{3x}{5} + \dfrac{x}{3} = \dfrac{3}{5}x + \dfrac{1}{3}x$

$\qquad\qquad = \left(\dfrac{3}{5} + \dfrac{1}{3}\right)x$ Distributive Property

$\qquad\qquad = \left(\dfrac{3}{5} \cdot \dfrac{3}{3} + \dfrac{1}{3} \cdot \dfrac{5}{5}\right)x$ Common denominator

$\qquad\qquad = \left(\dfrac{9}{15} + \dfrac{5}{15}\right)x = \dfrac{14}{15}x$

■

Starter Exercise 3   *Fill in the blanks.*

Simplify the following expressions.

(a) $(5x)(-2x^3) + (4x^2)^2 = 5(x \cdot (-2))x^3 + (4x^2)^2$

$\qquad\qquad\qquad\qquad\quad = 5(-2 \cdot x)x^3 + (4x^2)^2$

$\qquad\qquad\qquad\qquad\quad = (5 \cdot \boxed{})(x \cdot x^3) + (4x^2)^2$

$\qquad\qquad\qquad\qquad\quad = -10x^{\boxed{}+\boxed{}} + 4^2(x^2)^{\boxed{}}$

$\qquad\qquad\qquad\qquad\quad = -10x^{\boxed{}} + 16x^{2\cdot\boxed{}}$

$\qquad\qquad\qquad\qquad\quad = -10x^{\boxed{}} + 16x^{\boxed{}}$

$\qquad\qquad\qquad\qquad\quad = (-10 + 16)\boxed{} = \boxed{}$

(b) $\dfrac{5x}{9} + \dfrac{2x}{3} = \dfrac{5}{9}x + \dfrac{2}{3}x = (\boxed{} + \boxed{})x = \left(\boxed{} + \boxed{} \cdot \dfrac{3}{3}\right)x$

$\qquad\qquad = (\boxed{} + \boxed{})x = \boxed{}x$

EXAMPLE 4 ■   **Evaluating Algebraic Expressions**

Evaluate the following expressions when $x = -2$, $y = 5$, and $z = -1$.

(a) $5x + 3y$ (b) $4y^2 - 2xy + z$ (c) $|z^2 - y|$

(d) $\dfrac{x - 2z}{xy - z^2}$ (e) $(y - z)(2x + y)$

Solution

(a) When $x = -2$ and $y = 5$, the value of $5x + 3y$ is

$$5(-2) + 3(5) = -10 + 15 = 5.$$

(b) When $x = -2$, $y = 5$ and $z = -1$, the value of $4y^2 - 2xy + z$ is

$$4(5)^2 - 2(-2)(5) + (-1) = 4(25) - 2(-2)(5) + (-1)$$
$$= 100 - (-20) + (-1) = 119.$$

(c) When $y = 5$ and $z = -1$, the value of $|z^2 - y|$ is

$$|(-1)^2 - 5| = |1 - 5| = |-4| = 4.$$

(d) When $x = -2$, $y = 5$, and $z = -1$, the value of $(x - 2z)/(xy - z^2)$ is

$$\frac{-2 - 2(-1)}{-2(5) - (-1)^2} = \frac{-2 - (-2)}{-2(5) - (-1)^2} = \frac{0}{-11} = 0.$$

(e) When $x = -2$, $y = 5$, and $z = -1$, the value of $(y - z)(2x + y)$ is

$$(5 - (-1))(2(-2) - 5) = (6)(-9) = -54.$$

■

■ **Starter Exercise 4** | *Fill in the blanks.*

Evaluate the following expressions when $x = -2$ and $y = 0$.

(a) $6x^2 - 2y = 6(-2)^2 - 2(0) = 6 \cdot \boxed{} - \boxed{} = \boxed{}$

(b) $\dfrac{x^2 - 5xy}{2x - y} = \dfrac{(\boxed{})^2 - 5(-2)(0)}{2(\boxed{}) - \boxed{}} = \dfrac{\boxed{} - 0}{-4 - \boxed{}} = \cdots$

(c) $(3x + y)(x - 7y) = (3(\boxed{}) + \boxed{})(\boxed{} - 7 \cdot \boxed{})$

$$= (\boxed{} + \boxed{})(\boxed{} - \boxed{}) = \cdots$$

EXAMPLE 5 ■ **Evaluating Algebraic Expressions**

Evaluate $-6x + 3[x - 2(2x - 1)]$ when (a) $x = 2$, (b) $x = -1$, and (c) $x = 0$.

Solution

It may make our work shorter to simplify the algebraic expression before evaluating it.

$$
\begin{aligned}
-6x + 3[x - 2(2x - 1)] &= -6x + 3[x + (-2)(2x + (-1))] \\
&= -6x + 3[x + (-2)(2x) + (-2)(-1)] \quad \text{Distributive Property} \\
&= -6x + 3[x + (-2 \cdot 2)x + (-2)(-1)] \quad \text{Associative Property} \\
&= -6x + 3[1x + (-4x) + 2] \\
&= -6x + 3[(1 + (-4))x + 2] \quad \text{Distributive Property} \\
&= -6x + 3[-3x + 2] \\
&= -6x + 3(-3x) + 3(2) \quad \text{Distributive Property} \\
&= -6x + (3 \cdot (-3))x + 3(2) \quad \text{Associative Property} \\
&= -6x + (-9x) + 6 \\
&= (-6 + (-9))x + 6 \quad \text{Distributive Property} \\
&= -15x + 6
\end{aligned}
$$

(a) $-15x + 6 = -15(2) + 6 = -30 + 6 = -24$

(b) $-15x + 6 = -15(-1) + 6 = 15 + 6 = 21$

(c) $-15x + 6 = -15(0) + 6 = 0 + 6 = 6$

■ **Solutions to Starter Exercises** ■

1. (a) $= \boxed{-4z} + \boxed{(-4)} \, 6 = \boxed{-4z} + \boxed{(-24)} = \boxed{-4z} - \boxed{24}$

(b) $= \boxed{2} (7x) + \boxed{2} \cdot 1 + (-3x) + \boxed{(-3)} (-2)$

$= (\boxed{2} \cdot \boxed{7})x + \boxed{2} \cdot 1 + (-3x) + \boxed{(-3)} (-2)$

$= \boxed{14} x + 2 + (-3x) + \boxed{6} = \boxed{14} x + (-3x) + 2 + \boxed{6}$

$= (\boxed{14} + (-3)) \boxed{x} + 2 + \boxed{6} = \boxed{11} x + \boxed{8}$

■   **Solutions to Starter Exercises**   ■

2. (a) $= 3[x + (-2) \boxed{(5x)} + (-2) \boxed{(7)}] = 3[x + (-2 \cdot \boxed{5}) \boxed{x} + (-2) \boxed{(7)}]$

$= 3[1x + (\boxed{-10} x) + \boxed{(-14)}] = 3[(1 + \boxed{(-10)})x + \boxed{(-14)}]$

$= 3[\boxed{-9} x + \boxed{(-14)}] = \boxed{3}(\boxed{-9} x) + \boxed{3} \cdot \boxed{(-14)}$

$= (\boxed{3} \cdot (\boxed{-9}))x + \boxed{3} \cdot \boxed{(-14)} = \boxed{-27x - 42}$

(b) $= -2[4y + 4 \cdot \boxed{6}] + (-6y) + 14 = -2[4y + \boxed{24}] + (-6y) + 14$

$= -2(\boxed{4y}) + \boxed{(-2)} \boxed{(24)} + (-6y) + 14.$

$= (-2 \cdot \boxed{4})y + \boxed{(-2)} \boxed{(24)} + (-6y) + 14$

$= -8y + (-48) + (-6y) + 14 = -8y + (-6y) + (-48) + 14$

$= [-8 + (-6)]y + (-48) + 14 = \boxed{-14y - 34}$

3. (a) $= (5 \cdot \boxed{(-2)})(x \cdot x^3) + (4x^2)^2 = -10x^{\boxed{1+3}} + 4^2(x^2)^{\boxed{2}}$

$= -10x^{\boxed{4}} + 16x^{2 \cdot \boxed{2}} = -10x^{\boxed{4}} + 16x^{\boxed{4}}$

$= (-10 + 16)\boxed{x^4} = \boxed{6x^4}$

(b) $= \left(\boxed{\dfrac{5}{9}} + \boxed{\dfrac{2}{3}}\right) x = \left(\boxed{\dfrac{5}{9}} + \boxed{\dfrac{2}{3}} \cdot \dfrac{3}{3}\right) x$

$= \left(\boxed{\dfrac{5}{9}} + \boxed{\dfrac{6}{9}}\right) x = \boxed{\dfrac{11}{9}} x$

4. (a) $= 6 \cdot \boxed{4} - \boxed{0} = 24$

(b) $= \dfrac{(\boxed{-2})^2 - 5(-2)(0)}{2(\boxed{-2}) - \boxed{0}} = \dfrac{\boxed{4} - 0}{-4 - \boxed{0}} = \dfrac{4}{-4} = -1$

(c) $= (3(\boxed{-2}) + \boxed{0})(\boxed{-2} - 7 \cdot \boxed{(0)})$

$= (\boxed{-6} + \boxed{0})(\boxed{-2} - \boxed{0}) = (-6)(-2) = 12$

| 2.3 | **EXERCISES** |

In Exercises 1–20, simplify the variable expression.

1. $(2x)(-3x^2)$

2. $-4(-5y)^2$

3. $\left(\frac{1}{2}x^2\right)^3$

4. $(16x^2y^5)(2xy^3)$

5. $(2xy^2)^2(5x^3y^7)$

6. $4(x-2)+3x$

7. $-6y+2(y+1)$

8. $5(z+7)-2(3z-1)$

9. $8(r-6)-(2r-7)$

10. $\dfrac{2x}{5}+\dfrac{3x}{5}$

11. $\dfrac{x}{9}-\dfrac{7x}{9}$

12. $\dfrac{z}{2}+\dfrac{2z}{3}-\dfrac{3z}{5}$

13. $x-\frac{1}{2}(3x+1)$

14. $7y^2+y(5y+2)$

15. $(5x^2)(-2x)-(2x)^3$

16. $(-z^2y)^4+(7z)(z^3y)-16z^8y^4$

17. $-3x-[5x+2(x-1)]$

18. $2[-4y-(5y-7)]+16y-1$

19. $-z[12-2(z+4)]$

20. $13r+r[2r-5(3r-7)]$

In Exercises 21–28, evaluate the expressions when $x=2$, $y=0$, $z=-3$.

21. $|-z|$

22. $2x-y$

23. $\frac{9}{5}y+xz$

24. $14y^2-x^2z$

25. $4(x-5)+7y$

26. $(x-2z)(x^2+y^{10})$

27. $-6z(x-y)+11z$

28. $\dfrac{x-2y}{4x+yz}+z^2$

In Exercises 29–35, evaluate the expression for the specified values of the variables.

29. *Perimeter of a Rectangle:*
$2w+2l$ when $w=5, l=8$

30. *Area of a Rectangle:*
lw when $l=6, w=3$

31. *Area of a Triangle:*
$\frac{1}{2}bh$ when $b=12, h=9$

32. *Simple Interest:*
$P(1+r)$ when $P=\$1000, r=0.11$

33. *Area of a Trapezoid:* $\frac{1}{2}(b_1+b_2)h$ when $b_1=10$, $b_2=8, h=7$

34. *Heat Flow:* $\dfrac{KA(t_2-t_1)}{L}$ when $K=3$, $A=4$, $L=3$, $t_1=1$, $t_2=4$

35. *Freely-Falling Body:* $V_0t+\frac{1}{2}at^2$ when $V_0=0$, $t=3$, and $a=-32$.

2.4   Translating Expressions: Verbal to Algebraic

Section Highlights

1. Words or phrases that translate into *addition:* sum, plus, add, greater, increased by, more than, exceeds, total of.

2. Words or phrases that translate into *subtraction:* difference, minus, less, decreased by, subtract, reduced by, the remainder.

3. Words or phrases that translate into *multiplication:* product, multiplied by, twice, times, percent of.

4. Words or phrases that translate into *division:* quotient, divided by, ratio of, per.

5. If n is an integer, then $n + 1$, $n + 2$, $n + 3$, ... are the next consecutive integers.

6. If n is an even integer, then $n + 2$, $n + 4$, $n + 6$, ... are the next consecutive even integers.

7. If n is an odd integer, then $n + 2$, $n + 4$, $n + 6$, ... are the next consecutive odd integers.

8. If, for example, two numbers have a sum of 23 and one number is x, then the other number is $23 - x$.

EXAMPLE 1 ■   **Translating Verbal Phrases into Variable Expressions With a Specified Variable**

Translate the following into variable expressions.

(a) 6 less than x.

(b) The difference between 6 and x.

(c) The total of y and the square of x.

(d) Twice the sum of n and m.

(e) The quotient of x less than 13 and y.

(f) $\frac{1}{2}$ of the quantity m decreased by 12.

Solution

(a) $x - 6$;   Similarly, 6 less than 10 would be $10 - 6 = 4$.

(b) $6 - x$;   "The difference between" means subtract in the order given.

(c) $y + x^2$;   "Total of" means add.

(d) $2(n + m)$;   The parentheses are necessary since it says "twice the sum."

(e) $\dfrac{13 - x}{y}$;   Quotient means divide.

(f) $\frac{1}{2}(m - 12)$;   "The quantity" implies parentheses.

| **Starter Exercise 1** | *Fill in the blanks.* |

Translate the following into variable expressions.

(a) The difference between x and 12: $x \boxed{}$ 12

(b) Twice the sum of 5 and m: $2(\boxed{})$

(c) The ratio of x and x less than 12: $\dfrac{x}{\boxed{} - \boxed{}}$

(d) The product of 4 and the square of x, increased by 3: $\boxed{} + 3$

EXAMPLE 2 ■   Translating Verbal Phrases into Variable Expressions

Translate the following into variable expressions.

(a) A number increased by 4.

(b) The difference between a number and twice the number.

(c) The quotient of a number less than 3 and 5.

(d) The total of 6 times the square of a number and the number.

Solution

The first step is to identify the unknown and assign a variable to it. In all four of these, the unknown is "a number." We will let x be "a number."

(a) $x + 4$; "increased by" means add.

(b) $x - 2x$; Note the use of "a" and "the."

(c) $\dfrac{3 - x}{5}$; "and" separates the two quantities that are divided.

(d) $6x^2 + x$; "total of" means add.   ■

| **Starter Exercise 2** | *Fill in the blanks.* |

Translate the following into variable expressions. In each case identify the unknown and assign a variable to it.

(a) The total of 5 and a number. Let $x =$ "a number."

$\qquad 5 + \boxed{}$

(b) The difference between a number and the product of 6 and the number. Let $x =$ "a number."

$\qquad x \boxed{}$

(c) 8 increased by the ratio of 2 and a number. Let $x =$ "a number."

$\qquad \boxed{} + \boxed{}$

EXAMPLE 3 ■   **Translating Variable Expressions into Verbal Phrases**

Translate the following into verbal phrases.

(a) $y - 4$ (b) $4 - y$ (c) $6x + 1$ (d) $\dfrac{x + 2}{x}$

Solution

Note: There are many different correct answers for each one of these.

(a) The difference between y and 4.   (b) y less than 4.

(c) The product of 6 and x, increased by 1.   (d) The ratio of x increased by 2 and x.

EXAMPLE 4 ■   **Discovering Hidden Operations**

(a) A jar contains x nickels. Write a variable expression for the amount of money in the jar.

(b) A person worked x hours and gets paid $5.00 per hour. Write a variable expression for the amount of money earned by the person.

(c) A bookstore has a book on sale for 20% off the regular price p. Write a variable expression for the sale price of the book.

(d) A farmer has 100 boxes of fruit. He sells some of the boxes for $32 per box and some for $37 per box. Write a variable expression for the amount of money the farmer was paid for the fruit.

Solution

(a) *Verbal model:* (value of a nickel) · (number of nickels)
    Verbal expression: $0.05x$

(b) *Verbal model:* (pay per hour) · (number of hours)
    Verbal expression: $5.00x$

(c) *Verbal model:* (regular price) − (discount rate) · (regular price)
    Variable expression: $p - 0.20p$

(d) *Verbal model:*

        ($32) · (number of boxes sold at $32) + ($37) · (number of boxes sold at $37).

        If x boxes were sold at $32, then $100 - x$ boxes were sold at $37.

    Variable expression: $32x + 37(100 - x)$

| **Starter Exercise 3** | *Fill in the blanks.* |

Discover the hidden products.

(a) A wallet contains x dollars. Write a variable expression for the amount of money in the wallet.

        $1.00 ☐

(b) There are 15 coins in a box. If there are only dimes and quarters in the box, write a variable expression for the amount of money in the box. Let d = the number of dimes.

        Amount of money in the box = $0.1d + 0.25$ ☐

EXAMPLE 5 ■   Consecutive Integers

Translate the following into variable expressions.

(a) The sum of three consecutive integers.

(b) The product of two consecutive even integers.

(c) The ratio of two consecutive odd integers.

Solution

(a) Let n = an unknown integer. $n + 1$ and $n + 2$ will be the next two consecutive integers.

   Variable expression: $n + (n + 1) + (n + 2)$

(b) Let n = an even integer. $n + 2$ will be the next consecutive even integer.

   Variable expression: $n(n + 2)$

(c) Let n = an odd integer. $n + 2$ will be the next consecutive odd integer.

   Variable expression: $\dfrac{n}{n + 2}$

■

Starter Exercise 4   *Fill in the blanks.*

Translate the following into variable expressions.

(a) The sum of two consecutive integers increased by 5. Let n = an integer. $\boxed{}$ is the next consecutive integer.

   Variable expression: $n + (n + \boxed{}) + \boxed{}$

(b) The ratio of two consecutive even integers. Let n = an $\boxed{}$ integer. $\boxed{}$ is the next consecutive even integer.

   Variable expression: $\dfrac{n}{\boxed{}}$

(c) Twice the product of two consecutive odd integers. Let n = an $\boxed{}$ integer. $\boxed{}$ is the next consecutive odd integer.

   Variable expression: $\boxed{}\,n(\boxed{})$

■ **Solutions to Starter Exercises** ■

1. (a) $x \boxed{-} 12$

(b) $2(\boxed{5+m})$

(c) $\dfrac{x}{\boxed{12} \boxed{-} \boxed{x}}$

(d) $\boxed{4x^2} + 3$

2. (a) $5 + \boxed{x}$

(b) $x \boxed{-6x}$

(c) $\boxed{8} + \boxed{\dfrac{2}{x}}$

3. (a) $1.00 \boxed{x}$

(b) $0.1d + 0.25 \boxed{(15-d)}$

4. (a) $\boxed{n+1}$

$n + (n + \boxed{1}) + \boxed{5}$

(b) $\boxed{\text{even}}$, $\boxed{n+2}$

$\dfrac{n}{\boxed{n+2}}$

(c) $\boxed{\text{odd}}$, $\boxed{n+2}$

$\boxed{2} n(\boxed{n+2})$

2.4 EXERCISES

In Exercises 1–15, translate the following into variable expressions.

1. x increased by 4.

2. 17 less than y.

3. The product of 5 and the sum of a number and 10.

4. The difference between b and the product of 15 and b.

5. The ratio of x decreased by 9 and x increased by 2.

6. The total of twice x and 6.

7. The quotient of a number, and 8 decreased by the number.

8. 15 more than a number, decreased by 6.

9. The product of 16 and the square of a number, reduced by the number.

10. Increase 30% of a number by 11.

11. A number divided by the sum of the number and the square of the number.

12. Triple the absolute value of the difference between a number and 6.

13. 5 plus the quotient of a number increased by 4 and 17.

14. 19 decreased by the product of the cube of a number and 7.

15. The cube of the product of a number and 5, increased by 3.

In Exercises 16–22, translate the following into variable expressions and simplify the variable expression.

16. x increased by the product of 6 and x.

17. 5 less than a number decreased by 7.

18. The ratio of a number and 5 subtracted from the ratio of the number and 3.

19. The product of 5 and the sum of a number and 6.

20. The sum of three consecutive integers.

21. The product of two consecutive even integers.

22. The sum of two consecutive odd integers decreased by 5.

In Exercises 23–27, construct an algebraic expression that represents the indicated quantity.

23. *Total Amount of Money*   Mary has n nickels. How much money does Mary have in nickels?

24. *Total Amount of Money*   A wallet contains x $5 bills. How much money is in the wallet?

25. *Sales Tax*   A stereo power amplifier costs x dollars. How much does Bill pay for the stereo power amplifier including the 7% sales tax?

26. *Ticket Price*   Eight people go to a movie. Some of the people are adults, some are children. Children's tickets are $4.50, and adult tickets are $6.75. How much did it cost for the eight people to get into the movie?

27. *Revenue*   A farmer sells x boxes of fruit for $28 per box and twice as many boxes at $31 per box. How much money did the farmer get for his fruit?

2.5 Introduction to Equations

Section Highlights

1. An equation is a statement that two mathematical expressions are equal.
2. A *solution* of an equation is a number that makes the equation true when the variable is replaced by the number.
3. To *solve* an equation means to find all its solution set.
4. An equation can be thought of as having two sides that are in balance. Hence, if you add some quantity to one side of the equation, you must add the same quantity to the other side to maintain the balance. Similarly, if you are going to multiply one side of an equation by a quantity, you must also multiply the other side by the same quantity.

2.5   Introduction to Equations

EXAMPLE 1 ■   Trial Solution

Determine whether 2 is a solution to the following equations.

(a) $2(x + 4) = 12$ (b) $x^2 - 5x = x - 8$ (c) $2x^2 - 4 = x - 9$

Solution

(a) $2(x + 4) = 12$

     $2(2 + 4) \overset{?}{=} 12$ Replace x by 2.

       $2(6) \overset{?}{=} 12$ Simplify.

        $12 = 12$ 2 is a solution.

(b) $x^2 - 5x = x - 8$

     $2^2 - 5(2) \overset{?}{=} 2 - 8$ Replace x by 2.

       $4 - 10 \overset{?}{=} -6$ Simplify.

        $-6 = -6$ 2 is a solution.

(c) $2x^2 - 4 = x - 9$

     $2(2)^2 - 4 \overset{?}{=} 2 - 9$ Replace x by 2.

     $2(4) - 4 \overset{?}{=} -7$ Simplify.

       $4 \neq -7$ 2 is not a solution.

Starter Exercise 1   *Fill in the blanks.*

Determine whether 4 is a solution of the following equation.

(a) $3x^2 + 6x - 8 = x - 2x^2$

     $3(4)^2 + 6(4) - 8 \overset{?}{=} 4 - 2(4)^2$

     $3(16) + \boxed{24} - 8 \overset{?}{=} 4 - 2\boxed{16}$

     $48 + \boxed{24} - 8 \overset{?}{=} 4 - \boxed{32}$

     $\boxed{} \boxed{=} \boxed{}$

(b) $x^2 - x + 7 = x + 15$

      $\overset{?}{=}$

     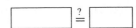 $\overset{?}{=}$

EXAMPLE 2 ■   Operations Used to Solve Equations

(a) $x + 4 = 6$

   $x + 4 + (-4) = 6 + (-4)$ Add -4 to both sides to get rid of the 4.

       $x + 0 = 2$ Additive Inverse Properties

         $x = 2$ Identity Property

(b) $5x = 4x + 17$

   $-4x + 5x = -4x + 4x + 17$ Add $-4x$ to both sides.

       $x = 17$ Solution

(c) $2x = 4$

$\frac{1}{2}(2x) = \frac{1}{2}(4)$ Multiply both sides by $\frac{1}{2}$.

$x = 2$ Solution

EXAMPLE 3 ■   Using Verbal Models to Construct Equations

Write an algebraic equation for the following.

(a) A number increased by 12 is twice the number.

(b) A car rental agency charges $35 per day and $0.25 per mile for their cars. Nancy rents a car and pays a total of $67.75 for the car. How many miles did Nancy drive?

Solution

(a)   *Verbal model:*   (a number) $+12 = 2\times$ (the number)

   Labels: $x =$ a number

   Albegraic equation: $x + 12 = 2x$

(b)   *Verbal model:*   (fixed cost) + (mileage cost) = total cost

   Labels: $35 =$ fixed cost
   $x =$ number of miles driven
   $0.25x =$ mileage cost
   $67.75 =$ total cost

   Algebraic equation: $35 + 0.25x = 67.75$

| **Starter Exercise 2** | *Fill in the blanks.* |

Write an algebraic equation for the following.

(a) The sum of three consecutive integers is 33.

   Verbal model: Sum = (integer) + (next integer) + (next integer)

   Labels: $n =$ the first integer

   $n + 1 = \boxed{}$

   $n + 2 = \boxed{}$

   Algebraic equation: $33 = \boxed{}$

(b) A pair of shoes are on sale for $22. This is 20% off the regular price. What is the regular price?

   Verbal model: $\boxed{}$

   Labels: $\$22 =$ sale price

   $R = \boxed{}$

   $0.20R = \boxed{}$

   Algebraic equation: $\boxed{}$

■ **Solutions to Starter Exercises** ■

1. (a) $3(16) + \boxed{24} - 8 \stackrel{?}{=} 4 - 2\boxed{(16)}$ (b) $\boxed{4^2 - 4 + 7} \stackrel{?}{=} \boxed{4 + 15}$

$48 + \boxed{24} - 8 \stackrel{?}{=} 4 - \boxed{32}$ $\boxed{16 - 4 + 7} \stackrel{?}{=} \boxed{4 + 15}$

$\boxed{64} \boxed{\neq} \boxed{-28}$ $\boxed{19} \boxed{=} \boxed{19}$

2. (a) *Labels:* $n + 1 = \boxed{\text{the next integer}}$
$n + 2 = \boxed{\text{the next integer}}$

Algebraic equation: $33 = \boxed{n + (n+1) + (n+2)}$

(b) *Verbal model:* $\boxed{\text{Sale price = regular price − discount}}$
Labels: $R = \boxed{\text{regular price}}$
$0.20R = \boxed{\text{discount}}$

Algebraic equation: $\boxed{22 = R - 0.20R}$

2.5 EXERCISES

In Exercises 1–10, determine whether the given value of x is a solution of the equation.

1. $2x + 4 = 6$
(a) $x = 1$
(b) $x = 6$

2. $x^2 + x = 0$
(a) $x = 0$
(b) $x = -1$

3. $-2x + 1 = 5x - 3$
(a) $x = 2$
(b) $x = \frac{1}{7}$

4. $\frac{1}{2}x - 3 = 7$
(a) $x = 20$
(b) $x = 2$

5. $x^2 - x = 6$
(a) $x = 3$
(b) $x = -2$

6. $2x^2 + 3 = x(x + 1)$
(a) $x = 1$
(b) $x = 6$

7. $\frac{1}{x} + 12 = 5$
(a) $x = 1$
(b) $x = -\frac{1}{7}$

8. $12x = x(x - 1)$
(a) $x = 0$
(b) $x = 3$

9. $x^3 - 1 = 0$
(a) $x = 1$
(b) $x = -1$

10. $\frac{1}{x} - \frac{4}{x+1} = 0$
(a) $x = 1$
(b) $x = \frac{1}{3}$

In Exercises 11–16, state the property used in each step of solving the equation.

11.
$$x + 2 = 14$$
$$x + 2 + (-2) = 14x(-2)$$
$$x + 0 = 12$$
$$x = 12$$

12.
$$2x = 6$$
$$\tfrac{1}{2}(2x) = \tfrac{1}{2}(6)$$
$$\left(\tfrac{1}{2} \cdot 2\right)x = \tfrac{1}{2}(6)$$
$$1x = 3$$
$$x = 3$$

13.
$$7x = 6x + 1$$
$$7x + (-6x) = 6x + (-6x) + 1$$
$$x = 1$$

14.
$$2(x + 1) = 13$$
$$2x + 2 = 13$$
$$2x + 2 + (-2) = 13 + (-2)$$
$$2x = 12$$
$$\tfrac{1}{2}(2x) = \tfrac{1}{2}(12)$$
$$x = 6$$

15.
$$5(x + 1) = x - 1$$
$$5x + 5 = x - 1$$
$$-x + 5x + 5 = -x + x - 1$$
$$4x + 5 = -1$$
$$4x + 5 + (-5) = -1 + (-5)$$
$$4x = -6$$
$$\tfrac{1}{4}(4x) = \tfrac{1}{4}(-6)$$
$$x = -\tfrac{3}{2}$$

16.
$$\tfrac{3}{4}x = \tfrac{5}{6}x + 1$$
$$-\tfrac{5}{6}x + \tfrac{3}{4}x = -\tfrac{5}{6}x + \tfrac{5}{6}x + 1$$
$$-\tfrac{1}{12}x = 1$$
$$-12\left(-\tfrac{1}{12}\right)x = -12(1)$$
$$x = -12$$

In Exercises 17–22, construct an equation for the given word problem. Do not solve the equation.

17. Loudspeaker cable costs $4.25 per foot. How many feet of loudspeaker cable can be bought for $68.

18. Five times the smaller of two consecutive integers is six times the larger. Find the two integers.

19. The difference between 6 and a number is 15. Find the number.

20. Twice a number exceeds the sum of the number and 3 by 5.

21. The sum of two numbers is 17. One number is two more than twice the other. What are the two numbers?

22. The volume of a box is 16 cubic feet. The length of the base of the box is 2 ft. by 4 ft. What is the height of the box?

Cumulative Practice Test for Chapters P–2

In Exercises 1–5, identify the Basic Rule of Algebra illustrated in each of the statements.

1. $2x + 3 + 5x = 2x + 5x + 3$

2. $4x + 7x = (4 + 7)x$

3. $x + 9x = 1x + 9x$

4. $\frac{1}{2}(2x) = \left(\frac{1}{2} \cdot 2\right)x$

5. $2(13x + 2) = 2(13x) + 2 \cdot 2$

In Exercises 6–10, use the given rule to complete the statement.

6. Additive Identity Property: $16y + 0 = \boxed{}$

7. Multiplicative Inverse Property: $\left(\frac{1}{3} \cdot 3\right)x = \boxed{}$

8. Distributive Property: $5(x + 2) = \boxed{}$

9. Additive Inverse Property: $-\frac{1}{2}x + 3 + (-3) = \boxed{}$

10. Distributive Property: $4x + 8x = \boxed{}$

In Exercises 11 and 12, determine whether the given value of x is a solution of the equation.

11. $5x - 3 = 7$
(a) $x = 2$
(b) $x = \frac{1}{2}$

12. $2x - 6 = 4x - 5$
(a) $x = 5$
(b) $x = -\frac{1}{2}$

In Exercises 13 and 14, state the property used in each step of solving the equation.

13.
$$5x = 4x + 3$$
$$-4x + 5x = -4x + 4x + 3$$
$$(-4 + 5)x = (-4 + 4)x + 3$$
$$1x = (-4 + 4)x + 3$$
$$1x = 0 + 3$$
$$x = 3$$

14.
$$3(x + 2) = 6$$
$$3x + 6 = 6$$
$$3x + 6 + (-6) = 6 + (-6)$$
$$3x + 0 = 0$$
$$3x = 0$$
$$\frac{1}{3}(3x) = \frac{1}{3}(0)$$
$$\left(\frac{1}{3} \cdot 3\right)x = \frac{1}{3}(0)$$
$$1x = 0$$
$$x = 0$$

In Exercises 15–20, evaluate the given variable expression for $x = 1$, $y = -2$, and $z = -1$.

15. $x^2 - 3y$

16. $y(x + z^2)^2 - z^4$

17. $xyz - \frac{1}{2}x^2$

18. $(14z^2 + 9x)y$

19. $x[x - y(z^2 + 3)]$

20. $(x^2y^4 - 3yz^2) + 6xy$

In Exercises 21–30, simplify the variable expression.

21. $2x + 3x$

22. $(-4x + 1) + 9x$

23. $2(x + 1) + 5x$

24. $-7x - (2x + 1)$

$5x + 1$

25. $-2(y - 6) - (-2y + 1)$

26. $2[z + (2z - 1)]$

27. $4[16 - (5x - 2)]$

28. $x(6x + y) - y(6x + y)$

29. $z(4z - 2(z + 2y)]$

30. $x(y^2 + 2z) - z(x + 2xy)$

In Exercises 31–36, translate the verbal phrase into a variable expression. Then simplify the variable expression.

31. The sum of three consecutive integers.

32. 7 less than the quotient of a number and 5.

33. A number increased by the product of 17 and the number.

34. Twice the sum of a number and 6 times the number.

35. 4 more than the ratio of a number and 2.

36. The difference between the square of a number and the product of the number 12.

37. Write two different variable expressions for the area of the figure.

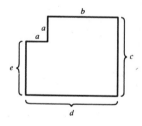

38. PVC pipe costs $0.22 per foot. Hector needs 306 ft. of PVC pipe. How much will the pipe cost Hector?

39. The area of a trapezoid is $\frac{1}{2}(b_1 + b_2)h$. Find the area of the trapezoid with $b_1 = 2$, $b_2 = 7$, and $h = 4$.

40. A box contains only nickels and dimes and there are 27 coins in the box. Write a variable expression for the total value of all the coins in the box.

2.1   **Answers to Exercises**

1. $-2x^2y^3$

2. x^3y^2

3. 6^3u^2v

4. a^6b^2

5. $(x-y)^3$

6. $2^2(2x+1)^2$

7. $2\cdot2\cdot2\cdot u\cdot u$

8. $(xy)(xy)(xy)(xy)$

9. $3ab\cdot b$

10. $7\cdot7\cdot x\cdot x\cdot x\cdot y\cdot y$

11. $(x+3)(x+3)$

12. $6(2a-1)(2a-1)(2a-1)$

13. x^7

14. a^6

15. $-17x^8$

16. u^4v^3

17. $-80a^8b^2$

18. $75s^4t^3$

19. $3456x^{13}y^{23}$

20. $16u^{39}v^{42}$

21. $-128a^9b^{28}$

22. $(x-y)^2$

23. $(2x+3)^7$

24. $2(a+b)^6$

25. $x^3y^3-u^2v^2$

26. $-8x^{15}y^{18}+2x^9y^4$

27. $3a^7b^3+8a^9b^{18}$

28. $4,290$

29. $700,650$

30. 3

31. 5

32. $P(1+r)^4$

33. x^2y

2.2   **Answers to Exercises**

1. Commutative Property of Addition

2. Distributive Property

3. Associative Property of Addition

4. Additive Identity Property

5. $1y^2$

6. $1x$

7. $3x+3y$

8. $-3(9x)$

9. $5x+5y$

10. $(-3+8)y$

11. $(4+9)x^2$

12. $2(3x)+2x+2z$

13. $(6a)(-5)+2b(-5)$

14. $2x(x+4)+3(x+4)$ or $(2x+3)x+(2x+3)4$

15. $22x$

16. $6y^2$

17. $3a+1$

18. $-5z-9$

19. $14x^2+9y$

20. $-6a+12b$

21. $-\frac{1}{4}z+13a$

22. $\frac{14}{5}b-\frac{3}{20}c$

23. $2xy+2z$

24. $5a+8b$

25. False

26. False

27. True **28.** False **29.** True **30.** True

31. $2 + 1 = 3$ **32.** $5 + 2.5 = 7.5$ **33.** $144 + 24 = 168$

34. $400 - 16 = 384$ **35.** $60 - 0.2 = 59.8$ **36.** $240 - 6 - 3 = 231$

2.3 Answers to Exercises

1. $-6x^3$ **2.** $-100y^2$ **3.** $\frac{1}{8}x^6$ **4.** $32x^3y^8$

5. $20x^5y^{11}$ **6.** $7x - 8$ **7.** $-4y + 2$ **8.** $-z + 37$

9. $6r - 41$ **10.** x **11.** $-\frac{2x}{3}$ **12.** $\frac{17z}{30}$

13. $-\frac{1}{2}x - \frac{1}{2}$ **14.** $12y^2 + 2y$ **15.** $-18x^3$ **16.** $-15z^8y^4 + 7z^4y$

17. $-10x + 2$ **18.** $-2y + 13$ **19.** $2z^2 - 4z$ **20.** $-13r^2 + 48r$

21. 3 **22.** 4 **23.** -6 **24.** 12

25. -12 **26.** 32 **27.** 3 **28.** $9\frac{1}{4}$

29. 26 **30.** 18 **31.** 54 **32.** 1,110

33. 63 **34.** 12 **35.** -144

2.4 Answers to Exercises

1. $x + 4$ **2.** $y - 17$ **3.** $5(x + 10)$ **4.** $b - 15b$

5. $\frac{x - 9}{x + 2}$ **6.** $2x + 6$ **7.** $\frac{x}{8 - x}$ **8.** $(15 + x) - 6$

9. $16x^2 - x$ **10.** $0.30x + 11$ **11.** $\frac{x}{x + x^2}$ **12.** $3|x - 6|$

13. $5 + \frac{x + 4}{17}$ **14.** $19 - 7x^3$ **15.** $(5x)^3 + 3$ **16.** $x + 6x;\ 7x$

17. $(x - 5) - 7;\ x - 12$ **18.** $\frac{x}{3} - \frac{x}{5};\ \frac{2x}{15}$ **19.** $5(x + 6);\ 5x + 30$

20. $n + (n + 1) + (n + 2);\ 3n + 3$ **21.** $n(n + 2);\ n^2 + 2n$

22. $n + (n + 2) - 5$; $2n - 3$ **23.** $0.05n$ **24.** $5x$

25. $x + 0.07x$ **26.** $4.50c + 6.75(8 - c)$ **27.** $28x + 31(2x)$

2.5	**Answers to Exercises**

1. (a) 1 is a solution.
   (b) 6 is not a solution.

2. (a) 0 is a solution.
   (b) -1 is a solution.

3. (a) 2 is not a solution.
   (b) $\frac{1}{7}$ is not a solution.

4. (a) 20 is a solution.
   (b) 2 is not a solution.

5. (a) 3 is a solution.
   (b) -2 is a solution.

6. (a) 1 is not a solution.
   (b) 6 is not a solution.

7. (a) 1 is not a solution.
   (b) $-\frac{1}{7}$ is a solution.

8. (a) 0 is a solution.
   (b) 3 is not a solution.

9. (a) 1 is a solution.
   (b) -1 is not a solution.

10. (a) 1 is not a solution.
   (b) $\frac{1}{3}$ is a solution.

11. Add -2 to both sides.
   Inverse Property
   Identity Property

12. Multiply both sides by $\frac{1}{2}$.
   Associative Property
   Inverse Property
   Identity Property

13. Add $-6x$ to both sides.
   Simplify.

14. Distributive Property
   Add -2 to both sides.
   Simplify.
   Multiply both sides by $\frac{1}{2}$.
   Simplify.

15. Distributive Property
   Add $-x$ to both sides.
   Simplify.
   Add -5 to both sides.
   Simplify.
   Multiply both sides by $\frac{1}{4}$.
   Simplify.

16. Add $-\frac{5}{6}x$ to both sides.
   Simplify.
   Multiply both sides by -12.
   Simplify.

17. $4.25x = 68$

18. $5x = 6(x + 1)$ **19.** $6 - x = 15$ **20.** $2x = (x + 3) + 5$

21. $x = 2 + 2(17 - x)$ **22.** $2(4)h = 16$

Answers to Cumulative Practice Test P–2

1. Commutative Property of Addition

2. Distributive Property

3. Multiplicative Identity Property

4. Associative Property of Multiplication

5. Distributive Property

6. $16y$

7. $1x$

8. $5x + 5 \cdot 2$

9. $-\frac{1}{2}x + 0$

10. $(4 + 8)x$

11. (a) 2 is a solution.
    (b) $\frac{1}{2}$ is not a solution.

12. (a) 5 is not a solution.
    (b) $-\frac{1}{2}$ is not a solution.

13. Add $-4x$ to both sides.
    Distributive Property.
    Simplify.
    Additive Inverse Property
    Multiplication by Zero
    Identity Property

14. Distributive Property
    Add -6 to both sides.
    Additive Inverse Property
    Identity Property
    Multiply both sides by $\frac{1}{3}$.
    Associative Property
    Simplify.
    Identity Property

15. 7

16. -9

17. $\frac{3}{2}$

18. -46

19. 9

20. 10

21. $5x$

22. $5x + 1$

23. $7x + 2$

24. $-9x - 1$

25. 11

26. $6z - 2$

27. $-20x + 72$

28. $6x^2 - 5xy - y^2$

29. $2z^2 - 4yz$

30. $xy^2 + xz - 2xyz$

31. $n + (n + 1) + (n + 2)$; $3n + 3$

32. $\frac{x}{5} - 7$

33. $x + 17x$; $18x$

34. $2(x + 6x)$; $14x$

35. $4 + \frac{x}{2}$

36. $x^2 - 12x$

37. $cd - a^2$ and $ab + de$

38. $67.32

39. 18

40. $0.05n + 0.10(27 - n)$

CHAPTER THREE
Linear Equations and Problem Solving

| 3.1 | Solving Linear Equations |

Section Highlights

1. Any equation, in one variable, that can be equivalently written in the form $ax + b = c$ (where a, b, and c are real numbers with $a \neq 0$) is a *linear equation* in one variable.
2. Steps for solving linear equations in one variable:
   i) Simplify both sides of the equation.
   ii) Get one variable term in the equation.
   iii) If there is a constant term added to the variable term, add the opposite of this constant term to both sides of the equation (recall that subtracting is adding the opposite).
   iv) Multiply both sides of the equation by the reciprocal of the coefficient of the variable term. At this point the variable should be isolated.

EXAMPLE 1 ■ Solving Linear Equations in Standard Form

Solve the following linear equations.

(a) $2x - 1 = 5$

(b) $-3x - 7 = 1$

(c) $\dfrac{x}{4} + 6 = 5$

Solution

(a)

$2x - 1 = 5$	Given equation
$2x - 1 + 1 = 5 + 1$	Add 1 to both sides.
$2x = 6$	Simplify.
$\frac{1}{2}(2x) = \frac{1}{2}(6)$	Multiply both sides by $\frac{1}{2}$.
$x = 3$	

Check:

$2x - 1 = 5$	Given equation
$2(3) - 1 \overset{?}{=} 5$	Replace x by 3.
$6 - 1 \overset{?}{=} 5$	
$5 = 5$	

Thus, the solution is 3.

(handwritten:) $-3x - 7 + 7 = 1 + 7$
$\dfrac{-3x}{-3} = \dfrac{8}{-3}$

(b)

$-3x - 7 = 1$	Given equation
$-3x - 7 + 7 = 1 + 7$	Add 7 to both sides.
$-3x = 8$	Simplify.
$-\frac{1}{3}(-3x) = -\frac{1}{3}(8)$	Multiply both sides by $-\frac{1}{3}$.
$x = -\frac{8}{3}$	

Check:

$-3x - 7 = 1$	Given equation
$-3\left(-\frac{8}{3}\right) - 7 \overset{?}{=} 1$	Replace x by $-\frac{8}{3}$.
$8 - 7 \overset{?}{=} 1$	
$1 = 1$	

Thus, the solution is $-\frac{8}{3}$.

(c) $\dfrac{x}{4} + 6 = 5$ Given equation

$\dfrac{x}{4} + 6 + (-6) = 5 + (-6)$ Add −6 to both sides.

$\dfrac{x}{4} = -1$ Simplify.

$4\left(\dfrac{x}{4}\right) = 4(-1)$ Multiply both sides by 4.

$x = -4$

Check:

$\dfrac{x}{4} + 6 = 5$ Given equation

$-\dfrac{4}{4} + 6 \overset{?}{=} 5$ Replace x by −4.

$-1 + 6 \overset{?}{=} 5$

$5 = 5$

Thus, the solution is −4.

| **Starter Exercise 1** | *Fill in the blanks.* |

Solve the following linear equations.

(a) $4x = 8$ ⟨handwritten: $\frac{4x}{4} = \frac{8}{4}$ $x = 2$⟩

$\boxed{\tfrac{1}{4}}(4x) = \boxed{\tfrac{1}{4}} \cdot 8$

$x = \boxed{}$

Check:

(b)   ⟨handwritten: $-2x + 5 - 5 = 0 - 5$ $-2x = 0$⟩

$-2x + 5 = 0$

$-2x + 5 + \left(\boxed{-5}\right) = 0 + \left(\boxed{5}\right)$

$-2x = \boxed{}$

$\boxed{}(-2x) = \boxed{}\,\boxed{}$

$x = \boxed{}$

Check:

(c) $\dfrac{x}{2} - 8 = -3$

$\dfrac{x}{2} - 8 + \left(\boxed{8}\right) = -3 + \boxed{8}$

$\boxed{\tfrac{x}{2}} = \boxed{5}$

$\boxed{} = \boxed{}$

$\boxed{} = \boxed{}$

Check:

EXAMPLE 2 ■   Solving Linear Equations in Nonstandard Form

Solve the following linear equations.

(a) $17x + 2 - 15x = 6$ (b) $-2x - 6 = -3x + 1$ (c) $x - 1 = 2x + 6 - x$

Solution

(a) $17x + 2 - 15x = 6$ Given equation

$2x + 2 = 6$ Simplify.

$2x + 2 + (-2) = 6 + (-2)$ Add −2 to both sides.

$2x = 4$ Simplify.

$\tfrac{1}{2}(2x) = \tfrac{1}{2}(4)$ Multiply both sides by $\tfrac{1}{2}$.

$x = 2$

Check:

$17x + 2 - 15x = 6$ Given equation

$17(2) + 2 - 15(2) \overset{?}{=} 6$ Replace x by 2.

$34 + 2 - 30 \overset{?}{=} 6$

$6 = 6$

Thus, the solution is 2.

(handwritten at top of page)
$-2y-6 = -3x+1$
$-6 = 3$
$3x-2x$

(b) $-2x - 6 = -3x + 1$ Given equation

$3x - 2x - 6 = 3x - 3x + 1$ Add $3x$ to both sides.

$x - 6 = 1$ Simplify.

$x - 6 + 6 = 1 + 6$ Add 6 to both sides.

$x = 7$

Check:

$-2x - 6 = -3x + 1$ Given equation

$-2(7) - 6 \overset{?}{=} -3(7) + 1$ Replace x by 7.

$-14 - 6 \overset{?}{=} -21 + 1$

$-20 = -20$

Thus, the solution is 7.

(c) $x - 1 = 2x + 6 - x$ Given equation

$x - 1 = x + 6$ Simplify.

$-x + x - 1 = -x + x + 6$ Add $-x$ to both sides.

$-1 = 6$ Simplify.

Since there is no x that satisfies this equation, there is no solution.

Starter Exercise 2   *Fill in the blanks.*

Solve the following linear equations.

(a) $\frac{1}{2}x - 2 + \frac{1}{3}x = 6$

$\frac{1}{2}x + \frac{1}{3}x - 2 = 6$

$\left(\Box + \Box\right)x - 2 = 6$

$\Box\, x - 2 = 6$

$\Box\, x - 2 + \Box = 6 + \Box$

$\Box\, x = \Box$

$\Box\left(\Box\, x\right) = \Box \cdot \Box$

$x = \Box$

Check:

(b) $5x - 7 = 2x + 1$

$\boxed{2x} + 5x - 7 = \Box + 2x + 1$

$\Box\, x - 7 = 1$

$\boxed{} = \boxed{}$

$\Box = \Box$

$\Box = \Box$

$\Box = \Box$

(handwritten on right)
$5x-7 = 2x+1$
$5x-2x=$
$3x-7 = 7+1$
$+7$
$3x = 8$
$\frac{3}{3} \quad \frac{8}{3}$

EXAMPLE 3 ■   An Application: Discount

A compact disc is on sale for $12.75, which is 15% off the regular price. What is the regular price?

Solution

Verbal model: Sale price = (Regular price) − Discount.

Labels: Sale price = $12.75
Regular price = x
Discount = $0.15x$

Algebraic equation: $12.75 = x - 0.15x$

Solve:

$$12.75 = x - 0.15x$$
$$12.75 = 1x - 0.15x$$
$$12.75 = 0.85x \qquad \text{Simplify.}$$
$$\frac{1}{0.85}(12.75) = \frac{1}{0.85}(0.85x)$$
$$15 = x$$

Check: $\qquad 15 - 0.15(15) = 15 - 2.25 = 12.75$

Thus, the regular price of the compact disc is $15.

■ **Solutions to Starter Exercises** ■

1. (a) $\boxed{\frac{1}{4}}(4x) = \boxed{\frac{1}{4}} \cdot 8$ $\qquad$ (b) $-2x + 5 + \left(\boxed{-5}\right) = 0 + \left(\boxed{-5}\right)$

$\qquad\qquad x = \boxed{2}$ $\qquad\qquad\qquad -2x = \boxed{-5}$

$\qquad\qquad\qquad\qquad\qquad\qquad \boxed{-\frac{1}{2}}(-2x) = \boxed{-\frac{1}{2}}\boxed{(-5)}$

$\qquad\qquad\qquad\qquad\qquad\qquad\qquad x = \boxed{\frac{5}{2}}$

(c) $\dfrac{x}{2} - 8 + \left(\boxed{8}\right) = -3 + \boxed{8}$

$\qquad\qquad \boxed{\dfrac{x}{2}} = \boxed{5}$

$\qquad\qquad \boxed{2\left(\dfrac{x}{2}\right)} = \boxed{2(5)}$

$\qquad\qquad\quad \boxed{x} = \boxed{10}$

2. (a) $\left(\boxed{\frac{1}{2}} + \boxed{\frac{1}{3}}\right)x - 2 = 6$ $\qquad\qquad$ (b) $\boxed{-2x} + 5x - 7 = \boxed{-2x} + 2x + 1$

$\qquad\quad \boxed{\frac{5}{6}}x - 2 = 6$ $\qquad\qquad\qquad\qquad\qquad \boxed{3}\,x - 7 = 1$

$\quad \boxed{\frac{5}{6}}x - 2 + \boxed{2} = 6 + \boxed{2}$ $\qquad\qquad \boxed{3x - 7 + 7} = \boxed{1 + 7}$

$\qquad\qquad \boxed{\frac{5}{6}}x = \boxed{8}$ $\qquad\qquad\qquad\qquad\qquad \boxed{3x} = \boxed{8}$

$\boxed{\frac{6}{5}}\left(\boxed{\frac{5}{6}}x\right) = \boxed{\frac{6}{5}} \cdot \boxed{8}$ $\qquad\qquad \boxed{\frac{1}{3}(3x)} = \boxed{\frac{1}{3}(8)}$

$\qquad\qquad x = \boxed{\frac{48}{5}}$ $\qquad\qquad\qquad\qquad\qquad \boxed{x} = \boxed{\frac{8}{3}}$

| **3.1** | **EXERCISES** |

In Exercises 1–4, verify that the solution of the last equation is a solution of all the preceding equations in the solving process.

1. $\qquad 5x = 10$

$\qquad \frac{1}{5}(5x) = \frac{1}{5}(10)$

$\qquad\qquad x = 2$

2. $\qquad x - 7 = 10$

$\qquad x - 7 + 7 = 10 + 7$

$\qquad\qquad x = 17$

3.
$$-2x + 1 = 5$$
$$-2x + 1 + (-1) = 5 + (-1)$$
$$-2x = 4$$
$$-\tfrac{1}{2}(-2x) = -\tfrac{1}{2}(4)$$
$$x = -2$$

4.
$$6 - 2x = 7$$
$$-6 + 6 - 2x = -6 + 7$$
$$-2x = 1$$
$$-\tfrac{1}{2}(-2x) = -\tfrac{1}{2}(1)$$
$$x = -\tfrac{1}{2}$$

In Exercises 5–8, solve the linear equation and state the algebraic property used in each step of the solving process.

5. $2x = 5$ **6.** $x - 7 = -11$ **7.** $-2x - 1 = 7$ **8.** $3x + 6 = 15$

In Exercises 9–30, solve the given equation and check your solution.

9. $x - 1 = 6$ *(handwritten: $x - 1 + 1 = 6 + 1$ $x = 7$)* **10.** $-5x = 15$ **11.** $2x - 1 = 4$

12. $6 - 11x = 28$ *(handwritten: $6 - 11x = 34$ $\frac{11}{11}$)* **13.** $12 = 2x - 8$ **14.** $-8x - 11 = 5$

15. $15 = 6x + 12$ *(handwritten: $15 = 6x + 12 - 12$ -12)* **16.** $2x = 2x + 1$ **17.** $12x + 16 = -32$

18. $-\dfrac{x}{3} + 1 = 7$ **19.** $\dfrac{x}{8} + 10 = 11$ **20.** $\dfrac{x}{2} - \dfrac{1}{3} = \dfrac{5}{6}$

21. $2x + 1 = x - 6$ **22.** $5x + 6 - x = 14$ **23.** $\dfrac{x}{4} - 2 = 2x - 1$

24. $x + 7 = 2x + 6 - x$ **25.** $7x - \tfrac{1}{2} = 6x - \tfrac{1}{6}$ **26.** $1 - 0.4x = 15$

27. $\dfrac{x}{3} - \dfrac{1}{4} = \dfrac{2x}{5} + \dfrac{1}{6}$ **28.** $2x + 6 - 5x = 3x + 2$

29. $15x - \dfrac{1}{2} = \dfrac{5x}{2} + 17$ **30.** $-1.1 - 0.3x = 0.4x + 1$

In Exercises 31–35, solve the given word problem.

31. The sum of the three consecutive integers is 24. Find the three integers.

32. The sum of three consecutive even integers is 48. Find the three integers.

33. Seven times the smaller of two consecutive odd integers is five times the larger. Find the two integers.

34. The sale price of a television set is $450, which is 20% off the regular price. What is the regular price of the television?

35. The perimeter of a basketball court is 288 ft. The length of the basketball court is 6 less than twice its width. What are the dimensions of the court?

3.2   Percents and the Percent Equation

Section Highlights

1. **Percent** means *per hundred* or *parts of 100*.
2. **Percents** are fractions with an implied denominator of 100.
3. When using percents, usually you must convert them into equivalent fraction or decimal form.
4. $a = p \cdot b$, where
   $b =$ base number,
   $p =$ percent (in fraction or decimal form), and
   $a =$ number being compared to b.

EXAMPLE 1 ■   Converting Decimals and Fractions to Percents

Convert (a) $\frac{7}{15}$ and (b) 2.4 to percents.

Solution

(a) *Verbal model:*   (Fraction) $\cdot$ (100%) $=$ percent

     Equation: $\frac{7}{15} \cdot 100\% = \frac{700}{15}\% = 46\frac{2}{3}\%$

(b) *Verbal model:*   (Decimal) $\cdot$ (100%) $=$ percent

     Equation: $(2.4)(100\%) = 240\%$

■ **Starter Exercise 1**   *Fill in the blanks.*

Convert (a) $\frac{5}{8}$ and (b) 0.7 to percents.

(a) *Verbal model:*   (Fraction) $\cdot$ (100%) $=$ percent

     Equation: $\frac{5}{8} \cdot 100\% = \frac{\boxed{500}}{8}\% = \boxed{}$

(b) *Verbal model:* $\boxed{}$

     Equation: $(0.7) \cdot \boxed{} = \boxed{}$

EXAMPLE 2 ■ Converting Percents to Decimals and Fractions

(a) Convert $40\frac{1}{2}\%$ to a fraction.

(b) Convert 17% to a decimal.

Solution

(a) *Verbal model:* (Fraction) · (100%) = percent

    Label: x = fraction

    Equation: $x \cdot (100\%) = 40\frac{1}{2}\%$

$$x = \frac{40\frac{1}{2}\%}{100\%}$$

$$x = \frac{81}{200}$$

Thus, $40\frac{1}{2}\%$ corresponds to the fraction $\frac{81}{200}$.

(b) *Verbal model:* (Decimal) · (100%) = percent

    Label: x = decimal

    Equation: $x \cdot (100\%) = 17\%$

$$x = \frac{17\%}{100\%}$$

$$x = 0.17$$

Thus, 17% corresponds to the the the decimal. ■

Starter Exercise 2   *Fill in the blanks.*

(a) Convert 17.3% to a decimal.

    Verbal model: (Decimal) · (100%) = percent

    Label: x = ☐

    Equation: $\boxed{X} \cdot \boxed{100} = 17.3\%$

$$\boxed{} = \frac{17.3\%}{\boxed{}}$$

$$\boxed{} = \boxed{0.173}$$

(b) Convert 45% to a fraction.

    Verbal model: ☐

    Label: x = ☐

    Equation: ☐ · ☐ = ☐

EXAMPLE 3 ■ Solving Percent Equations

(a) What percent of 27 is 9?

(b) What is 25% of 64?

(c) 15 is 20% of what number?

Solution

(a) *Verbal model:* 9 = what percent of 27

    Label: p = percent in fraction form

    Percent equation: $9 = p \cdot 27$

$$\frac{9}{27} = p$$

$$\frac{1}{3} = p$$

Thus, $33\frac{1}{3}\%$ of 27 is 9.

(c) *Verbal model:* 15 = 20% of what number

    Label: b = unknown number

    Percent equation: $15 = 0.20b$

$$\frac{15}{0.20} = b$$

$$75 = b$$

Thus, 15 is 20% of 75.

(b) *Verbal model:* What number = 25% of 64

    Label: a = unknown number

    Percent equation: $a = 0.25(64)$

$$a = 16$$

Thus, 16 is 25% of 64.

■

▮ Starter Exercise 3 | *Fill in the blanks.*

(a) What is 15% of 9?

Verbal model:	What number = 15% of 9
Label:	$a = \boxed{}$
Percent equation:	$a = 0.15\left(\boxed{}\right)$
	$a = \boxed{}$

(b) 30% of what is 27?

Verbal model:	27 = 30% of what number
Label:	$b = \boxed{}$
Percent equation:	$27 = \boxed{}\left(\boxed{}\right)$
	$\boxed{} = \boxed{}$
	$\boxed{} = \boxed{}$

(c) What percent of 42 is 7?

Verbal model:	$7 = \boxed{}$
Label:	$p = \boxed{}$
Percent equation:	$\boxed{} = \boxed{} \cdot \boxed{}$
	$\boxed{} = \boxed{}$
	$\boxed{} = \boxed{}$

EXAMPLE 4 ▦ An Application

A business executive received a $3,000 bonus last year. Her bonus was 5% of her yearly salary. What was her yearly salary last year?

Solution

Verbal model: Bonus = 5% of salary

Labels: $a = \text{bonus} = \$3000$
 $b = \text{salary} = \text{unknown}$
 $p = 0.05$

Equation: $3000 = 0.05b$

$$\frac{3000}{0.05} = b$$

$$60{,}000 = b$$

Thus, the executive's yearly salary last year was $60,000. ▪

| **Starter Exercise 4** | *Fill in the blanks.*

Mark will receive a 5.3% cost of living salary increase this year. His salary last year was $48,250. Find his salary this year. We must first find the amount of the increase, then add it to Mark's old salary.

Verbal model: Amount of increase = 5.3% of old salary

Label: a = unknown amount of increase

$p = \boxed{}$

b = old salary = $\boxed{}$

Equation: $a = \left(\boxed{}\right)\left(\boxed{}\right)$

$\boxed{} = \boxed{}$

Thus, his new salary is $48,250 + \boxed{} = \boxed{}$.

EXAMPLE 5 ■ An Application

A coffee maker is on sale for $45. Its regular price is $60. Find the discount rate (i.e., find the percent the price reduced).

Solution

Verbal model: Amount of discount = percent of price

Labels: a = amount of discount = $15
b = regular price = $60
p = unknown percent

Equation: $15 = p \cdot 60$

$\dfrac{15}{60} = p$

$0.25 = p$

Thus, the discount rate is 25%.

■

■ **Solutions to Starter Exercises** ■

1. (a) $\dfrac{5}{8} \cdot 100\% = \dfrac{\boxed{500}}{8}\%$

$= \boxed{62.5\%}$

(b) $\boxed{(\text{Fraction}) \cdot (100\%) = \text{percent}}$

$(0.7) \cdot \boxed{100\%} = \boxed{70\%}$

■ **Solutions to Starter Exercises** ■

2. (a) $x = \boxed{\text{decimal}}$

$\boxed{x} \cdot \boxed{100\%} = 17.3\%$

$\boxed{x} = \dfrac{17.3\%}{\boxed{100\%}}$

$\boxed{x} = \boxed{0.173}$

(b) $\boxed{(\text{Fraction}) \cdot (100\%) = \text{percent}}$

$x = \boxed{\text{fraction}}$

$\boxed{x} \cdot \boxed{100\%} = \boxed{45\%}$

$\boxed{x} = \boxed{\dfrac{45\%}{100\%}}$

$\boxed{x} = \boxed{\dfrac{45}{100}}$

$\boxed{x} = \boxed{\dfrac{9}{20}}$

3. (a) $a = \boxed{\text{unknown number}}$

$a = 0.15\left(\boxed{9}\right)$

$a = \boxed{1.35}$

(b) $b = \boxed{\text{unknown number}}$

$27 = \boxed{0.3}\left(\boxed{b}\right)$

$\boxed{\dfrac{27}{0.3}} = \boxed{b}$

$\boxed{90} = \boxed{b}$

(c) $7 = \boxed{\text{what percent of 42}}$

$p = \boxed{\text{percent in fraction form}}$

$\boxed{7} = \boxed{p} \cdot \boxed{42}$

$\boxed{\dfrac{7}{42}} = \boxed{p}$

$\boxed{16\frac{2}{3}\%} = \boxed{p}$

4. $p = \boxed{0.053}$

$b = \text{old salary} = \boxed{\$48,250}$

$a = \left(\boxed{0.053}\right)\left(\boxed{48,250}\right)$

$\boxed{a} = \boxed{\$2557.25}$

Thus, his new salary is $\$48,250 + \boxed{\$2557.25} = \boxed{\$50,807.25}$.

3.2 EXERCISES

In Exercises 1–4, change the percent to a decimal.

1. 50%   **2.** 4.3%   **3.** 0.01%   **4.** $\frac{2}{5}\%$

In Exercises 5–8, change the decimal to a percent.

5. 0.46   **6.** 0.97   **7.** 3.81   **8.** 0.0024

In Exercises 9–12, change the percent to a fraction in reduced form.

9. 50%   **10.** $33\frac{1}{3}\%$ **11.** 2.2%   **12.** 100%

In Exercises 13–16, change the fraction to a percent.

13. $\frac{1}{2}$ **14.** $\frac{3}{4}$ **15.** $\frac{2}{3}$ **16.** $\frac{5}{8}$

17. What is 40% of 24?   **18.** 40 is what percent of 50?

19. 9 is $33\frac{1}{3}\%$ of what number?   **20.** 15% of what number is 60?

21. What is 300% of 12?   **22.** 30 is what percent of 45?

23. $\frac{1}{3}\%$ of what number is 2?   **24.** 20 is what percent of 8?

25. A tire that regularly sells for $65 is on sale for $55.25. Find the discount rate.

26. My son had a 50% increase in his weight in one year. At the beginning of that year he weighed 20 pounds. How much did he weigh at the end of that year?

27. Your car is worth $8,000 and your license fee was $180 for the car. What percent of your car's value is the license fee?

28. An appliance store claims that 12% of the washing machines they sell need service in the first year. At this rate, how many washing machines were sold in a year in which 6 washing machines were serviced?

29. In an algebra class, 15% of the students received a grade of A. If there were 20 students in this class, how many A's were given?

30. A toothpaste manufacturer claims that their toothpaste will reduce cavities by 30%. If you have 10 cavities now, how many would you have if you had used this toothpaste?

31. A forest fire burned 7,000 acres of a 90,000 acre forest. What percent of the forest was burned?

32. Each month, Joan deposits 14% of her salary into her savings account. If Joan's monthly salary is $2,400, what is her monthly deposit into her savings account?

33. Because of a drought, you must reduce your monthly water usage from 15 units to 9 units. What is the reduction percentage?

3.3 More About Solving Linear Equations

Section Highlights

1. If a linear equation in one variable contains fractions, we may multiply both sides of the equation by the least common denominator of all the fractions, producing an equivalent equation without fractions.
2. Recall all steps for solving linear equations in one variable from section 3.1.

EXAMPLE 1 ■ Solving Linear Equations Involving Parentheses

Solve the following equations.

(a) $4(x + 6) = -4$ 　　　(b) $-2(2x - 6) + x = 17$ 　　　(c) $3(2x - 1) - 5x = 6(x - 11)$

Solution

(a)

$$4(x + 6) = -4 \quad \text{Given equation}$$
$$4x + 24 = -4 \quad \text{Distributive Property}$$
$$4x + 24 + (-24) = -4 + (-24) \quad \text{Add } -24 \text{ to both sides.}$$
$$4x = -28$$
$$\tfrac{1}{4}(4x) = \tfrac{1}{4}(-28) \quad \text{Multiply both sides by } \tfrac{1}{4}.$$
$$x = -7$$

Thus, -7 is the solution. Check this in the original equation.

(b)

$$-2(2x - 6) + x = 17 \quad \text{Given equation}$$
$$-4x + 12 + x = 17 \quad \text{Distributive Property}$$
$$-4x + x + 12 = 17 \quad \text{Commutative Property}$$
$$-3x + 12 = 17 \quad \text{Simplify}$$
$$-3x + 12 + (-12) = 17 + (-12) \quad \text{Add } -12 \text{ to both sides.}$$
$$-3x = 5$$
$$-\tfrac{1}{3}(-3x) = -\tfrac{1}{3}(5) \quad \text{Multiply both sides by } -\tfrac{1}{3}.$$
$$x = -\tfrac{5}{3}$$

Thus, the solution is $-\tfrac{5}{3}$. Check this in the original equation.

(c) $3(2x - 1) - 5x = 6(x - 11)$ Given equation

$6x - 3 - 5x = 6x - 66$ Distributive Property

$6x - 5x - 3 = 6x - 66$ Commutative Property

$x - 3 = 6x - 66$

$-x + x - 3 = -x + 6x - 66$ Add $-x$ to both sides.

$-3 = 5x - 66$

$-3 + 66 = 5x - 66 + 66$ Add 66 to both sides.

$63 = 5x$

$\frac{1}{5}(63) = \frac{1}{5}(5x)$ Multiply both sides by $-\frac{1}{5}$.

$\frac{63}{5} = x$

Thus, the solution is $\frac{63}{5}$. Check this in the original equation.

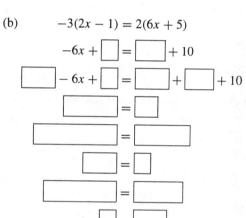

handwritten:
$2(x+5) - 7 = 3(x-2)$
$2x + 10 - 7 = 3x - 6$
$2x + 3x + 11$

■ Starter Exercise 1   *Fill in the blanks.*

Solve the following equations.

(a) $12 = 4(x - 2)$

$12 = \boxed{}x - \boxed{}$

$12 + \boxed{} = \boxed{}x - \boxed{} + \boxed{}$

$\boxed{} = \boxed{}x$

$\boxed{}\boxed{} = \boxed{}\boxed{}$

$\boxed{} = x$

(b) $-3(2x - 1) = 2(6x + 5)$

$-6x + \boxed{} = \boxed{} + 10$

$\boxed{} - 6x + \boxed{} = \boxed{} + \boxed{} + 10$

$\boxed{} = \boxed{}$

$\boxed{} = \boxed{}$

$\boxed{} = \boxed{}$

$\boxed{} = \boxed{}$

$\boxed{} = \boxed{}$

(c) $5(x - 1) - (2x + 4) = 6x$

$5x - 5 - 2x - 4 = 6x$

$5x - \boxed{} - \boxed{} - 4 = 6x$

$\boxed{} - \boxed{} = 6x$

$\boxed{} = \boxed{}$

$\boxed{} = \boxed{}$

$\boxed{} = \boxed{}$

$\boxed{} = \boxed{}$

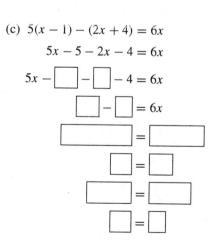

EXAMPLE 2 ■ **A Linear Equation Involving Nested Symbols of Grouping**

Solve $x + 2[3x - (2x + 1)] = 10x - 8$.

Solution

$x + 2[3x - (2x + 1)] = 10x - 8$	Given equation
$x + 2[3x - 2x - 1] = 10x - 8$	Distributive Property
$x + 2[x - 1] = 10x - 8$	
$x + 2x - 2 = 10x - 8$	Distributive Property
$3x - 2 = 10x - 8$	
$-10x + 3x - 2 = -10x + 10x - 8$	Add $-10x$ to both sides.
$-7x - 2 = -8$	
$-7x - 2 + 2 = -8 + 2$	Add 2 to both sides.
$-7x = -6$	
$-\frac{1}{7}(-7x) = -\frac{1}{7}(-6)$	Multiply both sides by $-\frac{1}{7}$.
$x = \frac{6}{7}$	

Thus, the solution set is $\frac{6}{7}$. Check this in the original equation. ■

Starter Exercise 2   *Fill in the blanks.*

Solve the following equation.

$$2[x - 3(2x - 1)] - x = -14$$

$$2\left[x - \boxed{} + \boxed{}\right] - x = -14$$

$$2\left[\boxed{} + \boxed{}\right] - x = -14$$

$$\boxed{} + \boxed{} - x = -14$$

$$\boxed{} = \boxed{}$$

$$\boxed{} = \boxed{}$$

$$\boxed{} = \boxed{}$$

$$\boxed{} = \boxed{}$$

$$\boxed{} = \boxed{}$$

$$\boxed{} = \boxed{}$$

EXAMPLE 3 ■   **Solving Linear Equations Involving Fractions**

Solve the following equations.

(a) $\dfrac{x}{4} + \dfrac{1}{2} = \dfrac{5}{6}$

(b) $\dfrac{2x}{5} + 1 = \dfrac{x}{3} + \dfrac{4}{15}$

(c) $\dfrac{2x-1}{3} = \dfrac{2x-1}{6}$

Solution

(a)

$\dfrac{x}{4} + \dfrac{1}{2} = \dfrac{5}{6}$	Given equation
$12\left(\dfrac{x}{4} + \dfrac{1}{2}\right) = 12\left(\dfrac{5}{6}\right)$	Multiply both sides by 12 (LCD).
$12 \cdot \dfrac{x}{4} + 12 \cdot \dfrac{1}{2} = 12\left(\dfrac{5}{6}\right)$	Distributive Property
$3x + 6 = 10$	Simplify.
$3x + 6 + (-6) = 10 + (-6)$	Add -6 to both sides.
$3x = 4$	
$\dfrac{1}{3}(3x) = \dfrac{1}{3}(4)$	Multiply both sides by $\frac{1}{3}$.
$x = \dfrac{4}{3}$	

Thus, the solution is $\frac{4}{3}$. Check this solution in the original equation.

(b)

$\dfrac{2x}{5} + 1 = \dfrac{x}{3} + \dfrac{4}{15}$	Given equation
$15\left(\dfrac{2x}{5} + 1\right) = 15\left(\dfrac{x}{3} + \dfrac{4}{15}\right)$	Multiply both sides by 15 (LCD).
$15 \cdot \dfrac{2x}{5} + 15 \cdot 1 = 15 \cdot \dfrac{x}{3} + 15 \cdot \dfrac{4}{15}$	Distributive Property
$6x + 15 = 5x + 4$	
$-5x + 6x + 15 = -5x + 5x + 4$	Add $-5x$ to both sides.
$x + 15 = 4$	
$x + 15 + (-15) = 4 + (-15)$	Add -15 to both sides.
$x = -11$	

Thus, the solution is -11. Check this solution in the original equation.

(c) $\dfrac{2x-1}{3} = \dfrac{2x-1}{6}$ Given equation

$6\left(\dfrac{2x-1}{3}\right) = 6\left(\dfrac{2x-1}{6}\right)$ Multiply both sides by 6 (LCD).

$2(2x-1) = 2x-1$ Simplify.

$4x-2 = 2x-1$ Distributive Property

$-2x+4x-2 = -2x+2x-1$ Add $-2x$ to both sides.

$2x-2 = -1$

$2x-2+2 = -1+2$ Add 2 to both sides.

$2x = 1$

$\dfrac{1}{2}(2x) = \dfrac{1}{2}(1)$ Multiply both sides by $\frac{1}{2}$.

$x = \dfrac{1}{2}$

Thus, the solution is $\frac{1}{2}$. Check this solution in the original equation. ∎

Starter Exercise 3   *Fill in the blanks.*

Solve the following equations.

(a) $\dfrac{5x-2}{3} = \dfrac{4x+3}{2}$

$6\left(\dfrac{5x-2}{3}\right) = 6\left(\dfrac{4x+3}{2}\right)$

$\boxed{}(5x-2) = \boxed{}(4x+3)$

$\boxed{} = \boxed{}$

$\boxed{} = \boxed{}$

$\boxed{} = \boxed{}$

$\boxed{} = \boxed{}$

$\boxed{} = \boxed{}$

$\boxed{} = \boxed{}$

$\boxed{} = \boxed{}$

(b) $\dfrac{3x}{5} + 3 = \dfrac{x}{3} + \dfrac{1}{5}$

$\boxed{}\left(\dfrac{3x}{5} + 3\right) = 15\left(\dfrac{x}{3} + \dfrac{1}{5}\right)$

$\boxed{}\dfrac{3x}{5} + \boxed{}3 = 15\boxed{} + 15\boxed{}$

$\boxed{} = \boxed{}$

$\boxed{} = \boxed{}$

$\boxed{} = \boxed{}$

$\boxed{} = \boxed{}$

$\boxed{} = \boxed{}$

$\boxed{} = \boxed{}$

$\boxed{} = \boxed{}$

■ **Solutions to Starter Exercises** ■

1. (a)

$$12 = \boxed{4}\, x - \boxed{8}$$

$$12 + \boxed{8} = \boxed{4}\, x - \boxed{8} + \boxed{8}$$

$$\boxed{20} = \boxed{4}\, x$$

$$\boxed{\tfrac{1}{4}}\ \boxed{(20)} = \boxed{\tfrac{1}{4}}\ \boxed{(4x)}$$

$$\boxed{5} = x$$

(b)

$$-6x + \boxed{3} = \boxed{12x} + 10$$

$$\boxed{-12x} - 6x + \boxed{3} = \boxed{-12x} + \boxed{12x} + 10$$

$$\boxed{-18x + 3} = \boxed{10}$$

$$\boxed{-18x + 3 + (-3)} = \boxed{10 + (-3)}$$

$$\boxed{-18x} = \boxed{7}$$

$$\boxed{-\tfrac{1}{18}(-18x)} = \boxed{-\tfrac{1}{18}(7)}$$

$$\boxed{x} = \boxed{-\tfrac{7}{18}}$$

(c)

$$5x - \boxed{2x} - \boxed{5} - 4 = 6x$$

$$\boxed{3x} - \boxed{9} = 6x$$

$$\boxed{-3x + 3x - 9} = \boxed{-3x + 6x}$$

$$\boxed{-9} = \boxed{3x}$$

$$\boxed{\tfrac{1}{3}(-9)} = \boxed{\tfrac{1}{3}(3x)}$$

$$\boxed{-3} = \boxed{x}$$

■ **Solutions to Starter Exercises** ■

2. $2\left[x - \boxed{6x} + \boxed{3}\right] - x = -14$

$\quad 2\left[\boxed{-5x} + \boxed{3}\right] - x = -14$

$\quad\quad \boxed{-10x} + \boxed{6} - x = -14$

$\quad\quad\quad \boxed{-10x - x + 6} = \boxed{-14}$

$\quad\quad\quad\quad \boxed{-11x + 6} = \boxed{-14}$

$\quad\quad\quad \boxed{-11x + 6 + (-6)} = \boxed{-14 + (-6)}$

$\quad\quad\quad\quad\quad \boxed{-11x} = \boxed{-20}$

$\quad\quad\quad \boxed{-\frac{1}{11}(-11x)} = \boxed{-\frac{1}{11}(-20)}$

$\quad\quad\quad\quad\quad\quad \boxed{x} = \boxed{\frac{20}{11}}$

3. (a) $\boxed{2}(5x - 2) = \boxed{3}(4x + 3)$

$\quad\quad\quad \boxed{10x - 4} = \boxed{12x + 9}$

$\quad\quad \boxed{-12x + 10x - 4} = \boxed{-12x + 12x + 9}$

$\quad\quad\quad\quad \boxed{-2x - 4} = \boxed{9}$

$\quad\quad\quad \boxed{-2x - 4 + 4} = \boxed{9 + 4}$

$\quad\quad\quad\quad\quad \boxed{-2x} = \boxed{13}$

$\quad\quad\quad \boxed{-\frac{1}{2}(-2x)} = \boxed{-\frac{1}{2}(13)}$

$\quad\quad\quad\quad\quad \boxed{x} = \boxed{-\frac{13}{2}}$

■ **Solutions to Starter Exercises** ■

3. —CONTINUED—

(b) $\boxed{15}\left(\dfrac{3x}{5}+3\right)=15\left(\dfrac{x}{3}+\dfrac{1}{5}\right)$

$\boxed{15}\dfrac{3x}{5}+\boxed{15}\,3=15\,\boxed{\dfrac{x}{3}}+15\,\boxed{\dfrac{1}{5}}$

$\boxed{9x+45}=\boxed{5x+3}$

$\boxed{-5x+9x+45}=\boxed{-5x+5x+3}$

$\boxed{4x+45}=\boxed{3}$

$\boxed{4x+45+(-45)}=\boxed{3+(-45)}$

$\boxed{4x}=\boxed{-42}$

$\boxed{\dfrac{1}{4}(4x)}=\boxed{\dfrac{1}{4}(-42)}$

$\boxed{x}=\boxed{-\dfrac{21}{2}}$

3.3 EXERCISES

In Exercises 1–20, solve the equation and check your solution.

1. $3(y+2)=-3$

2. $4=(2x-3)4$

3. $-4(-x+7)=0$

4. $10-(7z-4)=7$

5. $16x-(5x+6)=5$

6. $5(3x-5)=x$

7. $-2(x-1)=4(2x+1)$

8. $3(3x+6)=x-(4x+6)$

9. $5[4x-2(x+10)]=-10(x+10)$

10. $2[3(x-1)-(2x+2)]=10$

11. $\dfrac{x}{5}=\dfrac{2}{5}$

12. $\dfrac{2x}{7}=\dfrac{10}{7}$

13. $\dfrac{x}{2}+\dfrac{1}{3}=\dfrac{5}{6}$

14. $\dfrac{7x}{10}-\dfrac{2}{5}=\dfrac{x}{3}$

15. $\dfrac{1}{3}=\dfrac{z}{12}+\dfrac{5}{4}$

16. $\dfrac{2x+1}{2}=\dfrac{3x-1}{3}$

17. $\dfrac{z-6}{12}=\dfrac{3-4z}{8}$

18. $\dfrac{5x}{6}+\dfrac{2}{3}=\dfrac{1}{9}+\dfrac{5x}{18}$

19. $5[2x-4(x-1)]=4(x-1)+x$

20. $0.3x+1.7=0.2x-0.6$

21. The sum of two integers is 42. Four times one integer is twice the other. Find the two integers.

22. A 36-inch guitar string is tuned to a G note. To produce a B note divide the string so that four times the length of one piece is five times the length of the other piece. Find the length of each piece.

23. The time it takes two people working together to complete a job is t, where $(t/2) + (t/3) = 1$. Find the number of hours it takes for the two people to complete the job.

24. If the fulcrum that is supporting a 10-foot lever is located x feet from the end of the lever where a 25-pound force is applied, and a 100-pound force is applied to the other end, then the lever balances if $25x = 100(10 - x)$. Find the location of the fulcrum that makes the lever balance.

25. Jennifer has scores of 82, 96, and 85 on her first three algebra exams. What must Jennifer get on the fourth exam to have an average of 90?

3.4 | Ratio and Proportion

Section Highlights

1. The **ratio** of the real number a to the real number b is $\dfrac{a}{b}$.

2. A **proportion** is a statement of the equality of two ratios.

EXAMPLE 1 ■ Finding Ratios

(a) The ratio of 9 to 13 is given by $\frac{9}{13}$.

(b) The ratio of $2\frac{1}{2}$ to $1\frac{2}{3}$ is given by

$$\frac{2\frac{1}{2}}{1\frac{2}{3}} = \frac{\frac{5}{2}}{\frac{5}{3}} = \frac{5}{2} \cdot \frac{3}{5} = \frac{3}{2}.$$

(c) The ratio of 0.11 to 0.17 is given by $\dfrac{0.11}{0.17} = \dfrac{11}{17}$.

EXAMPLE 2 ■ Comparing Measurements

Find a ratio to compare the relative sizes of the following. Use the same unit of measurement in the numerator and denominator.

(a) 7 cups to 15 cups

(b) 5 feet to 2 yards

(c) 90 seconds to 2 minutes

Solution

(a) Since the units are the same, the ratio is $\frac{7}{15}$.

(b) Since 2 yards = 6 feet, the ratio is

$$\frac{5 \text{ feet}}{2 \text{ yards}} = \frac{5 \text{ feet}}{6 \text{ feet}} = \frac{5}{6}.$$

(c) Since 2 minutes = 120 seconds, the ratio is

$$\frac{90 \text{ seconds}}{2 \text{ minutes}} = \frac{90 \text{ seconds}}{120 \text{ seconds}} = \frac{90}{120} = \frac{3}{4}.$$

■ Starter Exercise 1   *Fill in the blanks*

Find a ratio to compare the relative sizes of the following. Use the same unit of measurement in the numerator and denominator.

(a) 4 meters to 9 meters

$$\frac{\boxed{}}{9}$$

(b) 5 pounds to 24 ounces

$$\frac{5 \text{ pounds}}{24 \text{ ounces}} = \frac{\boxed{}}{24 \text{ ounces}} = \frac{\boxed{}}{24} = \frac{\boxed{}}{3}$$

EXAMPLE 3 ■   Finding Unit Price

Find the unit price (in dollars per pound) for each of the following items.

(a) A 3-pound can of coffee for $6.22

(b) A 10-pound bag of fertilizer for $5.99

(c) A 14-ounce box of cereal for $2.80

Solution

(a)   *Verbal model:* $\text{Unit price} = \dfrac{\text{Total price}}{\text{Total units}}$

   Unit price: $\dfrac{\$6.32}{3} = \$2.10\overline{6} \approx \$2.11$ per pound

(b)   *Verbal model:* $\text{Unit price} = \dfrac{\text{Total price}}{\text{Total units}}$

   Unit price: $\dfrac{\$5.99}{10} = \$0.599$ per pound

(c)   We first convert ounces into pounds.

   $14 \text{ ounces} = \dfrac{14}{16} \text{ pounds} = 0.875 \text{ pounds}$

   Verbal model: $\text{Unit price} = \dfrac{\text{Total price}}{\text{Total units}}$

   Unit price: $\dfrac{\$2.80}{0.875} = \$3.20$ per pound

■ Starter Exercise 2   *Fill in the blanks.*

Find the unit price (in dollars per ounce) for each of the following items.

(a) A 12-ounce can of soda for $0.65

   Verbal model: $\text{Unit price} = \dfrac{\text{Total price}}{\text{Total units}}$

   Unit price: $\boxed{} = \$0.0541\overline{6}$ per ounce

(b) A 5.5-ounce bag of seed for $4.62

   Verbal model: $\text{Unit price} = \dfrac{\text{Total price}}{\text{Total units}}$

   Unit price: $\boxed{} = \boxed{}$ per ounce

EXAMPLE 4 ■ Comparing Unit Price

Which has the smaller unit price: a gallon of milk for $1.92 or a quart of milk for 59¢?

Solution

The unit price for the gallon of milk is:

$$\text{Unit price} = \frac{\text{Total price}}{\text{Total units}} = \frac{\$1.92}{1 \text{ gallon}} = \frac{\$1.92}{128 \text{ ounces}} = \$0.015 \text{ per ounce.}$$

The unit price for the quart of milk is:

$$\text{Unit price} = \frac{\text{Total price}}{\text{Total units}} = \frac{0.59¢}{1 \text{ quart}} = \frac{0.59¢}{32 \text{ ounces}} = 0.0184375 \text{ per ounce.}$$

Thus, the gallon of milk has the smaller unit price.

Starter Exercise 3   *Fill in the blanks.*

Which has the smaller unit price: a 50-foot spool of wire for $10.65 or a 20-foot spool of wire for $4.40?

The unit price for the larger spool is

$$\text{Unit price} = \frac{\text{Total price}}{\text{Total units}} = \frac{\boxed{}}{50} = \boxed{} \text{ per foot.}$$

The unit price for the smaller spool is

$$\text{Unit price} = \boxed{} = \boxed{} \text{ per foot.}$$

EXAMPLE 5 ■ Solving Proportions

Solve the following proportions.

(a) $\dfrac{x}{5} = \dfrac{4}{25}$ (b) $\dfrac{4}{x} = \dfrac{96}{15}$

Solution

(a)
$$\frac{x}{5} = \frac{4}{25} \qquad \text{Given}$$

$$5 \cdot \frac{x}{5} = 5 \cdot \frac{4}{25} \qquad \text{Multiply both sides by 5.}$$

$$x = \frac{4}{5} \qquad \text{Solution}$$

(b)
$$\frac{4}{x} = \frac{96}{15} \qquad \text{Given}$$

$$\frac{x}{4} = \frac{15}{96} \qquad \text{Invert both sides.}$$

$$4 \cdot \frac{x}{4} = 4 \cdot \frac{15}{96} \qquad \text{Multiply both sides by 4.}$$

$$x = \frac{15}{24} = \frac{5}{8} \qquad \text{Solution}$$

▮ **Starter Exercise 4**   *Fill in the blanks.*

Solve the following proportions.

(a)
$$\frac{9}{4} = \frac{36}{x}$$

$$\boxed{} = \frac{x}{36}$$

$$36 \cdot \left(\boxed{} \right) = 36 \cdot \frac{x}{36}$$

$$\boxed{} = x$$

(b)
$$\frac{23}{21} = \frac{z}{84}$$

$$\left(\boxed{} \right) \cdot \frac{23}{21} = \left(\boxed{} \right) \cdot \frac{z}{84}$$

$$\boxed{} = \boxed{}$$

EXAMPLE 6 ▮   Solving a Proportion

A number is to 15 as 6 is to 5. What is the number?

Solution

$$\frac{x}{15} = \frac{6}{5} \qquad \text{Equation}$$

$$15 \cdot \frac{x}{15} = 15 \cdot \frac{6}{5} \qquad \text{Multiply both sides by 15.}$$

$$x = 18$$

Thus, the number is 18.

▮

▮ **Starter Exercise 5**   *Fill in the blanks.*

8 is to 19 as a number is to 4. What is the number?

$$\frac{8}{19} = \frac{x}{\boxed{}}$$

$$\left(\boxed{} \right) \cdot \frac{8}{19} = \left(\boxed{} \right) \cdot \frac{x}{\boxed{}}$$

$$\boxed{} = x$$

EXAMPLE 7 ▮   An Application: Estimating the Amount of Tile

You are tiling your kitchen counter top with tiles that are 4 inches by 4 inches. You have 1488 square inches of counter top. How many tiles do you need?

Solution

Verbal model: $\dfrac{1 \text{ tile}}{\text{Area of 1 tile}} = \dfrac{\text{Total number of tiles}}{\text{Total area}}$

Labels: Area of 1 tile $= 4 \times 4 = 16$
Total area $= 1488$
Total number of tiles $= x$

Proportion:
$$\frac{1}{16} = \frac{x}{1488}$$

$$1488 \cdot \frac{1}{16} = \frac{x}{1488} \cdot 1488$$

$$93 = x$$

Thus, you need 93 tiles.

■ | **Starter Exercise 6** | *Fill in the blanks.*

Two ceiling lights are needed for every 130 square feet of floor space. How many lights are needed in an office with 3,120 square feet of floor space?

Verbal model: $$\frac{\text{Number of lights}}{\text{Floor space}} = \frac{\text{Total number of lights}}{\text{Total floor space}}$$

Label: ⬜

Proportion: $$\frac{2}{130} = \frac{\square}{\square}$$

EXAMPLE 8 ■   An Application: Population

The recent past voting records of a town of 25,000 registered voters shows that two out of every five registered voters vote in the elections. At this rate, how many people will vote in the next election?

Solution

Verbal model: $$\frac{\text{Total number of voters}}{\text{Total number of registered voters}} = \frac{\text{Voters}}{\text{Registered people}}$$

Labels: Total number of voters $= x$

Proportion: $$\frac{x}{25,000} = \frac{2}{5}$$

$$25,000\left(\frac{x}{25,000}\right) = 25,000\left(\frac{2}{5}\right)$$

$$x = 10,000$$

Thus, 10,000 people should vote in the next election.

■ Solutions to Starter Exercises ■

1. (a) $\dfrac{\boxed{4}}{9}$

(b) $\dfrac{5 \text{ pounds}}{24 \text{ ounces}} = \dfrac{\boxed{80 \text{ ounces}}}{24 \text{ ounces}}$

$= \dfrac{\boxed{80}}{24} = \dfrac{\boxed{10}}{3}$

2. (a) $\dfrac{\boxed{0.65}}{12} = \$0.0541\overline{6}$ per ounce

(b) *Unit price:*

$\dfrac{\boxed{4.62}}{5.5} = \boxed{\$0.84}$ per ounce

3. $\dfrac{\boxed{10.65}}{50} = \boxed{\$0.213}$ per foot

$\dfrac{\boxed{4.40}}{20} = \boxed{\$0.22}$ per foot

Thus, the larger spool has the smaller unit price.

4. (a) $\dfrac{9}{4} = \dfrac{36}{x}$

$\dfrac{\boxed{4}}{9} = \dfrac{x}{36}$

$36 \cdot \left(\dfrac{\boxed{4}}{9} \right) = 36 \cdot \dfrac{x}{36}$

$\boxed{16} = x$

(b) $\dfrac{23}{21} = \dfrac{z}{84}$

$\left(\boxed{84} \right) \cdot \dfrac{23}{21} = \left(\boxed{84} \right) \cdot \dfrac{z}{84}$

$\boxed{92} = \boxed{z}$

5. $\dfrac{8}{19} = \dfrac{x}{\boxed{4}}$

$\left(\boxed{4} \right) \cdot \dfrac{8}{19} = \left(\boxed{4} \right) \cdot \dfrac{x}{\boxed{4}}$

$\dfrac{\boxed{32}}{19} = x$

6. $\boxed{x = \text{total number of lights}}$

Proportion:

$\dfrac{2}{130} = \dfrac{\boxed{x}}{\boxed{3120}}$

$\boxed{3120 \cdot \dfrac{2}{130}} = \boxed{3120 \cdot \dfrac{x}{3120}}$

$\boxed{48} = \boxed{x}$

Thus, 48 lights are needed for the office.

| 3.4 | **EXERCISES** |

In Exercises 1–4, express the ratio as a fraction in reduced form. (Use the same units of measure for both quantities.)

1. Three ounces to 12 ounces

2. Four quarts to 96 ounces

3. Two feet to 10 inches

4. Two minutes to 90 seconds

In Exercises 5–7, express each as a ratio in reduced form. (Use the same units of measure for both quantities.)

5. In a town of 50,000 registered voters, 20,000 people voted in the last election. Find the ratio of voters to people registered.

6. You spent 20 hours on your algebra homework in the same week you spent 12 hours on your history homework. Find the ratio of algebra homework hours to history homework hours.

7. A half gallon of apple juice is mixed with 48 ounces of cranberry juice to make cranapple juice. Find the ratio of apple juice to cranberry juice in the cranapple juice.

In Exercises 8–10, find the unit price (in dollars per ounce) of the product.

8. A 12-ounce bag of corn chips for $1.80

9. A 1-pound can of coffee for $3.04

10. A gallon of water for 72¢

In Exercises 11–13, find which product has the smaller unit price.

11. A 3-pound bag of fertilizer for $4.80 or a 12-ounce bag for $1.32

12. A 12-foot piece of pipe for $1.44, or an 8-foot piece for 88¢

13. A 5-pound bag of nuts for $5.60, or a 24-ounce bag for $2.04

In Exercises 14–23, solve the given proportion.

14. $\dfrac{z}{3} = \dfrac{5}{9}$

15. $\dfrac{2}{5} = \dfrac{x}{15}$

16. $\dfrac{12}{t} = \dfrac{4}{3}$

17. $\dfrac{28}{15} = \dfrac{12}{y}$

18. $\dfrac{z+6}{14} = \dfrac{2}{7}$

19. $\dfrac{10}{27} = \dfrac{x-1}{36}$

20. $\dfrac{\frac{1}{2}}{6} = \dfrac{x}{8}$

21. $\dfrac{\frac{2}{3}}{10} = \dfrac{z}{9}$

22. $\dfrac{0.01}{0.12} = \dfrac{x}{6}$

23. $\dfrac{z}{0.15} = \dfrac{30}{225}$

24. A number is to six as 14 is to 3. Find the number.

25. The scale of a map is 1 inch equals 50 miles. Two cities are $2\frac{1}{2}$ inches apart on the map. What is the actual distance between the two cities?

26. A saline solution is made from 1 part salt for every 20 parts of water. How much salt is needed to mix with 10 ounces of water?

27. Two cups of flour are needed to make a batch of brownies. How much flour is needed to make $\frac{3}{4}$ of a batch of brownies?

28. Your car gets 38 miles to a gallon of gasoline. How much gasoline do you need for a 589 mile trip?

29. Two hundred fish in a lake are caught, tagged, and returned to the lake. Later 150 fish are caught and 30 of these had tags. Estimate the number of fish in the lake.

30. A saline solution is made from 1 part salt to 24 parts water. How much salt is needed to make 50 ounces of saline solution?

3.5 | Linear Inequalities and Applications

Section Highlights

1. Mechanically, the only difference between solving a linear inequality in one variable and solving a linear equation in one variable is, when multiplying (or dividing) both sides of an inequality by a negative quantity, we must reverse the inequality sign.

2. The difference between the solution set of a linear inequality in one variable and the solution set of a linear equation in one variable is the solution set of an inequality will have infinitely many elements and the solution set of a conditional linear equation has only one or zero elements.

EXAMPLE 1 ■   Graphs of Inequalities

Inequality	*Graph of Solution Set*	*Verbal Description*
(a) $x > 1$	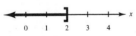	x is greater than 1.
(b) $x \le 2$		x is less than or equal to 2.
(c) $1 < x < 3$		x is greater than 1 *and x is less than 3.*
(d) $-1 \le x \le 2$		x is greater than or equal to -1 *and x is less than or equal to 2.*
(e) $-2 < x \le 0$		x is greater than -2 *and less than or equal to 0.*

EXAMPLE 2 ■   Solving a Linear Inequality

Solve the following inequalities and sketch the graph of its solution set.

(a) $x - 6 < -5$ (b) $-2x \geq -4$

(c) $3x - 1 > 5$ (d) $5x - 6 \leq 6x + 1$

Solution

(a) $x - 6 \; < -5$ Given inequality

   $x - 6 + 6 \; < -5 + 6$ Add 6 to both sides.

         $x \; < 1$

Thus, the solution set consists of all real numbers that are less than 1. The graph of the solution set is:

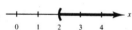

(b) $-2x \; \geq -4$ Given inequality

   $-\frac{1}{2}(-2x) \; \leq -\frac{1}{2}(-4)$ Multiply both sides by $-\frac{1}{2}$ and reverse the inequality sign.

         $x \; \leq 2$

Thus, the solution set consists of all real numbers less than or equal to 2. The graph of the solution set is:

(c) $3x - 1 \; > 5$ Given inequality

   $3x - 1 + 1 \; > 5 + 1$ Add 1 to both sides.

         $3x \; > 6$

   $\frac{1}{3}(3x) \; > \frac{1}{3}(6)$ Multiply both sides by $\frac{1}{3}$.

         $x \; > 2$

Thus, the solution set consists of all real numbers greater than 2. The graph of the solution set is:

(d) $5x - 6 \; \leq 6x + 1$ Given inequality

   $-6x + 5x - 6 \; \leq -6x + 6x + 1$ Add $-6x$ to both sides.

         $-x - 6 \; \leq 1$

   $-x - 6 + 6 \; \leq 1 + 6$ Add 6 to both sides.

         $-x \; \leq 7$

   $-1(-x) \; \geq -1(7)$ Multiply both sides by -1 and reverse the inequality sign.

         $x \; \geq -7$

Thus, the solution set consists of all real numbers greater than or equal to -7. The graph of the solution set is:

■ **Starter Exercise 1**   *Fill in the blanks.*

Solve the following inequalities and sketch the graph of its solution set.

(a) $x + 1 \geq -2$

$x + 1 + \boxed{} \geq -2 + \boxed{}$

$x \geq \boxed{}$

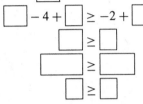

(b) $4 - 2x < 2$

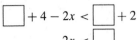

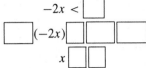

(c) $5x - 4 \geq 3x - 2$

$-3x + 5x - 4 \geq -3x + 3x - 2$

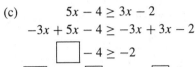

$\boxed{} - 4 \geq -2$

$\boxed{} - 4 + \boxed{} \geq -2 + \boxed{}$

$\boxed{} \geq \boxed{}$

$\boxed{} \geq \boxed{}$

$\boxed{} \geq \boxed{}$

EXAMPLE 3 ■   Solving a Double Inequality

Solve the following inequality.

$-2 < \tfrac{3}{2}x + 1 \leq 7$

Solution

$-2 < \tfrac{3}{2}x + 1 \leq 7$	Given inequality
$-2 + (-1) < \tfrac{3}{2}x + 1 + (-1) \leq 7 + (-1)$	Add -1 to all parts.
$-3 < \tfrac{3}{2}x \leq 6$	
$\tfrac{2}{3}(-3) < \tfrac{2}{3}\left(\tfrac{3}{2}x\right) \leq \tfrac{2}{3} \cdot 6$	Multiply all parts by $\tfrac{2}{3}$.
$-2 < x \leq 4$	

Thus, the solution set consists of all real numbers that are greater than -2 *and* less than or equal to 4. The graph of the solution set is:

■

■ **Starter Exercise 2**   *Fill in the blanks.*

Solve the inequality.

$7 \geq 4 - 2x \geq 0$

$\boxed{} + 7 \geq \boxed{} + 4 - 2x \geq \boxed{} + 0$

$\boxed{} \geq -2x \geq \boxed{}$

$\boxed{}\boxed{} \leq \boxed{}(-2x) \leq \boxed{}\boxed{}$

$\boxed{} \leq x \leq \boxed{}$

EXAMPLE 4 ■ An Application

Sam has $4.50 to spend on television antenna cable. The cable costs $0.15 per foot. How much cable can Sam buy?

Solution

Verbal model: (price per foot) · (number of feet) ≤ 4.50

Labels: $x =$ number of feet of cable

Inequality: $0.15x \leq 4.50$

$$\frac{1}{0.15}(0.15x) \leq \frac{1}{0.15}(4.50)$$

$$x \leq 30$$

Sam can buy at most 30 feet of cable.

■ **Solutions to Starter Exercises** ■

1. (a) $x + 1 + \boxed{(-1)} \geq -2 + \boxed{(-1)}$

$x \geq \boxed{-3}$

(b) $\boxed{-4} + 4 - 2x < \boxed{-4} + 2$

$-2x < \boxed{-2}$

$\boxed{-\frac{1}{2}}\boxed{(-2x)} \boxed{>} \boxed{-\frac{1}{2}}\boxed{(-2)}$

$x \boxed{>} \boxed{1}$

(c) $\boxed{2x} - 4 \geq -2$

$\boxed{2x} - 4 + \boxed{4} \geq -2 + \boxed{4}$

$\boxed{2x} \boxed{\geq} \boxed{2}$

$\boxed{\frac{1}{2}(2x)} \boxed{\geq} \boxed{\frac{1}{2}(2)}$

$\boxed{x} \boxed{\geq} \boxed{1}$

2. $\boxed{-4} + 7 \geq \boxed{-4} + 4 - 2x \geq \boxed{-4} + 0$

$\boxed{3} \geq -2x \geq \boxed{-4}$

$\boxed{-\frac{1}{2}}\boxed{(3)} \leq \boxed{-\frac{1}{2}}\boxed{(-2x)} \leq \boxed{-\frac{1}{2}}\boxed{(-4)}$

$\boxed{-\frac{3}{2}} \leq x \leq \boxed{2}$

3.5 | EXERCISES

In Exercises 1–20, solve the given inequalities.

1. $x + 2 < 7$

2. $3 + x \geq 14$

3. $x - 6 > 5$

4. $x - 1 \leq 8$

5. $2x \leq 4$

6. $3x < -9$

7. $-4x \geq 3$

8. $-\frac{1}{2}x > 4$

9. $\frac{3}{4}x < 9$

10. $-\frac{5}{6}z < \frac{15}{8}$

11. $3x + 1 \leq 10$

12. $1 - 2x > -5$

13. $5x - 2 < 7x + 4$

14. $6y - 1 \leq 2y + 7$

15. $2(4x - 1) - 4x < 6$

16. $-5(x + 2) > 4(3x - 2)$

17. $-3 < 2x + 1 < 5$

18. $-1 < -3(x + 4) \leq 5$

19. $\frac{1}{2} \leq \frac{1}{3}x + \frac{2}{3} < \frac{9}{16}$

20. $2 \leq -\frac{1}{3}(2x + 1) \leq 4$

In Exercises 21–30, solve the inequality and graph its solution set.

21. $5x - 1 \geq 9$

22. $1 - 6x < -5$

23. $x - 2(x + 2) < x$

24. $-2(3z + 1) > -4(z + 3)$

25. $\frac{1}{3}z + \frac{1}{2} \leq \frac{5}{6} + \frac{z}{6}$

26. $-1 < x + 3 < 5$

27. $0 \leq 4 - 2x \leq 4$

28. $12 \leq 3(2x + 4) < 14$

29. $-\frac{1}{2} < \frac{1}{2}x + 1 < \frac{2}{3}$

30. $-\frac{1}{6} \leq \frac{1}{2}(x - 1) - \frac{1}{3}x < \frac{1}{4}$

31. The sum of three consecutive integers is less than 42. What is the largest integer that could be the smallest of these three integers?

32. The perimeter of a rectangle is less than 24 ft. The length of the rectangle is twice the width. What is the longest width possible for this rectangle if the width is an integer?

33. To receive a grade of A in an algebra class, a student must have at least 90% of the total possible points on all four exams. Amy has 82, 97, and 94 on the first three exams. What is the smallest number of points Amy must get on the fourth exam to get an A if all exams are worth 100 points?

34. U-Drive car rental agency charges $42 per day for a car, regardless of the number of miles driven. The Wheels car rental agency charges $20 per day plus 10¢ per mile for their cars. How many miles can you drive a car from Wheels in a day and make the rental fee less than the fee at U-Drive?

35. The revenue from selling x units of a product is $R = 29.95x$. The cost of producing x units is $C = 21x + 500$. What is the smallest number of units needed to be produced to make a profit?

Cumulative Practice Test for Chapters P–3

In Exercises 1–6, match the correct graph of the solution set with each inequality.

1. $x \le 2$

2. $-1 < x \le 1$

3. $x - 1 \ge 2$

4. $-2x < 6$

5. $2x - 1 < -1$

6. $-5 < x - 2 < 0$

(a)

(b)

(c)

(d)

(e)

(f)

In Exercises 7–10, evaluate the given expressions.

7. $\frac{1}{2} - \left(\frac{1}{5} + \frac{3}{20}\right)$

8. $\frac{1}{3}(3)^2 - |-5^2|$

9. $2(6 - 2^2) - \frac{1}{3}$

10. $2^2 \cdot |-2| \div 4 - 2 \cdot 5$

In Exercises 11–16, simplify the variable expression.

11. $2(x - 1) - 2x$

12. $2x(3x - 1) - 4x^2 + 1$

13. $-3[2x - 4(2x - 1)]$

14. $x(2x - y) + y(5x + 2y)$

15. $\frac{1}{3}x - \frac{1}{2}\left(x - \frac{1}{2}\right)$

16. $2\left[\frac{1}{2}x - \left(\frac{1}{3}x + 1\right) + \frac{1}{6}x\right]$

In Exercises 17–30, solve the given equation or inequality.

17. $5x - 1 = 4$ $\quad 5x - 1 + 1 = 4 + 1$
$\quad 5x = -5$
$\quad \frac{5}{5} \quad \frac{5}{5}$

18. $6x + 12 = 9x - 3$

19. $4(2x - 7) = 8x - 7$

20. $\frac{1}{2}(2x - 3) = \frac{1}{3}x$

21. $5(2x + 1) = 3(3x - 2)$

22. $2[x - (2x + 6)] = 10$

23. $-2[2x + (6x - 1)] = x - (2x + 3)$

24. $1 - x \ge 7$

25. $3x + 12 > 0$

26. $\frac{4}{5}x - \frac{1}{3} \le \frac{1}{5} - \frac{1}{15}x$

27. $4(z - 1) \ge 3(2 - z)$

28. $1 \le 2x - 1 < 3$

29. $-2 < 2(3x - 1) < 5$

30. $-\frac{1}{6} \le \frac{1}{2}x - \frac{4}{3} \le \frac{2}{3}$

31. $\frac{1}{5} = \frac{x}{30}$

32. $\frac{x}{27} = \frac{5}{36}$

33. A number is 14 less than twice the number. What is the number?

34. The sum of three consecutive even integers is 48. Find the integers.

35. Two numbers have a sum of 26. Twice the smaller number is five less than the larger number. Find the two numbers.

36. What is the largest integer such that three times itself increased by six is less than 24?

37. The length of a rectangle is 6. What is the largest width so that the perimeter is less than 26 and the width is an integer?

38. A pair of shoes is on sale for $24. This is a 20% discount off the regular price. What is the regular price of the shoes?

39. If gasoline is $1.129 per gallon and your car gets 35 miles per gallon, how much does it cost you to drive your car 200 miles? (Round off to the nearest cent.)

40. If $1000 was put into an account earning 12% interest compounded monthly, then the amount in the account after one year is $A = 1000\left[1 + \left(\frac{0.12}{12}\right)\right]^{12}$. Use a calculator to calculate this amount.

3.1	**Answers to Exercises**

5. $2x = 5$

$\frac{1}{2}(2x) = \frac{1}{2}(5)$ Multiply both sides by $\frac{1}{2}$.

$x = \frac{5}{2}$ Solution

6. $x - 7 = -11$

$x - 7 + 7 = -11 + 7$ Add 7 to both sides.

$x = -4$ Solution

7. $-2x - 1 = 7$

$-2x - 1 + 1 = 7 + 1$ Add 1 to both sides.

$-2x = 8$ Simplify.

$-\frac{1}{2}(-2x) = -\frac{1}{2}(8)$ Multiply both sides by $-\frac{1}{2}$.

$x = -4$ Solution

8. $3x + 6 = 15$

$3x + 6 + (-6) = 15 + (-6)$ Add −6 to both sides.

$3x = 9$ Simplify.

$\frac{1}{3}(3x) = \frac{1}{3}(9)$ Multiply both sides by $\frac{1}{3}$.

$x = 3$ Solution

3.1	**Answers to Exercises**

9. $x = 7$ **10.** $x = -3$ **11.** $x = \frac{5}{2}$ **12.** $x = -2$

13. $x = 10$ **14.** $x = -2$ **15.** $x = \frac{1}{2}$ **16.** no solution

17. $x = -4$ **18.** $x = -18$ **19.** $x = 8$ **20.** $x = \frac{7}{3}$

21. $x = -7$ **22.** $x = 2$ **23.** $x = -\frac{4}{7}$ **24.** no solution

25. $x = \frac{1}{3}$ **26.** $x = -35$ **27.** $x = -\frac{25}{4}$ **28.** $x = \frac{2}{3}$

29. $x = \frac{7}{5}$ **30.** $x = -3$ **31.** 7, 8, 9   **32.** 14, 16, 18

33. 5, 7   **34.** $562.50   **35.** 94 ft by 50 ft

3.2	**Answers to Exercises**

1. 0.5   **2.** 0.043   **3.** 0.0001   **4.** 0.004   **5.** 46%

6. 97%   **7.** 381%   **8.** 0.24%   **9.** $\frac{1}{2}$ **10.** $\frac{1}{3}$

11. $\frac{11}{500}$ **12.** 1   **13.** 50%   **14.** 75%   **15.** $66\frac{2}{3}\%$

16. $62\frac{1}{2}\%$ **17.** 9.6   **18.** 80%   **19.** 27   **20.** 400

21. 36   **22.** $66\frac{2}{3}\%$ **23.** 600   **24.** 250%   **25.** 15%

26. 30   **27.** $2\frac{1}{4}\%$ **28.** 50   **29.** 3   **30.** 7

31. $7\frac{7}{9}\%$ **32.** 336   **33.** 40%

3.3	**Answers to Exercises**

1. $y = -3$ **2.** $x = 2$ **3.** $x = 7$ **4.** $z = 1$ **5.** $x = 1$

6. $x = \frac{25}{14}$ **7.** $x = -\frac{1}{5}$ **8.** $x = -2$ **9.** $x = 0$ **10.** $x = 10$

11. $x = 2$ **12.** $x = 5$ **13.** $x = 1$ **14.** $x = \frac{12}{11}$ **15.** $z = -11$

16. no solution   **17.** $z = \frac{3}{2}$ **18.** $x = -1$ **19.** $x = \frac{8}{5}$ **20.** $x = -23$

21. 14 and 28   **22.** 16 in. and 20 in.   **23.** 1 hr 12 min

24. 8 ft from the 25 lb force   **25.** 97

3.4   Answers to Exercises

1. $\frac{1}{4}$ **2.** $\frac{4}{3}$ **3.** $\frac{12}{5}$ **4.** $\frac{4}{3}$ **5.** $\frac{2}{5}$

6. $\frac{5}{3}$ **7.** $\frac{4}{3}$ **8.** $0.15   **9.** $0.19   **10.** $0.005625

11. The unit price of the larger bag is $0.10 per ounce, the unit price of the smaller bag is $0.11 per ounce.

12. The unit price of the longer pipe is $0.12 per foot, the unit price of the smaller pipe is $0.11 per foot.

13. The unit price of the larger bag is $0.07 per ounce, the unit price of the smaller bag is $0.085 per ounce.

14. $\frac{5}{3}$ **15.** 6   **16.** 9   **17.** $\frac{45}{7}$ **18.** -2

19. $\frac{43}{3}$ **20.** $\frac{2}{3}$ **21.** $\frac{3}{5}$ **22.** $\frac{1}{2}$ **23.** 0.02

24. 28   **25.** 125 mi   **26.** $\frac{1}{2}$ oz   **27.** $1\frac{1}{2}$ cups   **28.** 15.5 gal

29. 1000 fish   **30.** 2 oz

3.5   Answers to Exercises

1. $x < 5$ **2.** $x \geq 11$ **3.** $x > 11$ **4.** $x \leq 9$

5. $x \leq 2$ **6.** $x < -3$ **7.** $x \leq -\frac{3}{4}$ **8.** $x < -8$

9. $x < 12$ **10.** $z > -\frac{9}{4}$ **11.** $x \leq 3$ **12.** $x < 3$

13. $x > -3$ **14.** $y \leq 2$ **15.** $x < 2$ **16.** $x < -\frac{2}{17}$

17. $-2 < x < 2$ **18.** $-\frac{17}{3} \leq x < -\frac{11}{3}$ **19.** $-\frac{1}{2} \leq x < -\frac{5}{16}$

20. $-\frac{13}{2} \leq x \leq -\frac{7}{2}$ **21.** $x \geq 2$ **22.** $x > 1$

23. $x > -2$

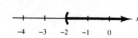

24. $z < 5$

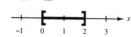

25. $z \leq 2$

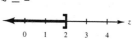

26. $-4 < x < 2$

27. $0 \leq x \leq 2$

28. $0 \leq x < \frac{1}{3}$

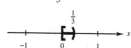

29. $-3 < x < -\frac{2}{3}$

30. $2 \leq x < \frac{9}{2}$

31. 12

32. 3 ft

33. at least 87 points

34. less than 220 miles

35. 56 units

Answers to Cumulative Practice Test P–3

1. (b)

2. (f)

3. (a)

4. (d)

5. (e)

6. (c)

7. $\frac{3}{20}$

8. -22

9. $\frac{11}{3}$

10. -8

11. -2

12. $2x^2 - 2x + 1$

13. $18x - 12$

14. $2x^2 + 4xy + 2y^2$

15. $-\frac{1}{6}x + \frac{1}{4}$

16. $\frac{2}{3}x - 2$

17. $x = 1$

18. $x = 5$

19. no solution

20. $x = \frac{9}{4}$

21. $x = -11$

22. $x = -11$

23. $-\frac{1}{3}$

24. $x \leq -6$

25. $x > -\frac{1}{4}$

26. $x \leq \frac{8}{13}$

27. $z \geq \frac{10}{7}$

28. $1 \leq x < 2$

29. $0 < x < \frac{7}{6}$

30. $\frac{7}{3} \leq x \leq 4$

31. $x = 6$

32. $x = \frac{15}{4}$

33. 14

34. 14, 16, 18

35. 7 and 19

36. 5

37. 6

38. $30

39. $6.45

40. $1126.83

C H A P T E R   F O U R
Graphs and Linear Applications

4.1 | Ordered Pairs and Graphs

Section Highlights

1. A **rectangular coordinate** system is formed by two real lines intersecting at right angles, at their origins. The horizontal line is usually called the x-axis and the vertical line is called the y-axis.
2. Every ordered pair of real numbers corresponds to a point in the rectangular coordinate system, and every point in the rectangular coordinate system corresponds to an ordered pair of real numbers.
3. An ordered pair (a, b) is a solution of an equation in x and y if replacing x by a and y by b makes the equation a true statement.

EXAMPLE 1 ■   Plotting Points in a Rectangular Coordinate System

Plot the points $(2, 4)$, $(1, 0)$, $(-1, 3)$, $(-2, -2)$, and $(4, -1)$ in a rectangular coordinate system.

Solution

The point $(2, 4)$ is 2 units to the right of the vertical axis and 4 units up from the horizontal axis. Similarly, the point $(1, 0)$ is one unit to the right of the vertical axis and on the horizontal axis. The other points are plotted similarly.

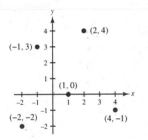

Starter Exercise 1 | *Fill in the blanks.*

Plot the points $(-1, 3)$, $(2, -3)$, $(5, 1)$, $(0, 0)$, and $(-3, -2)$ in a rectangular coordinate system.

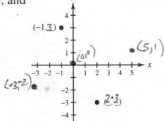

EXAMPLE 2 ■ Finding the Coordinates of Points

Find the coordinates of the points shown in the figure.

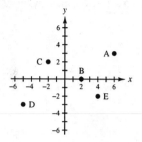

Solution

Point	Position	Coordinates
A	6 units right, 3 units up	(6, 3)
B	2 units right, 0 units up	(2, 0)
C	2 units left, 2 units up	(−2, 2)
D	5 units left, 3 units down	(−5, −3)
E	4 units right, 2 units down	(4, −2)

■

Starter Exercise 2   *Fill in the blanks.*

Find the coordinates of the points shown below.

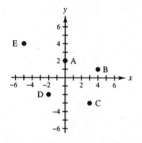

Point	Position	Coordinates
A	0 units right, 2 units up	(0, ☐)
B	4 units ☐, 1 unit up	(☐, ☐)
C	☐ units right, ☐ units down	(☐, ☐)
D	☐	(☐, ☐)
E	☐	(☐, ☐)

EXAMPLE 3 ■   An Application of the Rectangular Coordinate System

Every day for a two-week period, a meteorology student measured the temperature and recorded the information as the following ordered pairs (2, −1), (5, 3), (2, 0), (3, 1), and (2, 2), where the first coordinate is the number of days the temperature occurred and the second coordinate is the temperature. Plot these points on a rectangular coordinate system.

Solution

We plot the number of days on the x-axis and the temperature on the y-axis.

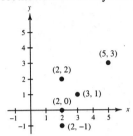

Starter Exercise 3   *An Application—Fill in the blanks.*

An algebra student recorded her study time in one week as follows.

Day	1	2	3	4	5
Hours	2	1	4	3	2

Plot the points on a rectangular coordinate system. Plot the day number on the x-axis and the hours on the y-axis.

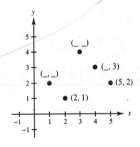

EXAMPLE 4 ■   Making Up a Table of Values

Make up a table of values showing five solution points for the equation $6x - 3y = 9$. Then plot the solution points in the rectangular coordinate system. (Choose x-values of −2, −1, 0, 1, and 2.)

Solution

We begin by solving the given equation for y.

$$6x - 3y = 9$$
$$-3y = -6x + 9$$
$$y = 2x - 3$$

Now we construct a table.

x	−2	−1	0	1	2
$y = 2x - 3$	−7	−5	−3	−1	1
Solution	(−2, −7)	(−1, −5)	(0, −3)	(1, −1)	(2, 1)

Now we plot these five points on the rectangular coordinate system.

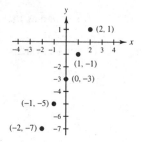

Starter Exercise 4 | *Fill in the blanks.*

Make up a table of values showing five solutions for the equation $8x + 2y = -2$. Then plot the solution points on the rectangular coordinate system. First solve the given equation for y.

$$8x + 2y = -2$$

$$2y = \boxed{} - 2$$

$$y = \boxed{} - \boxed{}$$

x	-2	-1	0	1	2
$y = \boxed{} - \boxed{}$	7	3	$\boxed{}$	-5	$\boxed{}$
Solution	$(-2, 7)$	$\left(-1, \boxed{}\right)$	$\left(0, \boxed{}\right)$	$\left(\boxed{}, \boxed{}\right)$	$\left(\boxed{}, \boxed{}\right)$

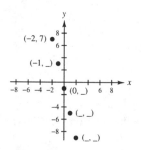

EXAMPLE 5 ■   Verifying Solutions of an Equation

Determine which of the following ordered pairs is a solution of the equation $2x - y = 3$.

(a) $(2, 1)$ 　　　　(b) $(4, -1)$ 　　　　(c) $\left(\frac{1}{3}, -\frac{7}{3}\right)$ 　　　　(d) $(-2, 1)$

Solution

(a) For the ordered pair (2, 1) we substitute 2 for x and 1 for y in the given equation.

$$2x - y = 3 \qquad \text{Given equation}$$
$$2(2) - 1 \stackrel{?}{=} 3 \qquad \text{Substitute.}$$
$$4 - 1 \stackrel{?}{=} 3$$
$$3 = 3 \qquad \text{(2, 1) is a solution.}$$

Since the substitution did satisfy the equation, we conclude that (2, 1) is a solution of the equation.

(b) The ordered pair (4, −1) is not a solution of the given equation since
$$2(4) - (-1) = 8 + 1 = 9 \neq 3.$$

(c) The ordered pair $\left(\frac{1}{3}, -\frac{7}{3}\right)$ is a solution of the equation since
$$2\left(\frac{1}{3}\right) - \left(-\frac{7}{3}\right) = \frac{2}{3} + \frac{7}{3} = \frac{9}{3} = 3.$$

(d) The ordered pair (−2, 1) is not a solution of the equation since
$$2(-2) - 1 = -4 - 1 = -5 \neq 3.$$

■

Starter Exercise 5 *Fill in the blanks.*

Determine which of the following ordered pairs is a solution of the equation $2x + 4y = 4$.

Test		*Conclusion*

(a) (5, 2)
$$2(5) + 4(2) = 10 + 8 = \boxed{} \qquad \text{No}$$

(b) (4, −1)
$$2(4) + 4(-1) = \boxed{} + \boxed{} = 4 \qquad \boxed{}$$

(c) (2, 0)
$$2(2) + 4(0) = 4 + \boxed{} = \boxed{} \qquad \boxed{}$$

(d) (−1, −1)
$$2\boxed{} + 4\boxed{} = \boxed{} + \boxed{} = \boxed{} \qquad \boxed{}$$

EXAMPLE 6 ■ An Application

The cost y of producing x units is given by $y = 10x + 1000$. Plot the cost of producing 100, 150, 200, and 250 units.

Solution

Calculation *Solution*

$y = 10(100) + 1000 = 1000 + 1000 = 2000$ (100, 2000)
$y = 10(150) + 1000 = 1500 + 1000 = 2500$ (150, 2500)
$y = 10(200) + 1000 = 2000 + 1000 = 3000$ (200, 3000)
$y = 10(250) + 1000 = 2500 + 1000 = 3500$ (250, 3500)

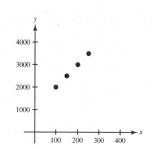

■

◼ **Starter Exercise 6** | *An Application—Fill in the blanks.*

When an employee produces x units per hour, the hourly wage is given by $y = x + 5$. Plot the wages for producing 2, 3, 4, and 5 units.

Calculation   *Solution*

$y = 2 + 5 = 7$ $(2, 7)$

$y = 3 + 5 = 8$ $\left(3, \boxed{}\right)$

$y = \boxed{} + 5 = \boxed{}$ $\left(4, \boxed{}\right)$

$y = \boxed{} + 5 = \boxed{}$ $\left(\boxed{}, \boxed{}\right)$

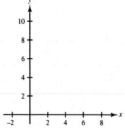

◼ **Solutions to Starter Exercises** ◼

1.

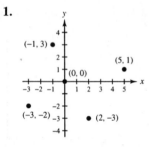

2. *Point*   *Position*   *Coordinates*

A   0 units right, 2 units up $\left(0, \boxed{2}\right)$

B   4 units $\boxed{\text{right}}$, 1 unit up $\left(\boxed{4}, \boxed{1}\right)$

C $\boxed{3}$ units right, $\boxed{3}$ units down $\left(\boxed{3}, \boxed{-3}\right)$

D $\boxed{\text{2 units left, 2 units down}}$ $\left(\boxed{-2}, \boxed{-2}\right)$

E $\boxed{\text{5 units left, 4 units up}}$ $\left(\boxed{-5}, \boxed{4}\right)$

3.

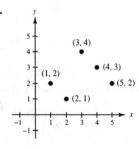

■ **Solutions to Starter Exercises** ■

4. $8x + 2y = -2$

$$2y = \boxed{-8x} - 2$$

$$y = \boxed{-4x} - \boxed{1}$$

x	-2	-1	0	1	2
$y = -4x - 1$	7	3	-1	-5	-9
Solution	$(-2, 7)$	$(-1, 3)$	$(0, -1)$	$(1, -5)$	$(2, -9)$

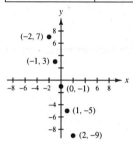

5.   *Test*   *Conclusion*

(a) $(5, 2)$

$2(5) + 4(2) = 10 + 8 = \boxed{18}$ No

(b) $(4, -1)$

$2(4) + 4(-1) = \boxed{8} + \boxed{(-4)} = 4$ $\boxed{\text{Yes}}$

(c) $(2, 0)$

$2(2) + 4(0) = 4 + \boxed{0} = \boxed{4}$ $\boxed{\text{Yes}}$

(d) $(-1, -1)$

$2\boxed{(-1)} + 4\boxed{(-1)} = \boxed{-2} + \boxed{(-4)} = \boxed{-6}$ $\boxed{\text{No}}$

6. *Calculation*   *Solution*

$y = 2 + 5 = 7$ $(2, 7)$

$y = 3 + 5 = 8$ $\left(3, \boxed{8}\right)$

$y = \boxed{4} + 5 = \boxed{9}$ $\left(4, \boxed{9}\right)$

$y = \boxed{5} + 5 = \boxed{10}$ $\left(\boxed{5}, \boxed{10}\right)$

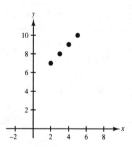

4.1 | EXERCISES

In Exercises 1–4, plot the points in the rectangular coordinate system.

1. $(2, -2)$, $(-3, 1)$, $(0, 0)$ **2.** $(0, -5)$, $(3, 2)$, $(-1, -3)$

3. $(6, -1)$, $(1, 1)$, $(2, -7)$ **4.** $(6, 0)$, $(-1, -3)$, $(5, 1)$

In Exercise 5, determine the quadrant in which the point is located.

5. (a) $\left(-1, -\frac{3}{4}\right)$ (b) $\left(\frac{7}{8}, -\frac{4}{3}\right)$

In Exercise 6, label the approximate coordinate of the points on the coordinate system.

6.

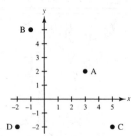

In Exercises 7–10, plot the points and connect them with line segments to form the figure.

7. Triangle:
$(1, 1)$, $(3, 4)$, $(-1, 2)$

8. Triangle:
$(0, 3)$, $(-4, 1)$, $(-2, -3)$

9. Square:
$(0, 0)$, $(2, 2)$, $(4, 0)$, $(2, -2)$

10. Parallelogram:
$(-1, 3)$, $(1, 5)$, $(3, 0)$, $(5, 2)$

In Exercises 11–14, determine whether the ordered pairs are solutions of the given equation.

Equation	*Ordered Pair*		*Equation*	*Ordered Pair*
11. $y = 5x - 1$	(a) $(0, 0)$	**12.**	$y = \frac{1}{2}x + 3$	(a) $(2, 4)$
	(b) $(2, 9)$			(b) $(0, 3)$
	(c) $(-1, -6)$			(c) $\left(\frac{1}{2}, 3\right)$
	(d) $\left(\frac{1}{5}, 0\right)$			(d) $(-4, 5)$
13. $2x - 3y = 6$	(a) $(0, 2)$	**14.**	$x - \frac{1}{2}y = 2$	(a) $(1, 2)$
	(b) $(3, 0)$			(b) $(3, -2)$
	(c) $(0, -2)$			(c) $(0, 0)$
	(d) $\left(\frac{1}{2}, \frac{1}{3}\right)$			(d) $(4, 5)$

In Exercises 15–24, plot the solution points, in the rectangular coordinate system, that correspond to the given values of x.

	Equation	*Values of x*		*Equation*	*Values of x*
15.	$y = 2x - 3$	$-2, -1, 0, 1, 2$	**16.**	$y = -\frac{1}{2}x + 1$	$-4, -2, 0, 2, 4$
17.	$y = \frac{3}{2}x$	$-4, -2, 0, 2, 4$	**18.**	$y = 3x - 4$	$1, 2, 3, 4$
19.	$y = x - 3$	$-5, -3, 0, 3, 5$	**20.**	$2x + 3y = 6$	$-6, -3, 0, 3, 6$
21.	$6x - 4y = 12$	$-2, 0, 2, 4, 6$	**22.**	$x - y = 1$	$-2, -1, 0, 1, 2$
23.	$15x - 5y = 25$	$-1, 0, 1, 2, 3$	**24.**	$3x + 9y = 18$	$-6, -3, 0, 3$

25. *Cost* The cost y of producing x units of a product is given by $y = 2x + 50$. Complete the following table to determine the cost of producing the specified number of units. Plot your results on the rectangular coordinate system.

x	50	100	150	200
$y = 2x + 50$				

26. *Hourly Wages* When an employee produces x units per hour, the hourly wage is given by $y = 0.5x + 8$. Complete the following table to determine the hourly wages for producing the specified number of units. Plot your results on the rectangular coordinate system.

x	2	6	10	14
$y = 0.5x + 8$				

4.2 Graphs of Equations

Section Highlights

1. The **graph** of an equation in two variables is the set of all points in the rectangular coordinate system that corresponds to an element of the solution set of the equation.

2. A point at which a graph of an equation meets the x-axis is called an **x-intercept**. To find x-intercepts, replace y by zero and solve the equation for x. Note that x-intercepts are of the form (a, 0).

3. A point at which a graph of an equation meets the y-axis is called a **y-intercept**. To find y-intercepts, replace x by zero and solve the equation for y. Note that y- intercepts are of the form (0, b).

EXAMPLE 1 ■ Sketching the Graph of an Equation

Sketch the graph of $4x + y = 1$.

Solution

Start by constructing the table below. To help with this, solve the equation for y.

$y = -4x + 1$

x	-2	-1	0	1	2
$y = -4x + 1$	9	5	1	-3	-7
Solution	$(-2, 9)$	$(-1, 5)$	$(0, 1)$	$(1, -3)$	$(2, -7)$

Now plot these points on the rectangular coordinate system. Complete the sketch by drawing a line through the points.

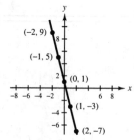

Starter Exercise 1   *Fill in the blanks.*

Complete the following table and sketch the graph of $6x - 3y = 9$.

x	-2	-1	0	1	2
$y = 2x - 3$	-7	$\square$	$\square$	$\square$	$\square$
Solution	$(\square, \square)$	$(-1, -5)$	$(\square, \square)$	$(\square, \square)$	$(\square, \square)$

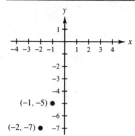

EXAMPLE 2 ■ Sketching the Graph of a Nonlinear Equation

Sketch the graph of $x^2 - y = 3$.

Solution

First solve the equation for y, then construct a table of values.

$$y = x^2 - 3$$

x	-3	-2	-1	0	1	2	3
$y = x^2 - 3$	6	1	-2	-3	-2	1	6
Solution	$(-3, 6)$	$(-2, 1)$	$(-1, -2)$	$(0, -3)$	$(1, -2)$	$(2, 1)$	$(3, 6)$

Now plot the points and connect them with a smooth curve.

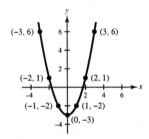

Starter Exercise 2 | *Fill in the blanks.*

Sketch the graph of $y = |x|$ by completing the table, plotting the points, and connecting the points with a smooth curve.

x	-2	-1	0	1	2
$y = |x|$	2	☐	☐	☐	☐
Solution	(☐, ☐)	$(-1, 1)$	(☐, ☐)	(☐, ☐)	(☐, ☐)

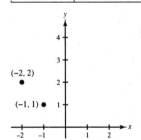

EXAMPLE 3 ■   Finding the Intercepts of a Graph

Find the intercepts and sketch the graph of $4x - 2y = 8$.

Solution

To find the x-intercept, replace y by zero and solve for x.

$$4x - 2y = 8 \qquad \text{Given equation}$$
$$4x - 2(0) = 8 \qquad \text{Let } y = 0.$$
$$x = 2 \qquad \text{Solve for } x.$$

Thus, we conclude the graph has an x-intercept at $(2, 0)$. To find the y-intercept, replace x by zero and solve for y.

$$4x - 2y = 8 \qquad \text{Given equation}$$
$$4(0) - 2y = 8 \qquad \text{Let } x = 0.$$
$$y = -4 \qquad \text{Solve for } y.$$

Thus, we conclude the graph has a y-intercept at $(0, -4)$. Now, construct a table of values and sketch the graph. (Solve the equation for y.)

x	-2	-1	0	1	2
$y = 2x - 4$	-8	-6	-4	-2	0
Solution	$(-2, -8)$	$(-1, -6)$	$(0, -4)$	$(1, -2)$	$(2, 0)$

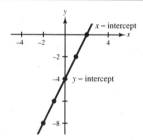

Starter Exercise 3 | *Fill in the blanks.*

Find the intercepts of the graph of $y = \frac{1}{2}x - 4$.

x-intercept

$$y = \frac{1}{2}x - 4$$
$$\boxed{0} = \frac{1}{2}x - 4$$
$$\boxed{4} = \frac{1}{2}x$$
$$\boxed{8} = x$$
$$\left(\boxed{8}, \boxed{0}\right)$$

y-intercept

$$y = \frac{1}{2}x - 4$$
$$y = \frac{1}{2}\boxed{0} - 4$$
$$y = \boxed{-4}$$
$$\left(\boxed{0}, \boxed{-4}\right)$$

EXAMPLE 4 ■ A Graph That has Two *x*-intercepts

Find the intercepts and sketch the graph of $y = x^2 - x - 6$.

Solution

To find any *x*-intercepts, replace *y* by zero and solve for *x*.

$y = x^2 - x - 6$	Given equation
$0 = x^2 - x - 6$	Let $y = 0$.
$0 = (x - 3)(x + 2)$	Factor.
$x - 3 = 0$ or $x + 2 = 0$	Set both factors equal to zero.
$x = 3$ $x = -2$	Solve for *x*.

Thus, the graph has two *x*-intercepts at $(3, 0)$ and $(-2, 0)$. To find any *y*-intercepts, replace *x* by zero and solve for *y*.

$y = x^2 - x - 6$	Given equation
$y = 0^2 - 0 - 6$	Let $x = 0$.
$y = -6$	Solve for *y*.

Thus, the graph has a *y*-intercept at $(0, -6)$. Now, construct a table of values and sketch the graph.

x	-2	-1	0	$\frac{1}{2}$	1	2	3
$y = x^2 - x - 6$	0	-4	-6	$-6\frac{1}{4}$	-6	-4	0
Solution	$(-2, 0)$	$(-1, -4)$	$(0, -6)$	$(\frac{1}{2}, -6\frac{1}{4})$	$(1, -6)$	$(2, -4)$	$(3, 0)$

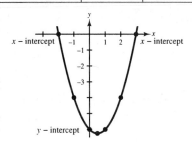

Starter Exercise 4 | *Fill in the blanks.*

Find the intercepts of the graph of $y = x^2 + 4x + 3$.

x-intercepts

$y = x^2 + 4x + 3$

$\boxed{} = x^2 + 4x + 3$

$\boxed{} = \left(x + \boxed{}\right)\left(x + \boxed{}\right)$

$x + \boxed{} = \boxed{}$ or $x + \boxed{} = \boxed{}$

$x = \boxed{}$ $x = \boxed{}$

$\left(\boxed{}, \boxed{}\right)$ and $\left(\boxed{}, \boxed{}\right)$

y-intercept

$y = x^2 + 4x + 3$

$y = 0^2 + 4\boxed{} + 3$

$y = \boxed{}$

$\left(\boxed{}, \boxed{}\right)$

■ **Solutions to Starter Exercises** ■

1.

x	-2	-1	0	1	2
$y = 2x - 3$	-7	-5	-3	-1	1
Solution	$(-2, -7)$	$(-1, -5)$	$(0 - 3)$	$(1, -1)$	$(2, 1)$

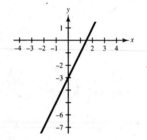

2.

x	-2	-1	0	1	2
$y = |x|$	2	1	0	1	2
Solution	$(-2, 2)$	$(-1, 1)$	$(0, 0)$	$(1, 1)$	$(2, 2)$

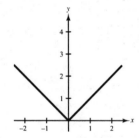

3.

x-intercept

$y = \frac{1}{2}x - 4$

$\boxed{0} = \frac{1}{2}x - 4$

$\boxed{4} = \frac{1}{2}x$

$\boxed{8} = x$

$\left(\boxed{8}, \boxed{0} \right)$

y-intercept

$y = \frac{1}{2}x - 4$

$y = \frac{1}{2} \boxed{(0)} - 4$

$y = \boxed{-4}$

$\left(\boxed{0}, \boxed{-4} \right)$

■ Solutions to Starter Exercises ■

4.

x-intercepts	y-intercept

x-intercepts:

$$y = x^2 + 4x + 3$$

$$\boxed{0} = x^2 + 4x + 3$$

$$\boxed{0} = \left(x + \boxed{1}\right)\left(x + \boxed{3}\right)$$

$$x + \boxed{1} = \boxed{0} \quad \text{or} \quad x + \boxed{3} = \boxed{0}$$

$$x = \boxed{-1} \qquad\qquad x = \boxed{-3}$$

$$\left(\boxed{-1}, \boxed{0}\right) \text{ and } \left(\boxed{-3}, \boxed{0}\right)$$

y-intercept:

$$y = x^2 + 4x + 3$$

$$y = 0^2 + 4\boxed{(0)} + 3$$

$$y = \boxed{3}$$

$$\left(\boxed{0}, \boxed{3}\right)$$

4.2 | EXERCISES

In Exercises 1–4, complete the table and use the resulting solution points to sketch the graph of the equation.

1. $y = 2x - 1$

x	-2	-1	0	1	2
y					
(x, y)					

2. $3x + 2y = 4$

x	-4	-2	0	2	4
y					
(x, y)					

3. $y = |x - 12|$

x	10	11	12	13	14
y					
(x, y)					

4. $y = x^2 - 5$

x	-2	-1	0	1	2
y					
(x, y)					

In Exercises 5–8, solve the equation for y.

5. $x - y = 6$

6. $2x + 3y = 12$

7. $x^2 + 3x + y = 8$

8. $5x^2 - 2y = 17$

In Exercises 9–16, find x- and y-intercepts (if any) of the graph of the equation.

9. $x - y = 4$

10. $x - 2y + 4 = 0$

11. $y = \frac{3}{5}x + 9$

12. $y = x^2 - 4$

13. $x = 6$

14. $y = x^2 + 4$

15. $y = x^2 - 6x + 8$

16. $y = x(x + 1) - 5(x + 1)$

In Exercises 17–24, sketch the graph of the equation and show the coordinates of at least two solution points on the graph.

17. $y = 3 - x$ **18.** $y = \frac{1}{2}x - 1$ **19.** $x = 4$

20. $y - 3 = 0$ **21.** $4x - 2y = 6$ **22.** $3x + 2y = 4$

23. $y = x^2 - 2$ **24.** $y = |x - 2|$

25. *Distances Traveled*   Let y represent the distance traveled by a car that is moving at a constant rate of 55 mph. Let x represent the number of hours that the car has been traveling. Write an equation that gives the distance y in terms of x and sketch the graph of the equation.

4.3 | Graphs and Graphing Utilities

Section Highlights

Basic Graphing Steps for a Graphing Calculator

To sketch the graph of an equation involving x and y, use the following steps. (Before performing these steps, you should set your calculator so that all of the standard defaults are active. Press ⎡ZOOM⎤ ⎡6⎤ ⎡ENTER⎤ .)

1. Rewrite the equation so that y is written as a function of x. In other words, rewrite the equation so that y is isolated on the left side of the equation.

2. Press the ⎡Y =⎤ key. Then enter the right side of the equation on the first line of the display. (The first line is labeled as Y_1=.)

3. Press the ⎡GRAPH⎤ key.

⊞ EXAMPLE 1 ■   Sketching the Graph of a Linear Equation

Sketch the graph of $3x - 2y = 4$.

Solution:

Start by solving the equation for y.

$3x - 2y = 4$ Given equation

$-2y = -3x + 4$ Add $-3x$ to both sides.

$y = \frac{3}{2}x - 2$ Multiply both sides by $-\frac{1}{2}$.

Next, after pressing the ⎡Y =⎤ key, enter the following keystrokes.

The screen should look like the figure at the right.

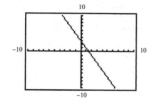 **Starter Exercise 1** *Fill in the blanks.*

Sketch the graph of $2x + y = 3$.

$$2x + y = 3$$

$$y = \boxed{} + 3$$

Keystrokes: $\boxed{Y=}$ $\boxed{(-)}$ 2 $\boxed{}$ $\boxed{+}$ 3 $\boxed{GRAPH}$

Screen:

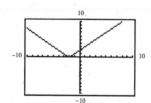

EXAMPLE 2 ■ Sketching the Graph of an Equation Involving Absolute Value

Sketch the graph of $y = |x + 2|$.

Solution

Use the following keystrokes.

$\boxed{Y=}$ $\boxed{2nd}$ $\boxed{ABS}$ $\boxed{(}$ $\boxed{X, T, \theta}$ $\boxed{+}$ 2 $\boxed{)}$ $\boxed{GRAPH}$

Screen:

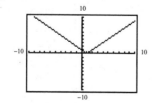 **Starter Exercise 2** *Fill in the blanks.*

Sketch the graph of $y = |x - 1|$.

Keystrokes: $\boxed{Y=}$ $\boxed{2nd}$ $\boxed{}$ $\boxed{(}$ $\boxed{}$ $\boxed{}$ 1 $\boxed{)}$ $\boxed{GRAPH}$

Screen:

▦ EXAMPLE 3 ■ Resetting the Scales on an Axis

Sketch the graph of $y = |x| + 12$.

Solution

If you press [Y=] [2nd] [ABS] [X, T, θ] [+] 12 [GRAPH], you will not see any graph on your screen. The reason for this is that the lowest point on the graph of $y = |x| + 12$ occurs at (0, 12). Using the standard range settings, we obtain a screen whose largest y-value is 10.

To change the settings, press [RANGE] and change the Ymax=10 to Ymax=30. Then change Yscl=1 to Yscl=5. Now press [GRAPH] and you will obtain the graph.

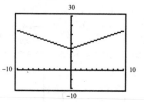

▦ Starter Exercise 3 | *Fill in the blanks.*

Sketch the graph of $y = (x - 15)^2$. First try to graph it as usual with the following keystrokes.

[Y=] [( ] [ ] [−] 15 [ ] [∧] 2

Note that you only get half of the parabola. Change Xmin=−15 to Xmin=−10, Xmax=15 to Xmax= 20, and Ymin=−10 to Ymin=−2. Now press [GRAPH].

Note: Return your calculator to standard setting by pressing [ZOOM], moving cursor to "Standard," and pressing [ENTER].

▦ EXAMPLE 4 ■ Using the Zoom Key

Sketch the graph of $y = |x|$. Do the two rays make 45° angles with the x-axis and y-axis?

Solution

First press [Y=] [2nd] [ABS] [X, T, θ] [GRAPH]. Your screen should look like this:

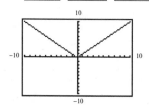

It does not appear to have the 45° angles desired. To change this, press [ZOOM] [5], [ENTER]. Now press [GRAPH] to obtain this graph:

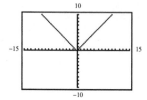

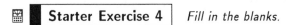 **Starter Exercise 4** *Fill in the blanks.*

With your calculator set to "Standard," graph $y = x$. Then change the settings of your calculator so that the line makes a 45° angle with the x-axis and y-axis. Press [＿＿] [＿＿] [＿＿] .

▦ EXAMPLE 5 ■   Sketching More than One Graph on the Same Screen

Sketch the graphs of $y = x + 3$, $y = x$, and $y = x - 3$.

Solution

Press [Y=] and enter all three equations.

Keystrokes: [X, T, θ] [+] 3   Move cursor down
[X, T, θ]   Move cursor down
[X, T, θ] [−] 3

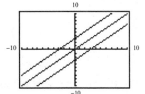

Now press [GRAPH] . You should have three parallel lines.

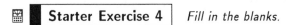 **Starter Exercise 5** *Fill in the blanks.*

Sketch the graphs of $y = \sqrt{81 - x^2}$ and $y = -\sqrt{81 - x^2}$ on the same screen.

Keystrokes: [Y=] [2nd] [√] [( ] 81 [−] [X, T, θ] [∧] 2 [ )]

Move cursor down, then

[(−)] [2nd] [＿＿] [＿＿] 81 [＿＿] [X, T, θ] [∧] 2 [＿＿] [GRAPH] .

■ **Solutions to Starter Exercises** ■

1. $y =$ [$-2x$] $+ 3$

Keystrokes: [Y=] [(−)] 2 [X, T, θ] [+] 3 [GRAPH]

2. [Y=] [2nd] [ABS] [( ] [X, T, θ] [−] 1 [ )] [GRAPH]

3. [Y=] [( ] [X, T, θ] [−] 15 [ )] [∧] 2

4. Press [ZOOM] [5] [ENTER] .

5. [(−)] [2nd] [√] [( ] 81 [−] [X, T, θ] [∧] 2 [ )] [GRAPH] .

The graph should be a circle, if not, set your calculator to "Square."

4.3 | EXERCISES

In Exercises 1–10, use a graphing calculator to sketch the graphs of the following equations. (Use standard settings on each graph.)

1. $y = 2x - 1$ **2.** $y = -4x$ **3.** $y = \frac{1}{2}x$ **4.** $y = -\frac{2}{3}x + 2$

5. $y = -\frac{1}{2}x^2$ **6.** $y = -x^2 + 4$ **7.** $y = (x - 1)^2$

8. $y = x^2 + 4x + 1$ **9.** $y = |x - 2|$ **10.** $y = -|x + 2| - 2$

In Exercises 11–14, solve each equation for y and use a graphing calculator to sketch the graph of the resulting equations. (Use standard settings.)

11. $-2x - y = 3$ **12.** $-3x + 2y = 8$

13. $x^2 + y = 4$ **14.** $x^2 + 2y = 2$

In Exercises 15 and 16, use a graphing calculator to sketch the graph of the given equation. (Use the indicated setting.)

15. $y = 10x^2 - 12$

```
RANGE
Xmin= -5
Xmax= 5
Xscl= 1
Ymin= -15
Ymax= 10
Yscl= 1
Xres= 1
```

16. $y = 3x^2 + 9x + 2$

```
RANGE
Xmin= -10
Xmax= 5
Xscl= 1
Ymin= -6
Ymax= 5
Yscl= 3
Xres= 1
```

In Exercises 17 and 18, use a graphing calculator to determine the number of x-intercepts of the graph of the equation.

17. $y = -x^2 + 16$ **18.** $y = x^3 - x^2 - 9x$

In Exercises 19–22, use a graphing calculator to match the given equation with its graph.

19. $y = x^2$ **20.** $y = -x^2$

21. $y = x^2 - 4x + 4$ **22.** $y = -x^2 - 2x - 3$

(a)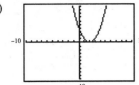

(b)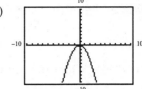

(c)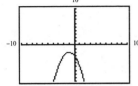

(d)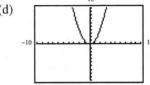

 In Exercises 23 and 24, use a graphing calculator to sketch the graphs of the given equations on the same screen.

23. $y = x^2$

$y = -x^2$

24. $y = |x| - 4$

$y = -|x| + 4$

 In Exercises 25 and 26, use the ⎡ TRACE ⎤ key to approximate the x-intercepts of the graphs of the following equations.

25. $y = x^2 - 10$

26. $y = -\frac{1}{2}x^2 - 2x + 3$

▌ 4.4 ▐ Business Applications and Graphs

┌───┐
│   Section Highlights   │
│   │
│ 1. ⎡Total amount⎤ = ⎡Base amount⎤ + ⎡Percent⎤ · ⎡Base amount⎤│
│   │
│ 2. ⎡Cost of taxable item⎤ = ⎡List price⎤ + ⎡Tax⎤   │
│   = ⎡x dollars⎤ + ⎡Percent⎤ · ⎡x dollars⎤│
│   │
│ 3. ⎡Selling price⎤ = ⎡Cost⎤ + ⎡Markup⎤   │
│   = ⎡Cost⎤ + ⎡Markup rate⎤ · ⎡Cost⎤   │
│   │
│ 4. ⎡Sale price⎤ = ⎡List price⎤ - ⎡Discount⎤   │
│   = ⎡List price⎤ - ⎡Discount rate⎤ · ⎡List price⎤│
└───┘

EXAMPLE 1 ■   **Finding Selling Price and Cost**

A stereo store uses 75% markup rate on the stereo components it sells.

(a) The cost of a cassette deck is $150. What is the selling price of the cassette deck?

(b) The selling price of a compact disc player is $350. What is the cost of the compact disc player?

Solution

(a) *Verbal model:* $\boxed{\text{Selling price}} = \boxed{\text{Cost}} + \boxed{\text{Markup rate}} \cdot \boxed{\text{Cost}}$

    Labels: Selling price $= x$
                    Cost $= \$150$
                    Markup rate $= 0.75$

    Equation: $x = 150 + (0.75)(150)$

$$= 150 + 112.5$$
$$= \$262.50$$

Thus the selling price of the cassette deck is $262.50.

(b) *Verbal model:* $\boxed{\text{Selling price}} = \boxed{\text{Cost}} + \boxed{\text{Markup rate}} \cdot \boxed{\text{Cost}}$

    Labels: Selling price $= \$350$
                    Cost $= x$
                    Markup rate $= 0.75$

    Equation: $350 = x + 0.75x$

$$350 = 1.75x$$
$$\frac{350}{1.75} = x$$
$$\$200 = x$$

Thus, the cost of the compact disc player is $200.　■

   ■ **Starter Exercise 1** | *Fill in the blanks.*

A bookstore uses a markup rate of 40%.

(a) If the selling price of a book is $28, find the cost of the book.

    Verbal model: $\boxed{\text{Selling price}} = \boxed{\text{Cost}} + \boxed{\text{Markup rate}} \cdot \boxed{\text{Cost}}$

      Labels: Selling price $= \boxed{}$
                   Cost $= x$
                   Markup rate $= \boxed{}$

      Equation: $\boxed{} = x + \boxed{}\,x$

$$\boxed{} = \boxed{}$$
$$\boxed{} = \boxed{}$$
$$\boxed{} = \boxed{}$$

(b) The cost of a book is \$17.50. What is the selling price?

Verbal model: [Selling price] = [ ] + [ ] · [ ]

Labels: Selling price = [ ]

Cost = [ ]

Markup rate = [ ]

Equation: $x =$ [ ] + [ ] · [ ]

[ ] = [ ]

[ ] = [ ]

EXAMPLE 2 ■ Finding the Markup Rate

A music store sells a compact disc for \$15. The cost of the compact disc is \$12. Find the markup rate used by the music store.

Solution

Verbal model: [Selling price] = [Cost] + [Markup rate] · [Cost]

Labels: Selling price = \$15
Cost = \$12
Markup rate = p (in decimal form)

Equation:
$$15 = 12 + p \cdot 12$$
$$-12 + 15 = -12 + 12 + 12p$$
$$3 = 12p$$
$$\tfrac{3}{12} = p$$
$$0.25 = p$$

Thus, the markup rate is 25%. ■

| **Starter Exercise 2** | *Fill in the blanks.*

A florist pays \$20 for a dozen roses. The florist sells the roses for \$45. What is the markup rate used by the florist?

Verbal model: [Selling price] = [Cost] + [Markup rate] · [Cost]

Labels: Selling price = [ ]

Cost = [ ]

Markup rate = [ ]

Equation: [ ] = [ ] + p · [ ]

⋮

EXAMPLE 3 ■ Finding the Discount Rate

A pair of shoes that regularly sells for $25 is on sale for $20. What is the discount rate?

Solution

Verbal model: $\boxed{\text{Sale price}} = \boxed{\text{List price}} - \boxed{\text{Discount rate}} \cdot \boxed{\text{List price}}$

Labels: Sale price = $20

                List price = $25

                Discount rate = p

Equation:

$$20 = 25 - p \cdot 25$$
$$-25 + 20 = -25 + 25 - p \cdot 25$$
$$-5 = -25p$$
$$\frac{-5}{-25} = p$$
$$0.2 = p$$

Thus, the discount rate is 20%.

■

Starter Exercise 3   *Fill in the blanks.*

A book is on sale for $10. This book regularly sells for $15. Find the discount rate.

Verbal model: $\boxed{\text{Sale price}} = \boxed{\text{List price}} - \boxed{\text{Discount rate}} \cdot \boxed{\text{List price}}$

Labels: Sale price = $\boxed{}$

              List price = $\boxed{}$

              Discount rate = p

Equation: $\boxed{} = \boxed{} - p \cdot \boxed{}$

              $\vdots$

EXAMPLE 4 ■ Finding the Sale Price

You have a coupon that is good for 10% off any dinner at Alice's Restaurant. You order a $11.90 dinner. What is the sale price of the dinner?

Solution

Verbal model: $\boxed{\text{Sale price}} = \boxed{\text{List price}} - \boxed{\text{Discount rate}} \cdot \boxed{\text{List price}}$

Labels: Sale price = x

              List price = $11.90

              Discount rate = 0.10

Equation: $x = 11.90 - 0.10(11.90) = 11.90 - 1.19 = \$10.71$

Thus, the sale price of the dinner is $10.71.

■

■ Starter Exercise 4 | *Fill in the blanks.*

A store has all its shirts on sale for 20% off the list price. The list price of a shirt is $30. What is the sale price of this shirt?

Verbal model: ⬚ Sale price ⬚ = ⬚ List price ⬚ − ⬚ Discount rate ⬚ · ⬚ List price ⬚

Labels: Sale price = ⬚

List price = ⬚

Discount rate = ⬚

Equation: ⬚ = ⬚ − ⬚ · ⬚

⋮

EXAMPLE 5 ■ Estimating the Amount of a Tip

Your dinner bill at a restaurant is $42.55. You decide to leave approximately a 15% tip. Approximately how much do you leave?

Solution

Verbal model: Amount left = Price of meal + 0.15 · Price of meal

Labels: Amount left = x
Price of meal = $42.55

Equation: $x = 42.55 + 0.15(42.55)$

$x = 42.55 + 6.3825$

$x = 48.9325$

$x \approx 48.93$

Thus, you leave $48.93. ■

■ Starter Exercise 5 | *Fill in the blanks.*

Alice's bill at a restaurant was $38.50. Alice left $45. Approximately what percent was the tip rate?

Verbal model: Amount left = Price of meal + Percent · Price of meal

Labels: Amount left = ⬚

Price of meal = ⬚

Percent = p

Equation: ⬚ = ⬚ + p · ⬚

⋮

EXAMPLE 6 ■ Commission

A salesperson makes a weekly base salary of $200 plus 3% of his sales. What was his pay in a week in which his sales were $4200?

Solution

Verbal model: | Total pay | = | Base salary | + | Commission rate | · | Total sales |

Labels: Total pay $= x$
                Base salary $= \$200$
                Commission rate $= 0.03$
                Total sales $= \$4200$

Equation: $x = 200 + 0.03 \cdot (4200) = 200 + 126 = \$326$

Thus, the pay was $326.

Starter Exercise 6   *Fill in the blanks.*

Lynn makes $300 plus a commission on her sales. One week her sales were $4800 and her pay was $396. What is Lynn's commission rate?

Verbal model: | Total pay | = | Base salary | + | Commission rate | · | Total sales |

Labels: Total pay $= \boxed{}$

             Base salary $= \boxed{}$

             Commission rate $= p$

             Total sales $= \boxed{}$

Equation: $396 = \boxed{} + p \cdot \boxed{}$
                     $\vdots$

■ Solutions to Starter Exercises ■

1. (a) Selling price $= \boxed{\$28}$

     Markup rate $= \boxed{0.4}$

     $\boxed{28} = x + \boxed{0.4}\,x$

     $\boxed{28} = \boxed{1.4x}$

     $\dfrac{\boxed{28}}{1.4} = \boxed{x}$

     $\boxed{\$20} = \boxed{x}$

Thus, the cost is $20.

(b) | Selling price | = | Cost | + | Markup rate | · | Cost |

     Selling price $= \boxed{x}$

     Cost $= \boxed{\$17.50}$

     Markup rate $= \boxed{0.4}$

     $x = \boxed{17.50} + \boxed{0.4} \cdot \boxed{(17.50)}$

     $\boxed{x} = \boxed{17.50 + 7}$

     $\boxed{x} = \boxed{\$24.50}$

Thus, the selling price of the book is $24.50.

Solutions to Starter Exercises

2. Selling price = $\boxed{\$45}$

Cost = $\boxed{\$20}$

Markup rate = $\boxed{p}$

$\boxed{45} = \boxed{20} + p \cdot \boxed{20}$

$\boxed{25} = \boxed{20p}$

$\boxed{\frac{25}{20}} = \boxed{p}$

$\boxed{1.25} = \boxed{p}$

Thus, the markup rate is 125%.

3. Sale price = $\boxed{\$10}$

List price = $\boxed{\$15}$

$\boxed{10} = \boxed{15} - p \cdot \boxed{15}$

$\boxed{-5} = \boxed{-15p}$

$\boxed{\frac{-5}{-15}} = \boxed{p}$

$\boxed{\frac{1}{3}} = \boxed{p}$

Thus, the discount rate is $33\frac{1}{3}\%$.

4. Sale price = $\boxed{x}$

List price = $\boxed{\$30}$

Discount rate = $\boxed{0.2}$

$\boxed{x} = \boxed{30} - \boxed{0.2} \cdot \boxed{(30)}$

$= \boxed{30 - 6}$

$= \boxed{\$24}$

Thus, the sale price is $24.

5. Amount left = $\boxed{\$45}$

Price of meal = $\boxed{\$38.50}$

$\boxed{45} = \boxed{38.50} + p \cdot \boxed{(38.50)}$

$\boxed{6.5} = \boxed{38.50p}$

$\boxed{\frac{6.5}{38.50}} = \boxed{p}$

$\boxed{0.17} \approx \boxed{p}$

Thus, the tip rate is approximately 17%.

6. Total pay = $\boxed{\$396}$

Base salary = $\boxed{\$300}$

Total sales = $\boxed{\$4800}$

$396 = \boxed{300} + p \cdot \boxed{4800}$

$\boxed{96} = \boxed{4800p}$

$\boxed{\frac{96}{4800}} = \boxed{p}$

$\boxed{0.02} = \boxed{p}$

Thus, Lynn's commission rate is 2%.

4.4 | EXERCISES

In Exercises 1–7, find the missing quantities.

	Merchandise	Cost	Selling Price	Markup Rate
1.	Socks	$1.75		40%
2.	Calculator		$45.00	50%
3.	Shoes	$30.00	$40.00	
4.	Necklace		$120.00	75%
5.	Milk	$1.80	$2.25	
6.	Book		$6.99	45%
7.	Freezer	$400.00	$540.00	

In Exercises 8–14, find the missing quantities.

	Merchandise	List Price	Sale Price	Discount Rate
8.	Clock	$80.00		20%
9.	Cassette Tape	$9.99	$7.99	
10.	Shirt		$20.00	30%
11.	Sofa	$799.99	$699.99	
12.	Stereo Receiver		$499.99	10%
13.	Television Set	$549.99	$500.00	
14.	Table		$450.00	$33\frac{1}{3}$%

15. A television repair shop charged $120 to repair a television set. $30 in new parts and $30 per hour of labor charged. How many hours did it take to repair the television set?

16. Your dinner bill was $52.75 and you want to leave a 15% tip. How much total money do you leave?

17. Gloria left $40 for a dinner bill of $33.25. What was Gloria's tip rate?

18. A rental car agency charges $25 per day and 27¢ per mile. Jack was charged $65.50 for a one-day car rental. How many miles did Jack drive?

19. You were charged $17.11, including 7% tax, for a compact disc. What was the retail price of the compact disc?

20. Your total price for a $24.99 pair of pants was $26.49, including tax. What was the tax rate?

21. Kay makes a base monthly salary of $2,000 plus 2% of her sales. If her pay was $2,231.36 in a month, what was her sales total?

▦ **22.** Dan gets paid $1750 per month plus commission on his total sales. If his pay one month was $2057.68 and his sales were $10,256, what is his commission rate?

23. Pete makes $14 per hour and time and a half for any hours over 40 per week. If his pay this week was $759.50, how many hours of overtime did Pete work?

▦ **24.** A toaster is on sale for $33\frac{1}{3}\%$ off the regular price. The discount is $13.33. Find the sale price.

▦ **25.** In San Diego County, the sales tax rate was 7% in 1994.
  (a) Write a linear equation giving the total amount paid y in terms of the price of the item x.
  (b) Use a graphing utility to graph the equation in part (a). Use the following settings.

   Xmin= 0   Ymin= 0
   Ymax= 1000   Ymax= 1100

  (c) Use the graph to estimate the amount paid for an item with a price of $500.

▎ 4.5 ▏ Formulas and Scientific Applications

Section Highlights

1. **Mixture Problems:**

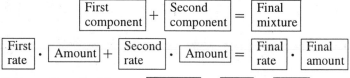

2. **Distance–Rate Problems:** | Distance | = | Rate | · | Time |

3. **Work–Rate Problems:**
   (a) The work rate is the reciprocal of the time needed to do an entire job.
   (b) | Work done | = | Work rate | · | Time |

EXAMPLE 1 ■   **Using a Geometric Formula**

A triangle has an area 30 square feet. The base of the triangle is 12 feet. What is the height of the triangle?

Solution

Common formula: $A=\frac{1}{2}bh$

Labels: $A = 30$ square feet
             $b = 12$ feet
             $h =$ height in feet

Equation: $30 = \frac{1}{2}(12)h$

             $30 = 6h$

             $\frac{30}{6} = h$

              $5 = h$

Thus, the height of the triangle is 5 feet.   ■

Starter Exercise 1   *Fill in the blanks.*

The perimeter of a square is 36 feet. How long is each side of the square?

Common formula: $P = 4s$

Labels: $P = $ ☐

$s = $ ☐

Equation: ☐ $= 4s$

$\vdots$

EXAMPLE 2 ■   Using a Geometric Formula

A rectangular box has a volume of 192 cubic inches. The width of the box is 6 inches, and the height is 4 inches. Find the length of the box.

Solution

Common formula: $V = lwh$

Labels: $V = 192$ cubic inches

$l = $ length in inches

$w = 6$ inches

$h = 4$ inches

Equation: $192 = l(6)(4)$

$192 = 24l$

$\frac{192}{24} = l$

$8 = l$

Thus, the length of the box is 8 inches.

Starter Exercise 2   *Fill in the blanks.*

A circular cylinder has a volume of 54π cubic inches, and radius 3 inches. Find the height of the circular cylinder.

Common formula: $P = \pi r^2 h$

Labels: $V = 54\pi$ cubic inches

$r = 3$ inches

$h = $ ☐

Equation: $54\pi = $ ☐

$\vdots$

EXAMPLE 3 ■ Simple Interest

(a) Solve for t in $I = Prt$.

(b) Solve for r in $A = P + Prt$.

Solution

(a) $\qquad I = Prt \qquad$ Given

$\qquad I = (Pr)t \qquad$ Group factors.

$\dfrac{1}{Pr}(I) = \dfrac{1}{Pr}(Pr)t \qquad$ Multiply both sides by $\frac{1}{Pr}$.

$\qquad \dfrac{I}{Pr} = t$

(b) $\qquad A = P + Prt \qquad$ Given

$\quad -P + A = -P + P + Prt \qquad$ Add $-P$ to both sides.

$\qquad A - P = (Pt)r \qquad$ Group factors.

$\qquad \dfrac{A - P}{Pt} = r$

■

Starter Exercise 3 | *Fill in the blanks.*

(a) Solve for P in $I = Prt$.

$I = Prt$

$I = P(rt)$

$I \cdot \dfrac{1}{\boxed{}} = P(rt)\dfrac{1}{\boxed{}}$

$\boxed{} = P$

(b) Solve for P in $A = P + Prt$.

$A = P + Prt$

$A = P\left(\boxed{} + \boxed{}\right)$

$\dfrac{A}{\boxed{}} = P$

EXAMPLE 4 ■ Simple Interest

You want to invest enough money at 6.5% simple annual interest so that in 6 months you have earned $162.50. How much money should you invest?

Solution

Common formula: $\quad I = Prt$

Labels: $\quad I = \$162.50$

$P = \text{principal}$

$r = 0.065$

$t = \dfrac{1}{2} \text{ year}$

Equation: $\quad 162.50 = P(0.065)\left(\dfrac{1}{2}\right)$

$162.50 = 0.0325P$

$\dfrac{162.50}{0.0325} = P$

$5000 = P$

Thus, $5000 should be invested.

■

Starter Exercise 4   *Fill in the blanks.*

You want to invest $1000 at 7% simple annual interest. How long should you leave your money in this account to earn $350?

Common formula: $I = Prt$

Labels: $I = \boxed{}$

$P = \boxed{}$

$r = \boxed{}$

$t = \boxed{}$

Equation: $350 = \boxed{}$

$\vdots$

EXAMPLE 5 ■   A Simple Interest Mixture Problem

A total of $30,000 is invested in two accounts, one paying 7% simple interest, and the other paying 8% simple interest. If at the end of one year the total interest earned on both accounts was $2280, how much was invested in each account?

Solution

Verbal model:

$$\boxed{\begin{array}{c}\text{Interest on}\\\text{first amount}\end{array}} + \boxed{\begin{array}{c}\text{Interest on}\\\text{second amount}\end{array}} = \boxed{\text{Total interest}} \text{ or}$$

$$P_1 r_1 \quad + \quad P_2 r_2 \quad = \quad 2280$$

Labels:
$$P_1 = x$$
$$P_2 = 30{,}000 - x$$
$$r_1 = 0.07$$
$$r_2 = 0.08$$

Equation:
$$0.07x + 0.08(30{,}000 - x) = 2280$$
$$0.07x + 2400 - 0.08x = 2280$$
$$2400 - 0.01x = 2280$$
$$-0.01x = -120$$
$$x = \frac{-120}{-0.01} = 12{,}000 \text{ and } 30{,}000 - x = 18{,}000$$

Thus, $12,000 was invested at 7% and $18,000 was invested at 8%.   ■

Starter Exercise 5 *Fill in the blanks.*

You invest \$10,000 at $9\frac{1}{2}\%$ simple interest. How much additional money must be invested at $8\frac{1}{2}\%$ so that the total interest earned on both accounts is \$1630?

Verbal model: $\boxed{\text{Interest on first amount}} + \boxed{\text{Interest on second amount}} = \boxed{\text{Total interest}}$ or

$$P_1 r_1 \quad + \quad P_2 r_2 \quad = \quad 1630$$

Labels: $P_1 = \boxed{}$

$P_2 = \boxed{}$

$r_1 = \boxed{}$

$r_2 = \boxed{}$

Equation: $0.095(10{,}000) + \boxed{} = 1630$

⋮

EXAMPLE 6 ■ A Coin Mixture Problem

A bank contains \$8.50 in dimes and quarters. If there are 43 coins in the bank, how many dimes and how many quarters are in the bank?

Solution

Verbal model: $\boxed{\text{Total value of dimes}} + \boxed{\text{Total value of quarters}} = \boxed{\text{Total value}}$

Labels: Dimes: value per coin = \$0.10, number of coins = x
Quarters: value per coin = \$0.25, number of coins = $43 - x$
Mixed coins: total value = \$8.50, number of coins = 43

Equation:
$$0.10x + 0.25(43 - x) = 8.50$$
$$0.10x + 10.75 - 0.25x = 8.50$$
$$-0.15x + 10.75 = 8.50$$
$$x = \frac{-2.25}{-0.15}$$
$$x = 15 \text{ dimes}$$
$$43 - x = 28 \text{ quarters}$$

Thus, the bank has 15 dimes and 28 quarters. ■

Starter Exercise 6 *Fill in the blanks.*

There is \$18.25 in a box of nickels and dimes. If there are twice as many dimes as nickels, find the number of each coin in the box.

Verbal model: $\boxed{\text{Total value of nickels}} + \boxed{\text{Total value of dimes}} = \boxed{\text{Total value}}$

Labels: Nickels: value per coin = \$0.05, number of coins = x

Dimes: value per coin = \$0.10, number of coins = $\boxed{}$
Mixed coins: total value = \$18.25

Equation: $\boxed{} + \boxed{} = 18.25$

⋮

EXAMPLE 7 ■ A Solution Mixture Problem

How many liters of a 10% saline solution must be mixed with 20 liters of a 15% saline solution to make a 12% saline solution?

Solution

Verbal model: | Amount of salt in 10% solution | + | Amount of salt in 15% solution | = | Total amount of salt in 12% solution |

Labels: 10% saline solution: percent of salt = 0.10, amount of solution = x
           15% saline solution: percent of salt = 0.15, amount of solution = 20
           Final solution: percent of salt = 0.12, amount of solution = $x + 20$

Equation: $0.10x + 0.15(20) = 0.12(x + 20)$

$$0.10x + 3 = 0.12x + 2.4$$

$$-0.02x + 3 = 2.4$$

$$-0.02x = -0.6$$

$$x = \frac{-0.6}{0.02} = 30$$

■

Starter Exercise 7   *Fill in the blanks.*

Some 7% hydrogen peroxide solution is mixed with 12% hydrogen peroxide solution to make 10 liters of a 10% hydrogen peroxide solution. How many liters of each was used?

Verbal model: | Amount of hydrogen peroxide in 7% solution | + | Amount of hydrogen peroxide in 12% solution | = | Total amount of hydrogen peroxide in 10% solution |

Labels: 7% solution: percent hydrogen peroxide = 0.07, amount = x

           12% solution: percent hydrogen peroxide = 0.12, amount = ⬚

           Final solution: percent hydrogen peroxide = 0.10, amount = ⬚

Equation: $0.07x +$ ⬚ $=$ ⬚ ⬚

⋮

EXAMPLE 8 ■ A Distance–Rate Problem

A family drove 55 miles per hour to a city and drove 45 miles per hour on the return trip. If the total time traveling was 4 hours, how long did it take to get to the city?

Solution

Verbal model: | Distance to the city | = | Distance of return trip |

Labels: We need rate and time for each trip since distance = rate · time.
           First rate = 55; first time = t
           Second rate = 45; second time = $4 - t$

Equation: $55t = 45(4 - t)$

$$55t = 180 - 45t$$

$$100t = 180$$

$$t = \frac{180}{100}$$

$$t = 1\frac{4}{5}$$

Thus, it took 1 hour 48 minutes to get to the city.

■

Starter Exercise 8   *Fill in the blanks.*

A hiker hiked for 2 hours, then reduced her speed by 2 miles per hour and returned. If her return trip took 3 hours, what was her rate on the first half of her trip?

Verbal model: $\boxed{\text{First distance}} = \boxed{\text{Second distance}}$

Labels: First rate $= r$; first time $= \boxed{}$

Second rate $= r - 2$; second time $= \boxed{}$

Equation: $\boxed{}\, r = \boxed{}(r - 2)$

$\vdots$

EXAMPLE 9 ■   A Distance–Rate Problem

One runner starts at one end of a 14-mile course running at 8 miles per hour. At the same time, another runner starts at the other end of the course at 6 miles per hour. How long will it take the two runners to meet?

Solution

Verbal model: $\boxed{\begin{array}{c}\text{First runner's}\\\text{distance}\end{array}} + \boxed{\begin{array}{c}\text{Second runner's}\\\text{distance}\end{array}} = \boxed{\begin{array}{c}\text{Total}\\\text{Distance}\end{array}}$

Labels: First runner: rate $= 8$, time $= t$
Second runner: rate $= 6$, time $= t$

Equation: $8t + 6t = 14$

$14t = 14$

$5 = \frac{14}{14}$

$t = 1$

Thus, the two runners meet after 1 hour.   ■

Starter Exercise 9   *Fill in the blanks*

Two cyclists start off in the same direction at the same time. One cyclist is traveling 15 miles per hour and the other 20 miles per hour. How long will it take for the two cyclists to be 8 miles apart?

Verbal model: $\boxed{\begin{array}{c}\text{Distance of}\\\text{faster cyclist}\end{array}} - \boxed{\begin{array}{c}\text{Distance of}\\\text{slower cyclist}\end{array}} = \boxed{\begin{array}{c}\text{Distance}\\\text{apart}\end{array}}$

Labels: Faster cyclist: rate $= 20$, time $= t$

Slower cyclist: rate $= \boxed{}$, time $= \boxed{}$

Equation: $20t - \boxed{} = 8$

$\vdots$

EXAMPLE 10 ■   **A Work–Rate Problem**

It takes one printing press 2 hours to print the daily edition of a newspaper. A smaller printing press takes 3 hours to print the same newspaper. How long would it take both printing presses working together to print the daily edition of the paper?

Solution

Verbal model: Work done = Portion done by large press + Portion done by small press   or

$$1 = \boxed{\text{Rate}} \cdot \boxed{\text{Time}} + \boxed{\text{Rate}} \cdot \boxed{\text{Time}}$$

Labels: Large press: rate $= \frac{1}{2}$, time $= t$

Small press: rate $= \frac{1}{3}$, time $= t$

Equation: $1 = \frac{1}{2}t + \frac{1}{3}t$

$1 = \left(\frac{1}{2} + \frac{1}{3}\right)t$

$1 = \frac{5}{6}t$

$\frac{6}{5} = t$

Thus, it would take 1 hour and 12 minutes ($\frac{6}{5}$ of an hour) for both presses to print the paper.   ■

■ **Starter Exercise 10**   *Fill in the blanks*

Mary can paint a room in 4 hours, while Beth can paint the room in 6 hours. How long would it take Mary and Beth to paint the room together?

Verbal model: Work Done = Portion done by Mary + Portion done by Beth   or

$$1 = \boxed{\text{Rate}} \cdot \boxed{\text{Time}} + \boxed{\text{Rate}} \cdot \boxed{\text{Time}}$$

Labels: Mary: rate $= \boxed{}$, time $= t$

Beth: rate $= \boxed{}$, time $= t$

Equation: $1 = \boxed{}t + \boxed{}t$

⋮

■   **Solutions to Starter Exercises**   ■

1. $P = \boxed{36}$

$s = \boxed{\text{length of a side}}$

$\boxed{36} = 4s$

$\boxed{\frac{36}{4}} = \boxed{s}$

$\boxed{9} = \boxed{s}$

2. $h = \boxed{\text{height}}$

$54\pi = \boxed{\pi(3)^2 h}$

$54\pi = \boxed{9\pi h}$

$\boxed{\frac{54\pi}{9\pi}} = \boxed{h}$

$\boxed{6} = \boxed{h}$

■ **Solutions to Starter Exercises** ■

3. (a) $I \cdot \dfrac{1}{\boxed{rt}} = P(rt)\dfrac{1}{\boxed{rt}}$

$\dfrac{I}{\boxed{rt}} = P$

(b) $A = P\left(\boxed{1} + \boxed{rt}\right)$

$\dfrac{A}{\boxed{1+rt}} = P$

4. $I = \boxed{\$350}$

$P = \boxed{\$1000}$

$r = \boxed{0.07}$

$t = \boxed{\text{time}}$

$350 = \boxed{1000(0.07)t}$

$\boxed{350} = \boxed{70t}$

$\boxed{\frac{350}{70}} = \boxed{t}$

$\boxed{5} = \boxed{t}$

5. $P_1 = \boxed{10{,}000}$

$P_2 = \boxed{x}$

$r_1 = \boxed{0.095}$

$r_2 = \boxed{0.085}$

$0.095(10{,}000) + \boxed{0.085x} = 1630$

$\boxed{950 + 0.085x} = \boxed{1630}$

$\boxed{0.085x} = \boxed{680}$

$\boxed{x} = \boxed{\dfrac{680}{0.085}}$

$= \boxed{\$8000}$

6. Number of dimes $= \boxed{2x}$

$\boxed{0.05x} + \boxed{0.10(2x)} = 18.25$

$\boxed{0.05x} + \boxed{0.2x} = \boxed{18.25}$

$\boxed{0.25x} = \boxed{18.25}$

$\boxed{x} = \boxed{73}$

Thus, there are 73 nickels and 146 dimes.

7. Amount of 12% solution $= \boxed{10 - x}$

Amount of final solution $= \boxed{10}$

$0.07x + \boxed{0.12(10 - x)} = \boxed{0.10}\ \boxed{(10)}$

$\boxed{0.07x + 1.2 - 0.12x} = \boxed{1}$

$\boxed{1.2 - 0.05x} = \boxed{1}$

$\boxed{-0.05x} = \boxed{-0.2}$

$\boxed{x} = \boxed{4}$

Thus, 4 liters of the 7% solution and 6 liters of the 12% solution were used.

8. First time $= \boxed{2}$

Second time $= \boxed{3}$

$\boxed{2}\ r = \boxed{3}\ \boxed{(r - 2)}$

$\boxed{2r} = \boxed{3r - 6}$

$\boxed{-r} = \boxed{-6}$

$\boxed{r} = \boxed{6}$

Thus, her rate on the first half of the trip was 6 mph.

■ **Solutions to Starter Exercises** ■

9. Slower cyclist:

rate = $\boxed{15 \text{ mph}}$, time = $\boxed{t}$

$20t - \boxed{15t} = 8$

$\boxed{5t} = \boxed{8}$

$\boxed{5} = \boxed{\frac{8}{5}}$

Thus, it takes 1 hour and 36 minutes (or $\frac{8}{5}$ hours).

10. Mary: rate = $\boxed{\frac{1}{4}}$

Beth: rate = $\boxed{\frac{1}{6}}$

$1 = \boxed{\frac{1}{4}}\, t + \boxed{\frac{1}{6}}\, t$

$\boxed{1} = \boxed{\frac{5}{12}t}$

$\boxed{\frac{12}{5}} = \boxed{t}$

Thus, it would take 2 hours and 24 minutes to paint the room together.

4.5 EXERCISES

In Exercises 1–4, evaluate the formula for the given values of the variables.

1. *Distance-Rate-Time Formula:* $d = rt$
$r = 60$ mph, $t = 3\frac{1}{2}$ hours

2. *Perimeter of a Rectangle:* $P = 2l + 2w$
$l = 8$ feet, $w = 6$ feet

3. *Simple Interest:* $A = P + Prt$
$P = \$500$, $r = 7\frac{1}{2}\%$, $t = 4$ years

4. *Power* (in amps): $I = \dfrac{P}{V}$
$P = 1000$ watts, $V = 110$

In Exercises 5–18, solve for the specified variable.

5. Solve for b.
Area of a Triangle: $A = \frac{1}{2}bh$

6. Solve for w.
Perimeter of a Rectangle: $P = 2l + 2w$

7. Solve for l.
Area of a Rectangle: $A = lw$

8. Solve for r.
Circumference of a Circle: $C = 2\pi r$

9. Solve for h.
Volume of a Circular Cylinder: $V = \pi r^2 h$

10. Solve for R.
Discount: $S = L - RL$

11. Solve for P.
Simple Interest Balance: $A = P + Prt$

12. Solve for a.
Area of a Trapezoid: $A = \frac{1}{2}(a + b)h$

13. Solve for V_0.
Freely-falling Body: $h = V_0 t + \frac{1}{2}at$

14. Solve for m_1.
Newton's Law of Universal Gravitation:
$F = \alpha \dfrac{m_1 m_2}{r^2}$

15. Solve for d.
Arithmetic Progression: $L = a + (n - 1)d$

16. Solve for a_n.
Arithmetic Progression: $S = \dfrac{n(a_1 + a_n)}{2}$

17. Solve for a_1.

Geometric Progression: $a_n = a_1 r^{n-1}$

18. Solve for h.

Surface Area: $S = 2\pi rh + 2\pi r^2$

19. A farmer has 600 feet of fence. He wants to use this fence to enclose a rectangular field in which the length is twice the width. If he uses all the fence, what are the dimensions of the field?

20. The circumference of a basketball hoop is 18π inches. What is the diameter of the hoop?

In Exercises 21–25, assume that the interest is simple annual interest.

21. Find the interest earned on $1000 invested at 9% for four years.

22. If an account paying 8% interest earned $120 in one year, what was the amount invested?

23. A total of $5800 is invested in two accounts, one paying $7\frac{3}{4}$% interest and the other paying 8% interest. How much was invested in each account if the total interest earned was $456.50?

24. A $7000 bond pays 9% annual interest. How much should be invested in an account paying 8% annual interest so that the interest earned in one year is the same as the interest earned on the bond in one year?

25. A $3000 bond paying 8% annual interest was purchased. How much must be invested in an account paying 6% annual interest so that the total interest earned on both in one year was 7.5% of the total investment?

26. How much candy that costs $1 per pound should be mixed with candy that costs $2 to make a 10-pound mixture that costs $1.60 per pound?

27. How many liters of apple juice that cost $1.80 per liter must be mixed with 4 liters of cranberry juice that cost $2.40 per liter to make a mixture that costs $2 per liter?

28. If 5 pounds of cashews that cost $7.50 per pound are mixed with 8 pounds of peanuts that cost $2.50 per pound, what is the cost per pound of the mixture?

29. A certain grade of hamburger is made by mixing 10 kilograms of hamburger that costs $1.90 per kilogram with some hamburger that costs $2.20 per kilogram. How many kilograms is in the mixture if the cost per kilogram is $2?

30. A box contains 210 coins, all dimes and quarters. If the box contains $30, how many dimes and how many quarters are in the box?

31. A bank contains $2.50 in nickels and dimes. If there are twice as many dimes as nickels, how many nickels and how many dimes are in the bank?

32. Ed spent $27.80 on 100 stamps. If he bought only 23¢ and 29¢ stamps, how many of each type did Ed buy?

33. Sue spent $15.98 on 19¢ and 29¢ stamps. The number of 29¢ stamps was two more than twice the number of 19¢ stamps. How many of each type did Sue buy?

34. 12,000 tickets were sold for a concert. Some tickets cost $8 each and others cost $10 each. If the total sales from the tickets were $104,000, how many of each type of ticket were sold?

35. A floral shop received an order for flowers that totaled $475. The prices per dozen of roses and carnations are $25 and $15, respectively. If there were 25 dozen flowers ordered, how many dozens of each type were ordered?

36. How many ounces of 15% saline solution must be mixed with 12 ounces of 10% saline solution to make a 12% saline solution?

37. How much of a 20% gold alloy must be mixed with a 28% gold alloy to make 10 grams of a 23% gold alloy?

38. What would be the percent concentration of hydrogen peroxide in a solution made by mixing 10 ounces of 8% hydrogen peroxide solution with 12 ounces of 14% hydrogen peroxide solution?

39. How much pure water must be added to a 20% saline solution to make 24 ounces of 12% saline solution?

40. A cyclist rode to the beach at 24 mph, then rode back at 20 mph. If the total trip took 11 hours, how much time was spent in each direction?

41. The first jogger started jogging at 8 mph. One hour later, a second jogger started jogging from the same point in the same direction at 10 mph. How long after the second jogger started will she catch the first jogger?

42. Bob drove for 2 hours, then reduced his speed by 10 mph, and drove 1 more hour. If he drove a total of 155 miles, what was his initial rate?

43. Jack starts at town A and drives toward town B at 45 mph. Patty starts at the same time at town B and drives the same road toward town A at 55 mph. If the two towns are 100 miles apart, how far from town A will Jack and Patty meet?

44. If pump A can empty a swimming pool in 7 hours, and pump B can empty the pool in 8 hours, how long will it take both pumps to empty the pool together?

45. Nancy can build a brick wall in 15 hours. Joan can build the wall in 12 hours. How long would it take Nancy and Joan to build the wall together?

46. Dan can paint a room in 4 hours, while Joe can paint the same room in 5 hours. Dan paints for 2 hours, then leaves to go surfing. How long does it take Joe to finish the job?

47. It takes you 2 hours to mow the lawn with your power mower, and 3 hours to mow the lawn with your push mower. You start mowing the lawn with your power mower, and in 30 minutes the power mower breaks down. How long does it take you to finish the job using your push mower?

48. Your age is twice your son's age. How old is your son if your combined ages total 60?

49. The difference in age of a mother and her daughter is 30 years. How old will the mother be when her age is twice her daughter's age?

50. Find two consecutive integers whose sum is 107.

51. Find three consecutive integers whose sum is 99.

52. Find two consecutive even integers whose sum is 42.

53. Find three consecutive odd integers whose sum is 99.

54. Find two consecutive odd integers such that 13 times the smaller is 11 times the larger.

Cumulative Practice Test for Chapters P–4

In Exercises 1–5, match the letter corresponding to the property of algebra that justifies each of the following statements.

(a) Multiplicative Identity Property
(b) Associative Property of Addition
(c) Distributive Property
(d) Associative Property of Multiplication
(e) Commutative Property of Addition

1. $2x + 3x = (2 + 3)x$

2. $5x + (7x - 3) = (7x - 3) + 5x$

3. $x = 1x$

4. $\frac{1}{4}(4x) = \left(\frac{1}{4} \cdot 4\right)x$

5. $5x + (7x - 3) = (5x + 7x) - 3$

In Exercises 6–10, solve the given linear equation.

6. $x + 6 = 10$

7. $2x = 7$

8. $5x - 6 = 9$

9. $2(x + 4) = 6x - 1$

10. $2(x + 2) = 2(x + 5) - x$

In Exercises 11–14, evaluate the given expression.

11. $\frac{1}{6} - \frac{3}{4}$

12. $|-2| + \frac{9 - 1}{2^2} - 4^2$

13. $\left(\frac{1}{2}\right)^2 + \frac{5 - 2^2}{6}$

14. $-|-3|^2 - \frac{2}{3} + \left(1\frac{2}{3}\right)^2$

In Exercises 15–18, solve the given inequality.

15. $x - 6 > 1$

16. $-2x \leq 8$

17. $2(2x - 1) < 6$

18. $0 \leq 5 - 3x < 8$

In Exercises 19–22, simplify the given expression.

19. $3(2 - x) + x$

20. $5z - (2x - 3z) + 4$

21. $4x^2 - 3x + x^2 - 2x$

22. $4x(4x + 3y) - 3y(4x + 3y)$

In Exercises 23–26, translate the given verbal statement into a variable expression.

23. Twice the sum of a number and three

24. The total of a number and six

25. The quotient of a number and seven less than that number

26. The difference between twice a number and four less than the number

27. The sum of two consecutive integers is 207. Find the integers.

28. The sum of three consecutive even integers is 102. Find the integers.

29. The sum of two consecutive odd integers is 32. Seventeen times the larger is fifteen times the smaller. Find the two integers.

30. A total of $13,000 is invested in two accounts. One account pays $7\frac{1}{2}$%, and the other pays 8.5%, simple annual interest. If the total interest earned on both accounts in one year was $1,035, how much was invested in each account?

31. A pair of pants regularly sell for $25. They are on sale for $20. What is the discount rate?

32. How many pounds of chocolates that cost $7 per pound must be mixed with three pounds of chocolates that cost $6 per pound to make a mixture of chocolates that cost $6.40 per pound?

33. A bookstore uses a markup rate of 40%. The store sells a book for $14. What is the cost of this book?

34. A box contains $2.50 in nickels and dimes. If there are twice as many dimes as nickels, how many of each are in the box?

35. Tommy rode his bicycle to the store in a half hour. On the return trip, he reduced his speed by three mph and it took him 36 minutes. What was Tommy's rate on the trip to the store?

36. How much pure water must be added to eight quarts of a 50% antifreeze solution to make a 40% antifreeze solution?

37. A man drove the first part of a 140 mile trip at 50 mph. He then reduced his speed by 10 mph and finished the trip. If the total trip took 3 hours, how far did he drive at each speed?

In Exercises 38–40, match the equation or inequality with the graphs below.

38. $y = 2x + 1$ **39.** $y = x^2 + 1$ **40.** $y = |x|$

(a)
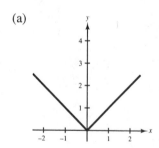

(b)
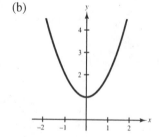

(c)
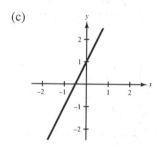

4.1 | **Answers to Exercises**

1.

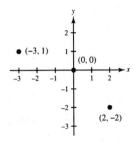

2.

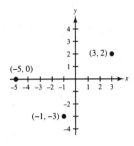

3.

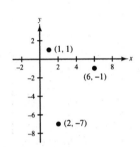

4.

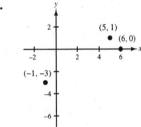

5. (a) Quadrant III
   (b) Quadrant IV

6. $A(3, \ 2)$
   $B(-1, \ 5)$
   $C(5, \ -2)$
   $D(-2, \ -2)$

7.

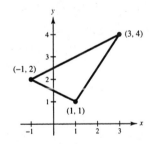

8.

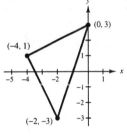

9.

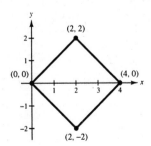

10.

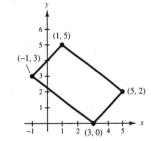

11. (a) No
  (b) Yes
  (c) Yes
  (d) Yes

12. (a) Yes
  (b) Yes
  (c) No
  (d) No

13. (a) No
  (b) Yes
  (c) Yes
  (d) No

14. (a) No
  (b) No
  (c) No
  (d) No

15.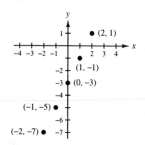

16.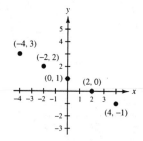

17.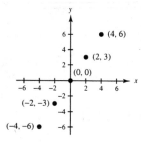

18.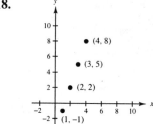

19.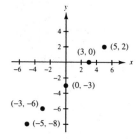

20.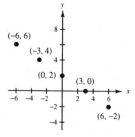

21.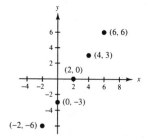

22.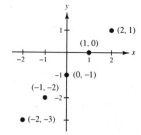

23.

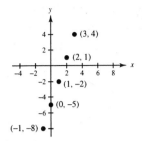

24.

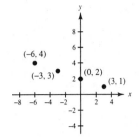

25.

x	50	100	150	200
$y = 2x + 50$	150	250	350	450

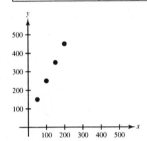

26.

x	2	6	10	14
$y = 0.5x + 8$	9	11	13	15

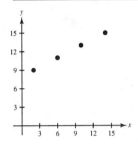

4.2	**Answers to Exercises**

1.

x	-2	-1	0	1	2
y	-5	-3	-1	1	3
$(x,\ y)$	$(-2, -5)$	$(-1, -3)$	$(0, -1)$	$(1, 1)$	$(2, 3)$

2.

x	-4	-2	0	2	4
y	8	5	2	-1	-4
$(x,\ y)$	$(-4, 8)$	$(-2, 5)$	$(0, 2)$	$(2, -1)$	$(4, -4)$

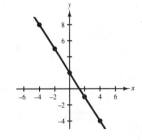

3.

x	10	11	12	13	14
y	2	1	0	1	2
(x, y)	$(10, 2)$	$(11, 1)$	$(12, 0)$	$(13, 1)$	$(14, 2)$

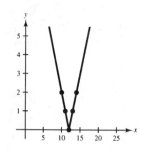

4.

x	-2	-1	0	1	2
y	-1	-4	-5	-4	-1
(x, y)	$(-2, -1)$	$(-1, -4)$	$(0, -5)$	$(1, -4)$	$(2, -1)$

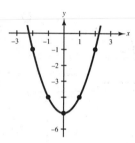

5. $y = x - 6$ **6.** $y = -\frac{2}{3}x + 4$ **7.** $y = -x^2 - 3x + 8$ **8.** $y = \frac{5}{2}x^2 - \frac{17}{2}$

9. x-intercept: $(4, 0)$
y-intercept: $(0, -4)$

10. x-intercept: $(-4, 0)$
y-intercept: $(0, 2)$

11. x-intercept: $(-15, 0)$
y-intercept: $(0, 9)$

12. x-intercepts: $(-2, 0)$, $(2, 0)$
y-intercept: $(0, -4)$

13. x-intercept: $(6, 0)$
y-intercept: None

14. x-intercept: None
y-intercept: $(0, 4)$

15. x-intercepts: $(2, 0)$, $(4, 0)$
y-intercept: $(0, 8)$

16. x-intercepts: $(5, 0)$, $(-1, 0)$
y-intercept: $(0, -5)$

17.

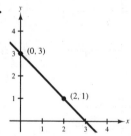

18.

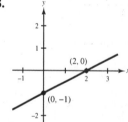

19.

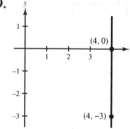

20.

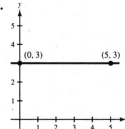

21.

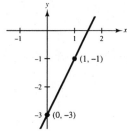

22.

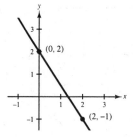

23.

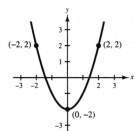

24.

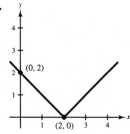

25. $y = 55x$

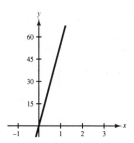

4.3   **Answers to Exercises**

1.

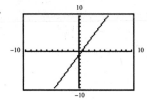

2.

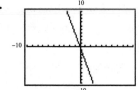

3.

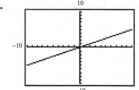

4.

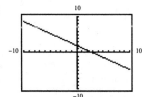

5.

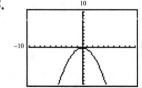

6.

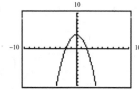

7.

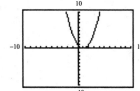

8.

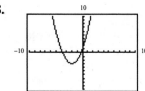

9.

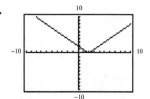

10.

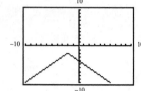

11. $y = -2x - 3$

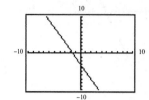

12. $y = \frac{3}{2}x + 4$

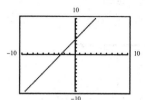

13. $y = -x^2 + 4$

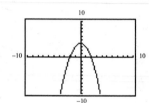

14. $y = -\frac{1}{2}x^2 + 1$

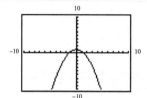

15.

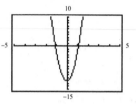

16.

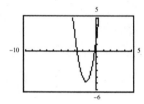

17. 2

18. 3

19. (d)

20. (b)

21. (a)

22. (c)

23.

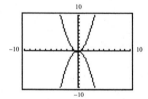

24.

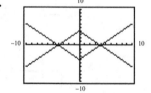

25. $(-3.15789475, 0)$ and $(3.15789475, 0)$

26. $(-5.210526, 0)$ and $(1.1052632, 0)$

4.4	**Answers to Exercises**

1. $2.45

2. $30

3. $33\frac{1}{3}\%$

4. $68.57

5. 25%

6. $4.82

7. 35%

8. $64

9. 20%

10. $28.57

11. 12.5%

12. $555.54

13. 9%

14. $675

15. 3 hrs

16. $60.66

17. 20.3%   **18.** 150 mi   **19.** $15.99   **20.** 6%

21. $11,568   **22.** 3%   **23.** 9.5 hrs   **24.** $26.66

25. (a) $y = 1.07x$

(b)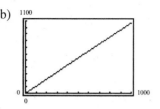

(c) $535

4.5	**Answers to Exercises**

1. 210 mi   **2.** 28 ft   **3.** $650   **4.** $9\frac{1}{11}$ amps   **5.** $b = \dfrac{2A}{h}$

6. $w = \dfrac{1}{2}(p - 2l)$ **7.** $l = \dfrac{A}{w}$ **8.** $r = \dfrac{c}{2\pi}$ **9.** $h = \dfrac{V}{\pi r^2}$

10. $R = \dfrac{L - S}{L}$ **11.** $P = \dfrac{A}{1 + rt}$ **12.** $a = \dfrac{2A - bh}{h}$ **13.** $\dfrac{h - \frac{1}{2}at}{t}$

14. $m_1 = \dfrac{Fr^2}{\alpha m_2}$ **15.** $d = \dfrac{L - a}{n - 1}$ **16.** $a_n = \dfrac{2s - na_1}{n}$ **17.** $a_1 = \dfrac{a_n}{r^{n-1}}$

18. $h = \dfrac{S - 2\pi r^2}{2\pi r}$ **19.** 100 ft by 200 ft   **20.** 18 in.   **21.** $360

22. $1500   **23.** $3000 at $7\frac{3}{4}$%   **24.** $7875   **25.** $1000
                  $2800 at 8%

26. 4 lbs   **27.** 8 liters   **28.** $\frac{115}{26} \approx 4.42$ lbs   **29.** 15 kg

30. 150 dimes and 60 quarters   **31.** 10 nickels and 20 dimes

32. 20–23¢ stamps; 80–29¢ stamps   **33.** 20–19¢ stamps; 42–29¢ stamps

34. 8000–$8 tickets; 4000–$10 tickets   **35.** 10 dozen roses; 15 dozen carnations

36. 8 oz   **37.** 6.25 gr   **38.** $11\frac{3}{11}$%   **39.** 16 oz

40. 5 hrs; 6 hrs   **41.** 4 hrs   **42.** 55 mph   **43.** 45 mi

44. 3 hrs 44 min   **45.** 6 hrs 40 min   **46.** $\frac{1}{2}$ hr   **47.** $2\frac{1}{4}$ hrs

48. 20 yrs old   **49.** 60 yrs old   **50.** 53 and 54   **51.** 32, 33, 34

52. 20, 22   **53.** 31, 33, 35   **54.** 11 and 13

Answers to Cumulative Practice Test P–4

1. (c)   **2.** (e)   **3.** (a)   **4.** (d)   **5.** (b)

6. $x = 4$ **7.** $x = \frac{7}{2}$ **8.** $x = 3$ **9.** $x = \frac{9}{4}$ **10.** $x = 6$

11. $-\frac{7}{12}$ **12.** -12 **13.** $\frac{5}{12}$ **14.** $-\frac{62}{9}$ **15.** $x > 7$

16. $x \geq -4$ **17.** $x < 2$ **18.** $-1 < x \leq \frac{5}{3}$ **19.** $-2x + 6$ **20.** $8z - 2x + 4$

21. $5x^2 - 5x$ **22.** $16x^2 - 9y^2$ **23.** $2(x + 3)$ **24.** $x + 6$ **25.** $\frac{x}{x - 7}$

26. $2x - (x - 4)$ **27.** 103 and 104   **28.** 32, 34, and 36   **29.** 15 and 17   **30.** $7000 at $7\frac{1}{2}$% $6000 at 8.5%

31. 20%   **32.** 2 lbs   **33.** $10   **34.** 10 nickels 20 dimes   **35.** 18 mph

36. 2 qts   **37.** 100 mi at 50 mph 40 mi at 40 mph   **38.** c   **39.** b   **40.** a

CHAPTER FIVE
Exponents and Polynomials

5.1 | Adding and Subtracting Polynomials

Section Highlights

1. Let $a_n, a_{n-1}, \ldots, a_1, a_0$ be real numbers, and let n be a nonnegative integer. A *polynomial in x* is an expression of the form

$$a_n x^n + a_{n-1} x^{n-1} + \cdots + a_1 x + a_0.$$

   Each a_i is called a *coefficient*. If $a_n \neq 0$, then the *degree* of the polynomial is n, and the *leading coefficient* is a_n. Each addend is called a *term*, and a_0 is the *constant term*.

2. A polynomial is in *standard form* if:
   (a) Each term is in standard form,
   (b) No two terms have the same variable parts, and
   (c) The terms are arranged in order of descending powers of the variable.

3. A polynomial with only one term is called a *monomial*.

4. A polynomial in standard form with two terms is called a *binomial*.

5. A polynomial with three terms, in standard form, is called a *trinomial*.

EXAMPLE 1 ■ Adding Polynomials Horizontally

Use a horizontal arrangement to find the following sums.

(a) $(x^2 + 2x - 1) + (2x^2 - 3x + 7)$

(b) $(x^3 - 2x - 6) + (4x^2 - 1) + (3x^2 + x + 5)$

Solution

(a) $(x^2 + 2x - 1) + (2x^2 - 3x + 7)$ Given

$\quad = (x^2 + 2x + (-1)) + (2x^2 + (-3x) + 7)$ Change subtraction to addition.

$\quad = (x^2 + 2x^2) + (2x + (-3x)) + (-1 + 7)$ Regroup.

$\quad = 3x^2 - x + 6$ Standard form

(b) $(x^3 - 2x - 6) + (4x^2 - 1) + (3x^2 + x + 5)$ Given

$\quad = (x^3 + (-2x) + (-6)) + (4x^2 + (-1)) + (3x^2 + x + 5)$ Change to addition.

$\quad = x^3 + (4x^2 + 3x^2) + (-2x + x) + (-6 + (-1) + 5)$ Regroup.

$\quad = x^3 + 7x^2 - x - 2$ Standard form

Starter Exercise 1 | *Fill in the blanks.*

Use a horizontal arrangement to find the following sums.

(a) $(5x^2 - x + 1) + (3x^3 + 2x^2)$ Given

$\quad = (5x^2 + (-x) + 1) + (3x^3 + 2x^2)$ Change to addition.

$\quad = 3x^3 + \left(5x^2 + \boxed{}\right) + \boxed{} + 1$ Regroup.

$\quad = \boxed{}$

(b) $(x^3 + 2x + 6) + [(3x^2 + 1) + (4x^3 - x + 2)]$ Given

$\quad = \boxed{}$ Change to addition.

$\quad = (x^3 + 2x + 6) + \left[4x^3 + 3x^2 + \left(\boxed{}\right)\right]$ Regroup.

$\quad = (x^3 + 4x^3) + 3x^2 + \boxed{} + \left(\boxed{}\right)$ Regroup.

$\quad = \boxed{}$

EXAMPLE 2 ■   Adding Polynomials Vertically

Use a vertical arrangement to find the following sums.

(a) $(4x^3 + 2x^2 - 3x + 6) + (x^3 - 4x + 1)$

(b) $(x^3 + 2x^2 + 7x + 3) + (2x^2 - 1) + (x^3 + 7x^2 - 6)$

Solution

(a)

$$
\begin{array}{r}
4x^3 + 2x^2 - 3x + 6 \\
x^3 \quad\;\; - 4x + 1 \\
\hline
5x^3 + 2x^2 - 7x + 1
\end{array}
$$

(b)

$$
\begin{array}{r}
x^3 + \;\; 2x^2 + 7x + 3 \\
2x^2 \quad\;\; - 1 \\
x^3 + \;\; 7x^2 \quad\;\; - 6 \\
\hline
2x^3 + 11x^2 + 7x - 4
\end{array}
$$

■

Starter Exercise 2 | *Fill in the blanks.*

Use a vertical arrangement to find the sum of $(7x^3 + 6x^2 - 5x + 1)$ and $(3x^2 - 4)$.

$$
\begin{array}{l}
7x^3 + 6x^2 - 5x + \;\; 1 \\
\boxed{} \\
\hline
7x^3 + \boxed{}
\end{array}
$$

EXAMPLE 3 ■ Subtracting Polynomials Horizontally

Perform the following operations.

(a) $(7x^2 + 5x + 1) - (5x^2 - 3x - 4)$

(b) $(x^3 + 7x - 3) - [(2x^2 + x - 1) + (4x^2 - 1)]$

Solution

(a) $(7x^2 + 5x + 1) - (5x^2 - 3x - 4)$ Given

$= (7x^2 + 5x + 1) + (-5x^2 + 3x + 4)$ Change signs and add.

$= (7x^2 - 5x^2) + (5x + 3x) + (1 + 4)$ Regroup.

$= 2x^2 + 8x + 5$

(b) $(x^3 + 7x - 3) - [(2x^2 + x - 1) + (4x^2 - 1)]$ Given

$= (x^3 + 7x - 3) - [(2x^2 + 4x^2) + x + (-1 - 1)]$ Regroup.

$= (x^3 + 7x - 3) - [6x^2 + x - 2]$

$= (x^3 + 7x - 3) + [-6x^2 - x + 2]$ Change signs.

$= x^3 - 6x^2 + (7x - x) + (-3 + 2)$ Regroup.

$= x^3 - 6x^2 + 6x - 1$

Starter Exercise 3   *Fill in the blanks.*

(a) Perform the following operations: $(16x^2 - x + 3) - (4x^2 + 2x + 4)$

$(16x^2 - x + 3) - (4x^2 + 2x + 4)$

$= (16x^2 - x + 3) + \left(\boxed{} \right)$

$= \left(16x^2 \boxed{}\right) + \left(-x \boxed{}\right) + \left(3 \boxed{}\right)$

$= \boxed{}$

(b) Perform the following operations: $(4x^3 + 2x^2 + 1) + [(4x^2 + 2x - 1) - (x^2 + 2x + 4)]$

$(4x^3 + 2x^2 + 1) + [(4x^2 + 2x - 1) - (x^2 + 2x + 4)]$

$= (4x^3 + 2x^2 + 1) + \left[(4x^2 + 2x - 1) + \left(\boxed{} \right) \right]$

$= (4x^3 + 2x^2 + 1) + \left[\left(4x^2 \boxed{}\right) + \left(2x \boxed{}\right) + \left(-1 \boxed{}\right) \right]$

$= (4x^3 + 2x^2 + 1) + \left[\boxed{} \right]$

$= 4x^3 + \left(2x^2 \boxed{}\right) + \boxed{}$

$= \boxed{}$

EXAMPLE 4 ■ Subtracting Polynomials Vertically

Use a vertical arrangement to perform the following operations.

(a) $(x^2 - 2x + 8) - (3x^2 + 4x + 1)$

(b) $(4x^3 + 2x^2 - 6) - (2x^3 - x + 8)$

Solution

(a)
$$
\begin{array}{ll}
(x^2 - 2x + 8) & \Rightarrow \\
-(3x^2 + 4x + 1) & \Rightarrow
\end{array}
\qquad
\begin{array}{l}
x^2 - 2x + 8 \\
\underline{-3x^2 - 4x - 1} \\
-2x^2 - 6x + 7
\end{array}
$$

(b)
$$
\begin{array}{ll}
(4x^3 + 2x^2 \quad\; - 6) & \Rightarrow \\
-(2x^3 \qquad - x + 8) & \Rightarrow
\end{array}
\qquad
\begin{array}{l}
4x^3 + 2x^2 \quad\; - 6 \\
\underline{-2x^3 \qquad + x - 8} \\
2x^3 + 2x^2 + x - 14
\end{array}
$$ ■

■ Starter Exercise 4   *Fill in the blanks.*

Use a vertical arrangement to perform the following operations.

(a) $(3x^2 - 7x) - (4x^2 + 6x + 3)$

$$
\begin{array}{ll}
(3x^2 - 7x) & \Rightarrow \qquad 3x^2 - 7x \\
-\left(4x^2 + \boxed{6x + 3}\,\right) & \Rightarrow \qquad -4x^2 - \boxed{} \\
& \qquad\qquad\;\; \boxed{}
\end{array}
$$

(b) $(5x^3 - 1) - (2x^2 + x - 6)$

$$
\begin{array}{ll}
5x^3 \boxed{} & \Rightarrow \quad \boxed{} \\
-(2x^2 + x - 6) & \Rightarrow \quad -2x^2 - x + 6 \\
& \qquad\quad\; \boxed{}
\end{array}
$$

■ Solutions to Starter Exercises ■

1. (a) $= 3x^3 + \left(5x^2 + \boxed{2x^2}\,\right) + \boxed{(-x)} + 1 = \boxed{3x^3 + 7x^2 - x + 1}$

(b) $= \boxed{(x^3 + 2x + 6) + [(3x^2 + 1) + (4x^3 + (-x) + 2)]}$

$= (x^3 + 2x + 6) + \left[4x^3 + 3x^2 + \left(\boxed{-x}) + (1 + \boxed{2}\,\right)\right]$

$= (x^3 + 4x^3) + 3x^2 + \boxed{(2x + (-x))} + \left(\boxed{6 + 3}\,\right)$

$= \boxed{5x^3 + 3x^2 + x + 9}$

2.
$$
\begin{array}{l}
7x^3 + \;6x^2 - 5x + 1 \\
\quad\;\; \boxed{3x^2 \qquad - 4} \\
\hline
7x^3 + \boxed{9x^2 - 5x - 3}
\end{array}
$$

■ **Solutions to Starter Exercises** ■

3. (a) $= (16x^2 - x + 3) + \left(\boxed{-4x^2 - 2x - 4}\right)$

$= \left(16x^2 \boxed{-4x^2}\right) + \left(-x \boxed{-2x}\right) + \left(3 \boxed{-4}\right) = \boxed{12x^2 - 3x - 1}$

(b) $= (4x^3 + 2x^2 + 1) + \left[(4x^2 + 2x - 1) + \left(\boxed{-x^2 - 2x - 4}\right)\right]$

$= (4x^3 + 2x^2 + 1) + \left[\left(4x^2 \boxed{-x^2}\right) + \left(2x \boxed{-2x}\right) + \left(-1 \boxed{-4}\right)\right]$

$= (4x^3 + 2x^2 + 1) + \left[\boxed{3x^2 - 5}\right]$

$= 4x^3 + \left(2x^2 \boxed{+3x^2}\right) + \boxed{(1-5)} = \boxed{4x^3 + 5x^2 - 4}$

4. (a) $\begin{array}{l}(3x^2 - 7x \quad) \\ -\left(4x^2 + \boxed{6x+3}\right)\end{array} \Rightarrow \begin{array}{l} 3x^2 - 7x \\ -4x^2 - \boxed{6x - 3} \\ \hline \boxed{-x^2 - 13x - 3}\end{array}$

(b) $\begin{array}{l} 5x^3 \boxed{\quad -1} \\ -(\quad 2x^2 + x - 6)\end{array} \Rightarrow \begin{array}{l} \boxed{5x^3 \quad\quad -1} \\ -2x^2 - x + 6 \\ \hline \boxed{5x^3 - 2x^2 - x + 5}\end{array}$

5.1 EXERCISES

In Exercises 1–4, write the polynomial in standard form. Then find its degree and leading coefficient.

Polynomial	Standard Form	Degree	Leading Coefficient
1. $2x - 1$		1	2
2. $1 - 3x^2$	$-3x^2 + 1$	2	-3
3. $x + 2x^2 - 3$	$2x^2 + x - 3$	2	2
4. $x^5 + x - 2x^3$	$x^5 - 2x^3 + x$	5	2

In Exercises 5–8, determine whether the polynomial is a monomial, a binomial, or a trinomial.

5. $2x + 1$ **6.** 14 **7.** $x^2 + 2x - 6$ **8.** $4 - x^2$

In Exercises 9–12, determine why the algebraic expression is not a polynomial.

9. $|x| - 4$ **10.** $\dfrac{1}{x}$ **11.** $x^{-1} + 5$ **12.** $\dfrac{x+1}{x-1}$

cannot have an absolute value of a variable

In Exercises 13–16, perform the addition using a horizontal arrangement.

13. $(2x^2 + 5x - 1) + (7x^2 - x - 13)$

14. $(5x^3 - 2x^2 + 6x + 1) + (6x^3 - 2x - 1)$

15. $(x + 1) + (x^3 + 2x - 6)$

16. $(x^2 + x + 7) + (x + 3) + (2x^2 - 3x - 7)$

In Exercises 17–20, perform the addition using a vertical arrangement.

17. $(2x^2 - 6x - 3) + (x^2 - x - 1)$

18. $(x^3 + 2x^2 - 3) + (4x^2 - x + 4)$

19. $(y^5 - 1) + (y^2 + 2y + 1)$

20. $(6t - 1) + (4t + 6)$

In Exercises 21–24, perform the subtraction using a horizontal arrangement.

21. $(2u - 1) - (5u + 6)$

22. $(5x^2 - 2x + 1) - (4x^2 + 2x - 7)$

23. $(t^3 - 1) - (t^3 + 1)$

24. $(4x^3 - 2x^2 + 1) - (5x^3 + 2x - 6)$

In Exercises 25–28, perform the subtractions using a vertical arrangement.

25. $(2y + 7) - (y - 1)$

26. $(x^2 + 3x + 1) - (4x^2 - 2x + 5)$

27. $(x^3 - 1) - (x^2 + x + 1)$

28. $(5y^3 - 2y^2 + 4y + 9) - (3y^3 + y - 1)$

In Exercises 29–40, perform the indicated operations.

29. $(7x - 5) - (19x + 7)$

30. $(2x^2 - 1) + (5x - 3)$

31. $(x^2 - 4x + 1) - [(x^2 + 3) + (2x^2 + x)]$

32. $(5y^2 + 3) - [(y^3 + 2y - 2) - (y^2 - 5y + 1)]$

33. $10x^2 - [4x^2 - (2x^2 + x - 3)]$

34. $(t^4 + t + 1) - [3t^4 - (t^4 - 3t + 16)]$

35. $2(u - 4) - 3(2u - 6)$

36. $-2(x^2 + 2x - 6) - (2x^2 + 4x + 9)$

37. $-4(u - 1) - 6(2u + 7) + 5(u - 3)$

38. $5x^6 - 4(x + 1) + 2(x - 1)$

39. $(4x^7 + 9x^5 + 3) - (2x^7 - 7x^5 - 5)$

40. $3(y^2 - 3y + 1) - 2(2y^2 - y + 2) - 3(y^2 - 5y + 2)$

5.2   Multiplying Polynomials: Special Products

Section Highlights

1. **Distributive Property**

$$a \cdot (b + c) = a \cdot c + a \cdot c$$

$$(a + b) \cdot c = a \cdot c + b \cdot c$$

2. **Sum and Difference of Two Terms**

$$(a + b)(a - b) = a^2 - b^2$$

3. **Square of a Binomial**

$$(a + b)^2 = a^2 + 2ab + b^2$$

$$(a - b)^2 = a^2 - 2ab + b^2$$

EXAMPLE 1 ■   Finding Products with Monomial Multipliers

Multiply the following.

(a) $-2x(x^2 - 6)$

(b) $(x^2 + 2x + 3)(x)$

Solution

(a) $-2x(x^2 - 6) = -2x(x^2) - (-2x)(6)$ Distributive Property

$\qquad\qquad\quad = -2x^3 + 12x$ Standard form

(b) $(x^2 + 2x + 3)(x) = x^2(x) + 2x(x) + 3(x)$ Distributive Property

$\qquad\qquad\qquad\quad = x^3 + 2x^2 + 3x$ Standard form

■ Starter Exercise 1   *Fill in the blanks.*

(a) Multiply: $(-x)(2x^2 - 4x + 1)$

$$(-x)(2x^2 - 4x + 1) = \left(\boxed{X} \right)(2x^2) - \left(\boxed{X} \right)(4x) + \left(\boxed{X} \right)(1)$$

$$= \boxed{3X^2 - 4X - 1}$$

(b) Multiply: $(5x^2 - 6x)(-2x)$

$$(5x^2 - 6x)(-2x) = 5x^2 \left(\boxed{X} \right) - 6x \left(\boxed{} \right)$$

$$= \boxed{}$$

EXAMPLE 2 ■   **Multiplying Binomials Using the Distributive Property**

(a) $(x + 1)(x + 2)$ (b) $(2x - 1)(x + 6)$

Solution

(a) $(x + 1)(x + 2) = (x + 1)(x) + (x + 1)(2)$ Distributive Property

$\qquad\qquad\qquad = x^2 + x + 2x + 2$ Distributive Property

$\qquad\qquad\qquad = x^2 + 3x + 2$ Standard form

(b) $(2x - 1)(x + 6) = (2x - 1)(x) + (2x - 1)(6)$ Distributive Property

$\qquad\qquad\qquad\quad = 2x^2 - x + 12x - 6$ Distributive Property

$\qquad\qquad\qquad\quad = 2x^2 + 11x - 6$ Standard form

Starter Exercise 2   *Fill in the blanks.*

(a) $(x - 7)(2x + 1) = (x - 7)\left(\boxed{}\right) + (x - 7)\left(\boxed{}\right)$

$\qquad\qquad\qquad = 2x^2 - \boxed{} + \boxed{} - 7 = \boxed{}$

(b) $(2x - 3)(2x - 5) = \left(\boxed{}\right)(2x) - \left(\boxed{}\right)(5)$

$\qquad\qquad\qquad = \boxed{} = \boxed{}$

EXAMPLE 3 ■   **Multiplying Binomials Using the FOIL Method**

Use the FOIL Method to multiply $(5x + 1)(x - 4)$.

Solution

$$\overset{\qquad\quad\text{F}\qquad\text{O}\quad\text{I}\quad\text{L}}{(5x + 1)(x - 4) = 5x^2 - 20x + x - 4}$$
$$\qquad\qquad\qquad = 5x^2 - 19x - 4$$

Starter Exercise 3   *Fill in the blanks.*

Use the FOIL Method to multiply $(x - 3)(4x + 3)$.

$$\overset{\qquad\qquad\quad\text{F}\qquad\text{O}\quad\text{I}\quad\text{L}}{(x - 3)(4x + 3) = 4x^2 + \boxed{} - \boxed{} - \boxed{}}$$
$$\qquad\qquad\quad = \boxed{}$$

EXAMPLE 4 ■ Simplifying Polynomial Expressions

Simplify the following expression and write the result in standard form.

(a) $(3x^2 + x)(x - 1) - (10x^2 + 6)$ (b) $(2x + 3)^2 - x^2$

Solution

(a) $(3x^2 + x)(x - 1) - (10x^2 + 6)$

$$= 3x^3 - 3x^2 + x^2 - x - 10x^2 - 6 \qquad \text{Multiply binomials.}$$

$$= 3x^3 - 12x^2 - x - 6 \qquad \text{Standard form}$$

(b) $(2x + 3)^2 - x^2 = (2x + 3)(2x + 3) - x^2 \qquad \text{Repeated multiplication}$

$$= 4x^2 + 6x + 6x + 9 - x^2 \qquad \text{Multiply binomials.}$$

$$= 3x^2 + 12x + 9 \qquad \text{Standard form}$$

■

Starter Exercise 4 | *Fill in the blanks.*

Simplify the following expressions and write the result in standard form.

(a) $(2x^2 - 3x + 1)(x) - (2x)^2 = 2x^3 \left(\boxed{} \right) - 3x \left(\boxed{} \right) + 1 \left(\boxed{} \right) - 4x^2$

$$= \boxed{} = \boxed{}$$

(b) $(2x - 3)(x - 4) - 7x^2 = 2x^2 - 8x - \boxed{} + \boxed{} - 7x^2 = \boxed{}$

EXAMPLE 5 ■ Multiplying Polynomials (Horizontal Arrangement)

Use a horizontal arrangement of factors to find the product $(x^2 - 3x + 2)(2x + 7)$.

Solution

$(x^2 - 3x + 2)(2x + 7) = (x^2 - 3x + 2)(2x) + (x^2 - 3x + 2)(7) \qquad \text{Distributive Property}$

$$= 2x^3 - 6x^2 + 4x + 7x^2 - 21x + 14 \qquad \text{Distributive Property}$$

$$= 2x^3 + x^2 - 17x + 14 \qquad \text{Standard form}$$

■

Starter Exercise 5 | *Fill in the blanks.*

Use a horizontal arrangement of factors to find the following product.

$$(2x + 1)(x^2 - 3x + 7) = (2x + 1)x^2 - \left(\boxed{} \right)(3x) + \left(\boxed{} \right)(7)$$

$$= \boxed{}$$

$$= \boxed{}$$

EXAMPLE 6 ■   Multiplying Polynomials (Vertical Arrangement)

Find the product $(2x^2 + x - 1)(3x + 2)$ using a vertical arrangement of factors.

Solution

$$
\begin{array}{r}
2x^2 + \ x - 1 \\
\times \qquad 3x + 2 \\
\hline
4x^2 + 2x - 2 \\
6x^3 + 3x^2 - 3x \\
\hline
6x^3 + 7x^2 - \ x - 2
\end{array}
$$

$\Leftarrow \quad 2(2x^2 + x - 1)$

$\Leftarrow \quad 3x(2x^2 + x - 1)$

Standard form   ■

Starter Exercise 6   *Fill in the blanks.*

Find the following product using a vertical arrangement of factors.

$$
\begin{array}{r}
3x^2 - \ x - 1 \\
x + 5 \\
\hline
15x^2 - 5x - 5 \\
3x^3 - \boxed{} \\
\hline
\boxed{}
\end{array}
$$

EXAMPLE 7 ■   Multiplying Polynomials

Multiply the following.

(a) $(x^2 + 5x - 1)(x + 2x^2 - 3)$

(b) $(x - 2)^3$

Solution

(a) $(x^2 + 5x - 1)(x + 2x^2 - 3)$

$= (x^2 + 5x - 1)(2x^2 + x - 3)$ 　　　　Factors in standard form

$= (x^2 + 5x - 1)(2x^2) + (x^2 + 5x - 1)(x) - (x^2 + 5x - 1)(3)$ 　Distributive Property

$= 2x^4 + 10x^3 - 2x^2 + x^3 + 5x^2 - x - 3x^2 - 15x + 3$ 　Distributive Property

$= 2x^4 + 11x^3 - 16x + 3$ 　　　　Standard form

(b) $(x - 2)^3 = [(x - 2)(x - 2)](x - 2)$ 　　　Repeated multiplication form

$= [x^2 - 2x - 2x + 4](x - 2)$ 　　　Multiply binomials.

$= [x^2 - 4x + 4](x - 2)$ 　　　Factors in standard form

$= [x^2 - 4x + 4](x) - [x^2 - 4x + 4](2)$ 　Distributive Property

$= x^3 - 4x^2 + 4x - 2x^2 + 8x - 8$ 　Distributive Property

$= x^3 - 6x^2 + 12x - 8$ 　　　Standard form   ■

Starter Exercise 7 | *Fill in the blanks.*

Multiply the following.

(a) $(2x + x^2 - 1)(-x + x^2 + 2) = (x^2 + 2x - 1)(x^2 - x + 2)$

$$= \left(\boxed{}\right)(x^2) - \left(\boxed{}\right)(x) + \left(\boxed{}\right)(2)$$

$$= \boxed{} = \boxed{}$$

(b) $(x + 3)^3 = [(x + 3)(x + 3)](x + 3) = [x^2 + 3x + \boxed{} + 9](x + 3)$

$$= [x^2 + \boxed{} + 9](x + 3) = x^2\left(\boxed{}\right) + \boxed{}(x + 3) + 9\left(\boxed{}\right)$$

$$= \boxed{} = \boxed{}$$

EXAMPLE 8 ■   Finding the Product of the Sum and Difference of Two Terms

Find the product $(2x + 5)(2x - 5)$.

Solution

$$(2x + 5)(2x - 5) = (2x)^2 - 5^2 = 4x^2 - 25$$

■

Starter Exercise 8 | *Fill in the blanks.*

Find the following product.

$$(3 - 4x)(3 + 4x) = (3)^2 - \left(\boxed{}\right)^2 = \boxed{}$$

EXAMPLE 9 ■   Squaring a Binomial

Find the following products.

(a) $(2x + 7)^2$ 　　　　　　　　　　　　(b) $(x - 6)^2$

Solution

(a) $(2x + 7)^2 = (2x)^2 + 2(2x)(7) + 7^2$ 　　　(b) $(x - 6)^2 = x^2 - 2(x)(6) + 6^2$

$\qquad = 4x^2 + 28x + 49$ 　　　　　　　　　　$= x^2 - 12x + 36$

■

Starter Exercise 9 | *Fill in the blanks.*

Find the following products.

(a) $(3x - 2)^2 = \left(\boxed{}\right)^2 - 2\left(\boxed{}\right)\left(\boxed{}\right) + \left(\boxed{}\right)^2 = \boxed{}$

(b) $(x + 4)^2 = \left(\boxed{}\right)^2 + 2\left(\boxed{}\right)\left(\boxed{}\right) + \left(\boxed{}\right)^2 = \boxed{}$

■ **Solutions to Starter Exercises** ■

1. (a) $(-x)(2x^2 - 4x + 1) = \left(\boxed{-x}\right)(2x^2) - \left(\boxed{-x}\right)(4x) + \left(\boxed{-x}\right)(1)$

$= \boxed{-2x^3 + 4x^2 - x}$

(b) $(5x^2 - 6x)(-2x) = 5x^2\left(\boxed{-2x}\right) - 6x\left(\boxed{-2x}\right) = \boxed{-10x^3 + 12x^2}$

2. (a) $(x - 7)(2x + 1) = (x - 7)\left(\boxed{2x}\right) + (x - 7)\left(\boxed{1}\right)$

$= 2x^2 - \boxed{14x} + \boxed{x} - 7 = \boxed{2x^2 - 13x - 7}$

(b) $(2x - 3)(2x - 5) = \left(\boxed{2x - 3}\right)(2x) - \left(\boxed{2x - 3}\right)(5)$

$= \boxed{4x^2 - 6x - 10x + 15} = \boxed{4x^2 - 16x + 15}$

3. $(x - 3)(4x + 3) = 4x^2 + \boxed{3x} - \boxed{12x} - \boxed{9} = \boxed{4x^2 - 9x - 9}$

4. (a) $(2x^2 - 3x + 1)(x) - (2x)^2 = 2x^3\left(\boxed{x}\right) - 3x\left(\boxed{x}\right) + 1\left(\boxed{x}\right) - 4x^2$

$= \boxed{2x^4 - 3x^2 + x - 4x^2} = \boxed{2x^4 - 7x^2 + x}$

(b) $(2x - 3)(x - 4) - 7x^2 = 2x^2 - 8x - \boxed{3x} + \boxed{12} - 7x^2$

$= \boxed{-5x^2 - 11x + 12}$

5. $(2x + 1)(x^2 - 3x + 7) = (2x + 1)x^2 - \left(\boxed{2x + 1}\right)(3x) + \left(\boxed{2x + 1}\right)(7)$

$= \boxed{2x^3 + x^2 - 6x^2 - 3x + 14x + 7}$

$= \boxed{2x^3 - 5x^2 + 11x + 7}$

6.

$$
\begin{array}{r}
3x^2 - x \;\; -1 \\
x \;\; +5 \\
\hline
15x^2 - 5x \;\; -5 \\
3x^3 - \boxed{x^2 - x} \\
\hline
\boxed{3x^3 + \;\; 14x^2 - 6x - 5}
\end{array}
$$

■ **Solutions to Starter Exercises** ■

7. (a) $= \left(\boxed{x^2 + 2x - 1} \right)(x^2) - \left(\boxed{x^2 + 2x - 1} \right)(x) + \left(\boxed{x^2 + 2x - 1} \right)(2)$

$= \boxed{x^4 + 2x^3 - x^2 - x^3 - 2x^2 + x + 2x^2 + 4x - 2}$

$= \boxed{x^4 + x^3 - x^2 + 5x - 2}$

(b) $= \left[x^2 + 3x + \boxed{3x} + 9 \right](x + 3)$

$= \left[x^2 + \boxed{6x} + 9 \right](x + 3)$

$= x^2 \left(\boxed{x + 3} \right) + \boxed{6x}(x + 3) + 9 \left(\boxed{x + 3} \right)$

$= \boxed{x^3 + 3x^2 + 6x^2 + 18x + 9x + 27}$

$= \boxed{x^3 + 9x^2 + 27x + 27}$

8. $(3 - 4x)(3 + 4x) = (3)^2 - \left(\boxed{4x} \right)^2 = \boxed{9 - 16x^2}$

9. (a) $(3x - 2)^2 = \left(\boxed{3x} \right)^2 - 2\left(\boxed{3x} \right)\left(\boxed{2} \right) + \left(\boxed{2} \right)^2 = \boxed{9x^2 - 12x + 4}$

(b) $(x + 4)^2 = \left(\boxed{x} \right)^2 + 2\left(\boxed{x} \right)\left(\boxed{4} \right) + \left(\boxed{4} \right)^2 = \boxed{x^2 + 8x + 16}$

5.2 | EXERCISES

In Exercises 1–32, perform the indicated operations and simplify.

1. $z(-3z^2)$

2. $(-2x)(-6x)$

3. $(4z^2)^3$

4. $x(2x)^2$

5. $\left(\dfrac{x}{2} \right)(10x)$

6. $\left(\dfrac{x}{3} \right)^2 (-9x)^2$

7. $2y(3y^2) - 5y(6)$

8. $5x(2x)^2 + (-6x)(-5x)$

9. $x(x + 2)$

10. $2x(3x - 1)$

11. $-3x(x^2 - 6)$

12. $-5t(7 - 2t^2)$

13. $x(3x^2 - 2x + 1)$

14. $-2x(x^2 - 4x + 3)$

15. $(x + 1)(x + 2)$

16. $(x - 1)(x + 3)$

17. $(2x + 1)(x - 6)$

18. $(2z - 1)(3z - 4)$

19. $(x + y)(2x - y)$

20. $(u - 2v)(4u + 7v)$

21. $(5 - 2t)(t + 7)$

22. $(4 + 2x)(3 - 2x)$

23. $(x - 3)(2x + 1) - (2x)^2$

24. $(2x - 3)(x + 1) - (2x - 1)(2x + 9)$

25. $(z + 1)(z^2 - 2z - 3)$

26. $(2x - 3)(4x^2 + x - 6)$

27. $(x^2 - 7x - 1)(2x + 1)$

28. $(2x^2 - 3x + 1)(x^2 + 6)$

29. $(x^2 - 2x + 4)(x^2 + 4x - 3)$

30. $(2x^2 + 2x + 3)(x^2 - 3x - 1)$

31. $(x + 1)(x + 2)(x + 3)$

32. $(2x - 1)(x - 4)(3x - 2)$

In Exercises 33–40, perform the indicated operation and simplify. (Use a special product formula whenever possible.)

33. $(x + 2)^2$

34. $(x - 3)^2$

35. $(2x - 4)^2$

36. $(x - y)(x + y)$

37. $(4x + 3)(4x - 3)$

38. $(2 - 3x)(2 + 3x)$

39. $(x + 1)^2 - (x - 1)^2$

40. $[(x + 2) + y]^2$

41. *Area of a Square* The sides of a square are of length $2x - 1$. Find a polynomial in standard form that represents the area of the square.

42. *Area of a Rectangle* The length of a rectangle is $3x + 7$, the width is $x + 2$. Find a polynomial in standard form that represents the area of the rectangle.

In Exercises 43–45, use a graphing utility to check your factoring in Exercises 33–35 by graphing both the given expression and your answer.

5.3 | Dividing Polynomials

Section Highlights

1. **Rule of Exponents**
   Let m and n be positive integers, and let a represent a real number, a variable, or an algebraic expression.

   i) $\dfrac{a^m}{a^n} = a^{m-n}$, if $m > n$;   ii) $\dfrac{a^m}{a^n} = \dfrac{1}{a^{n-m}}$, if $n > m$;   iii) $\dfrac{a^n}{a^n} = 1 = a^0$

   Note that $a \neq 0$ in all the above cases.

2. $\dfrac{a + b}{c} = \dfrac{a}{c} + \dfrac{b}{c}$,   if $c \neq 0$.

EXAMPLE 1 ■   Dividing a Monomial by a Monomial

Perform the following division. (Assume $x \neq 0$.)

(a) $4x^2 \div 2x$

(b) $14x^4 \div 4x^2$

(c) $27x^2 \div \dfrac{1}{3}x^2$

(d) $8x^2 \div 6x^4$

Solution

(a) $\dfrac{4x^2}{2x} = \dfrac{4}{2} \cdot \dfrac{x^2}{x} = 2x^{2-1} = 2x$

(b) $\dfrac{14x^4}{4x^2} = \dfrac{14}{4} \cdot \dfrac{x^4}{x^2} = \dfrac{7}{2} \cdot x^{4-2} = \dfrac{7}{2}x^2$

(c) $\dfrac{27x^2}{(1/3)x^2} = \dfrac{27}{1/3} \cdot \dfrac{x^2}{x^2} = 81 \cdot 1 = 81$

(d) $\dfrac{8x^2}{6x^4} = \dfrac{8}{6} \cdot \dfrac{1}{x^{4-2}} = \dfrac{4}{3} \cdot \dfrac{1}{x^2} = \dfrac{4}{3x^2}$

$\blacksquare$

Starter Exercise 1 | *Fill in the blanks.*

Perform the following divisions. (Assume $x \neq 0$.)

(a) $\dfrac{3x^3}{x^2} = 3 \cdot \dfrac{x^3}{x^2} = 3 \cdot x^{\boxed{} - \boxed{}} = 3x^{\boxed{}} = \boxed{}$

(b) $\dfrac{16x^4}{8x^4} = \dfrac{16}{8} \cdot \dfrac{x^4}{x^4} = \boxed{} \cdot \boxed{} = \boxed{}$

(c) $\dfrac{8x^3}{4x^5} = \dfrac{8}{4} \cdot \dfrac{x^3}{x^5} = 2 \cdot \boxed{} = \boxed{}$

EXAMPLE 2 ■ Dividing a Polynomial by a Monomial

Perform the following divisions. (Assume $x \neq 0$.)

(a) $\dfrac{4y+2}{2}$

(b) $\dfrac{2x^2+3x}{x}$

(c) $\dfrac{8x^3+6x^2+10x}{2x}$

(d) $\dfrac{4x^2+5x+1}{3x}$

Solution

(a) $\dfrac{4y+2}{2} = \dfrac{4y}{2} + \dfrac{2}{2} = 2y+1$

(b) $\dfrac{2x^2+3x}{x} = \dfrac{2x^2}{x} + \dfrac{3x}{x} = 2x+3$

(c) $\dfrac{8x^3+6x^2+10x}{2x} = \dfrac{8x^3}{2x} + \dfrac{6x^2}{2x} + \dfrac{10x}{2x} = 4x^2+3x+5$

(d) $\dfrac{4x^2+5x+1}{3x} = \dfrac{4x^2}{3x} + \dfrac{5x}{3x} + \dfrac{1}{3x} = \dfrac{4}{3}x + \dfrac{5}{3} + \dfrac{1}{3x}$

Note that in each example we wrote the division as the sum of fractions, then used the rules for dividing monomials. $\blacksquare$

Starter Exercise 2 | *Fill in the blanks.*

Perform the following division. (Assume $x \neq 0$.)

(a) $\dfrac{2x^2+2x}{x} = \dfrac{2x^2}{x} + \dfrac{2x}{x} = 2\boxed{} + 2\boxed{} = \boxed{}$

(b) $\dfrac{3x^3+9x^2+27x}{3x} = \dfrac{3x^3}{3x} + \dfrac{9x^2}{3x} + \dfrac{27x}{3x} = \boxed{}$

(c) $\dfrac{x^2-2x+1}{x} = \dfrac{x^2}{x} - \boxed{} + \dfrac{1}{x} = x - \boxed{} + \dfrac{1}{x}$

(d) $\dfrac{3x^3-2x^2-4x}{2x^2} = \dfrac{3x^3}{\boxed{}} - \dfrac{2x^2}{\boxed{}} - \dfrac{4x}{\boxed{}} = \boxed{}$

EXAMPLE 3 ■ Dividing a Polynomial by a Binomial

Perform the following divisions.

(a) Divide $x^2 + 5x + 6$ by $x + 2$

(b) $\dfrac{3x^2 + x + 1}{x + 2}$

(c) $(2x^3 + 3x^2 - x + 6) \div (2x + 1)$

(d) $(x^2 - x + 3x^3 - 4) \div (1 + x)$

(e) $(x^3 - 1) \div (x - 1)$

Solution

(a)
$$
\begin{array}{r}
x + 3 \\
x + 2 \overline{)\,x^2 + 5x + 6} \\
\underline{x^2 + 2x} \qquad\ \ \\
3x + 6 \\
\underline{3x + 6} \\
0
\end{array}
$$

Think $\dfrac{x^2}{x} = x$.

Think $\dfrac{3x}{x} = 3$.

$x(x + 2)$

$3x + 6$ Subtract; bring down the 6.

$3(x + 2)$

0 Subtract.

Thus, $\dfrac{x^2 + 5x + 6}{x + 2} = x + 3$.

(b)
$$
\begin{array}{r}
3x - 5 \\
x + 2 \overline{)\,3x^2 + x + 1} \\
\underline{3x^2 + 6x} \qquad\ \ \\
-5x + 1 \\
\underline{-5x - 10} \\
11
\end{array}
$$

Think $\dfrac{3x^2}{x} = 3x$.

Think $\dfrac{-5x}{x} = -5$.

$3x(x + 2)$

$-5x + 1$ Subtract; bring down the 1.

$-5(x + 2)$

11 Subtract.

Thus, $\dfrac{3x^2 + x + 1}{x + 2} = 3x - 5 + \dfrac{11}{x + 2}$.

(c)
$$
\begin{array}{r}
x^2 + x - 1 \\
2x + 1 \overline{)\,2x^3 + 3x^2 - x + 6} \\
\underline{2x^3 + x^2} \qquad\qquad\ \ \\
2x^2 - x \\
\underline{2x^2 + x} \\
-2x + 6 \\
\underline{-2x - 1} \\
7
\end{array}
$$

Thus,

$$\frac{2x^3 + 3x^2 - x + 6}{2x + 1} = x^2 + x - 1 + \frac{7}{2x + 1}.$$

(d) First, write the divisor and dividend in standard form.

$$
\begin{array}{r}
3x^2 - 2x + 1 \\
x + 1 \overline{)\,3x^3 + x^2 - x - 4} \\
\underline{3x^3 + 3x^2} \qquad\qquad\ \ \\
-2x^2 - x \\
\underline{-2x^2 - 2x} \\
x - 4 \\
\underline{x + 1} \\
-5
\end{array}
$$

Thus, $\dfrac{3x^3 - x^2 - x - 4}{x + 1} = 3x^2 - 2x + 1 - \dfrac{5}{x + 1}$.

(e)
$$
\begin{array}{r}
x^2 + x + 1 \\
x - 1 \overline{)\,x^3 + 0x^2 + 0x - 1} \\
\underline{x^3 - x^2} \qquad\qquad\ \ \\
x^2 + 0x \\
\underline{x^2 - x} \\
x - 1 \\
\underline{x - 1} \\
0
\end{array}
$$

$0x^2$ and $0x$ are place holders and used to line up subtraction.

Thus, $\dfrac{x^3 - 1}{x - 1} = x^2 + x + 1$.

■

Starter Exercise 3 | *Fill in the blanks.*

Perform the following divisions.

(a)

$$
\begin{array}{r}
2x - \boxed{} \\
x + 2 \overline{\smash{)}\, 2x^2 - 3x - 7} \\
\underline{\boxed{} + 4x} \\
-7x - 7 \\
\underline{\boxed{}} \\
\boxed{}
\end{array}
$$

$$\frac{2x^2 - 3x - 7}{x + 2} = \boxed{}$$

(b)

$$
\begin{array}{r}
\boxed{} + \boxed{} + \boxed{} \\
2x - 1 \overline{\smash{)}\, 4x^3 + 2x^2 + x + 6} \\
\underline{4x^3 - \boxed{}} \\
\boxed{} + x \\
\underline{\boxed{} - \boxed{}} \\
\boxed{} + \boxed{} \\
\underline{\boxed{} - \boxed{}} \\
\boxed{}
\end{array}
$$

$$\frac{4x^3 - 2x^2 + x + 6}{2x - 1} = \boxed{}$$

(c)

$$
\begin{array}{r}
x^2 - x + \boxed{} \\
x + 1 \overline{\smash{)}\, x^3 + \boxed{} + 1} \\
\underline{x^3 + x^2} \\
\boxed{} \\
-x^2 - x \\
\underline{\boxed{}} \\
\boxed{} \\
\underline{\boxed{}}
\end{array}
$$

$$\frac{x^3 + 1}{x + 1} = \boxed{}$$

◼ **Solutions to Starter Exercises** ◼

1. (a) $3 \cdot x^{\boxed{3}} {}^{-}{}^{\boxed{2}} = 3x^{\boxed{1}} = \boxed{3x}$

(b) $\boxed{2} \cdot \boxed{1} = \boxed{2}$

(c) $2 \cdot \boxed{\dfrac{1}{x^{5-3}}} = \boxed{\dfrac{2}{x^2}}$

2. (a) $2\boxed{x} + 2\boxed{\cdot 1} = \boxed{2x + 2}$

(b) $\boxed{x^2 + 3x + 9}$

(c) $\dfrac{x^2}{x} - \boxed{\dfrac{2x}{x}} + \dfrac{1}{x} = x - \boxed{2} + \dfrac{1}{x}$

(d) $\dfrac{3x^3}{\boxed{2x^2}} - \dfrac{2x^2}{\boxed{2x^2}} - \dfrac{4x}{\boxed{2x^2}} = \boxed{\dfrac{3}{2}x - 1 - \dfrac{2}{x}}$

■ **Solutions to Starter Exercises** ■

3. (a)

$$
\begin{array}{r}
2x - \boxed{7} \\
x + 2 \overline{\smash{\big)}\; 2x^2 - 3x - 7\;} \\
\underline{\boxed{2x^2} + 4x} \\
-7x - 7 \\
\underline{\boxed{-7x - 14}} \\
\boxed{7}
\end{array}
$$

$$
\frac{2x^2 - 3x - 7}{x + 2} = \boxed{2x - 7 + \dfrac{7}{x+2}}
$$

(b)

$$
\begin{array}{r}
\boxed{2x^2} + \boxed{2x} + \boxed{\tfrac{3}{2}} \\
2x - 1 \overline{\smash{\big)}\; 4x^3 + 2x^2 + x + 6\;} \\
\underline{4x^3 - \boxed{2x^2}} \\
\boxed{4x^2} + x \\
\underline{\boxed{4x^2} - \boxed{2x}} \\
\boxed{3x} + \boxed{6} \\
\underline{\boxed{3x} - \boxed{\tfrac{3}{2}}} \\
\boxed{\tfrac{15}{2}}
\end{array}
$$

$$
\frac{4x^3 - 2x^2 + x + 6}{2x - 1} = \boxed{2x^2 + 2x + \dfrac{3}{2} + \dfrac{15/2}{2x - 1}}
$$

(c)

$$
\begin{array}{r}
x^2 - x + \boxed{1} \\
x + 1 \overline{\smash{\big)}\; x^3 + \boxed{0x^2 + 0x} + 1\;} \\
\underline{x^3 + x^2} \\
\boxed{-x^2 + 0x} \\
\underline{-x^2 - x} \\
\boxed{x + 1} \\
\underline{\boxed{x + 1}} \\
\boxed{0}
\end{array}
$$

$$
\frac{x^3 + 1}{x + 1} = \boxed{x^2 - x + 1}
$$

| 5.3 | **EXERCISES** |

In Exercises 1–6, simplify the given expression. (Assume that each denominator is not zero.)

1. $\dfrac{4x^2}{2x}$

2. $\dfrac{-4x}{2x^2}$

3. $\dfrac{32z^4}{24z}$

4. $\dfrac{-28x^2}{-12x^4}$

5. $\dfrac{(4x^2)^2}{6x^7}$

6. $\dfrac{(a^2b)^3}{(3ab^2)^2}$

In Exercises 7–38, perform the indicated division and simplify. (Assume that each denominator is not zero.)

7. $\dfrac{6x+27}{3}$

8. $\dfrac{8x+4}{4}$

9. $\dfrac{2x^2+3x}{x}$

10. $(5x^2+17x)\div x$

11. $\dfrac{24x^2+16x}{4x}$

12. $\dfrac{15x^2-5x}{-3x}$

13. $\dfrac{n^3+2n^2+4n}{n}$

14. $\dfrac{2z^3-6z^2+8z}{2z}$

15. $\dfrac{x^2-3x+7}{2x}$

16. $\dfrac{2x^2(x-1)^2+3x(x-1)}{x-1}$

17. $\dfrac{x^2+7x+6}{x+1}$

18. $\dfrac{y^2-14y+24}{y-2}$

19. $(2x^2-x+1)\div(x+1)$

20. $(5y^2-4y-3)\div(y+2)$

21. $(4z^2+2z+9)\div(2z+3)$

22. $(15t^2-4t+1)\div(3t-1)$

23. $\dfrac{4x^2-9}{2x+3}$

24. $\dfrac{25-z^2}{5-z}$

25. $(x^3-64)\div(x-4)$

26. $(y^3+27)\div(y+3)$

27. $(z^3+2z+1)\div(z+2)$

28. $(2t^3-3t^2+6)\div(2t-1)$

29. $\dfrac{x^3+2x^2+3x-1}{x+1}$

30. $\dfrac{z^3-3z^2+6z-5}{z-3}$

31. $\dfrac{4x^3+2x^2-x+6}{2x-2}$

32. $\dfrac{2t^3-3t^2-4t+5}{2t-1}$

33. $(x+7)\div(x+2)$

34. $(6t-3)\div(2t+1)$

35. $(4x^3+2x^2)\div(2x+1)$

36. $(15z^2+2z)\div(3z+1)$

37. $(x^4-1)\div(x-1)$

38. $6x^3\div(x-1)$

In Exercises 39 and 40, simplify the given expression. (Assume that each denominator is not zero.)

39. $\dfrac{18x^2y^3}{3y}-(5xy)^2$

40. $3x-4-\dfrac{x^2-2x-8}{x+2}$

▦ In Exercises 41–44, use a graphing utility to check your answers to Exercises 7–10.

5.4 | Negative Exponents and Scientific Notation

Section Highlights

1. If n is an integer and a is a real number, a variable, or an algebraic expression such that $a \neq 0$, then a^{-n} is the reciprocal of a^n. That is, $a^{-n} = 1/a^n$, $(a \neq 0)$.
2. Very large or very small numbers can be written in a more convenient notation called *scientific notation*. Scientific notation has the form $c \times 10^n$, where $1 \leq c < 10$ and n is an integer.

EXAMPLE 1 ■ Monomials Involving Negative Exponents

Rewrite the following so that the expression has no negative exponents.

(a) x^{-4}

(b) $2x^{-2}$

(c) $\dfrac{1}{3x^{-3}}$

Solution

(a) $x^{-4} = \dfrac{1}{x^4}$ since $\dfrac{1}{x^4}$ is the reciprocal of x^{-4}.

(b) $2x^{-2} = 2 \cdot \dfrac{1}{x^2} = \dfrac{2}{x^2}$

(c) $\dfrac{1}{3x^{-3}} = \dfrac{1}{3 \cdot (1/x^3)} = \dfrac{1}{3/x^3} = 1 \cdot \dfrac{x^3}{3} = \dfrac{x^3}{3}$

Starter Exercise 1 *Fill in the blanks.*

Rewrite each of the following so that the expression has no negative exponents.

(a) $x^{-3} = \dfrac{1}{\boxed{}}$

(b) $x^2y^{-3} = x^2 \cdot \dfrac{1}{\boxed{}} = \dfrac{x^2}{\boxed{}}$

(c) $\dfrac{1}{x^{-2}} = \dfrac{1}{1/x^2} = 1 \cdot \dfrac{x^2}{\boxed{}} = \boxed{}$

EXAMPLE 2 ■ Using Rules of Exponents

Use rules of exponents to write each of the following without negative exponents. (Assume that each variable is not zero.)

(a) $(4xy^2)(3x^2y^{-2})$

(b) $\dfrac{2a^{-2}}{a^{-3}}$

(c) $\left(\dfrac{x^2y^{-3}}{xy^2}\right)^2$

(d) $\left(\dfrac{ab}{a^{-1}b^2}\right)^{-1}$

Solution

(a) $(4xy^2)(3x^2y^{-2}) = (4)(3)x \cdot x^2y^2y^{-2}$ Regroup factors.

$= 12x^{1+2}y^{2+(-2)}$ Apply rules of exponents.

$= 12x^3y^0$ Simplify.

$= 12x^3(1)$ Rules of exponents

$= 12x^3$

(b) $\dfrac{2a^{-2}}{a^{-3}} = 2 \cdot \dfrac{a^{-2}}{a^{-3}}$ Separate fraction.

$\phantom{\dfrac{2a^{-2}}{a^{-3}}} = 2 \cdot a^{-2-(-3)}$ Apply rules of exponents.

$\phantom{\dfrac{2a^{-2}}{a^{-3}}} = 2a^{1}$ Simplify.

$\phantom{\dfrac{2a^{-2}}{a^{-3}}} = 2a$

(c) $\left(\dfrac{x^2 y^{-3}}{xy^2}\right)^2 = \dfrac{(x^2 y^{-3})^2}{(xy^2)^2}$ Rules of exponents

$\phantom{\left(\dfrac{x^2 y^{-3}}{xy^2}\right)^2} = \dfrac{(x^2)^2(y^{-3})^2}{x^2(y^2)^2}$ Rules of exponents

$\phantom{\left(\dfrac{x^2 y^{-3}}{xy^2}\right)^2} = \dfrac{x^4 y^{-6}}{x^2 y^4}$ Rules of exponents

$\phantom{\left(\dfrac{x^2 y^{-3}}{xy^2}\right)^2} = \dfrac{x^4}{x^2} \cdot \dfrac{y^{-6}}{y^4}$ Separate fraction.

$\phantom{\left(\dfrac{x^2 y^{-3}}{xy^2}\right)^2} = x^{4-2} y^{-6-4}$ Rules of exponents

$\phantom{\left(\dfrac{x^2 y^{-3}}{xy^2}\right)^2} = x^2 y^{-10}$

$\phantom{\left(\dfrac{x^2 y^{-3}}{xy^2}\right)^2} = x^2 \cdot \dfrac{1}{y^{10}}$ Rules of exponents

$\phantom{\left(\dfrac{x^2 y^{-3}}{xy^2}\right)^2} = \dfrac{x^2}{y^{10}}$

(d) $\left(\dfrac{ab}{a^{-1}b^2}\right)^{-1} = \dfrac{a^{-1}b^2}{ab}$ Negative exponent means reciprocal.

$\phantom{\left(\dfrac{ab}{a^{-1}b^2}\right)^{-1}} = \dfrac{a^{-1}}{a} \cdot \dfrac{b^2}{b}$ Separate fraction.

$\phantom{\left(\dfrac{ab}{a^{-1}b^2}\right)^{-1}} = a^{-1-1} b^{2-1}$ Rules of exponents

$\phantom{\left(\dfrac{ab}{a^{-1}b^2}\right)^{-1}} = a^{-2} b^{1}$

$\phantom{\left(\dfrac{ab}{a^{-1}b^2}\right)^{-1}} = \dfrac{1}{a^2} \cdot b$ Reciprocal

$\phantom{\left(\dfrac{ab}{a^{-1}b^2}\right)^{-1}} = \dfrac{b}{a^2}$

■ Starter Exercise 2 | *Fill in the blanks.*

Use rules of exponents to write each of the following without negative exponents. (Assume that each variable is not zero.)

(a) $(2x^2)^2(3xy^2) = 2^2(x^2)^2(3xy^2) = 4 \cdot x^{\square}(3xy^2) = 4 \cdot \boxed{} \cdot x^{\square} \cdot xy^2$

$= \boxed{} x^{\square+\square} y^2 = \boxed{}$

(b) $\dfrac{x^{-2}y^4}{x^3 y} = \dfrac{x^{-2}}{x^3} \cdot \dfrac{y^4}{y} = x^{\square-\square} y^{\square-\square} = \dfrac{1}{x^{\square}} \cdot y^{\square} = \boxed{}$

(c) $\left(\dfrac{2xy^2}{x^{-2}y^3}\right)^3 = \dfrac{(2xy^2)^3}{(x^{-2}y^3)^3} = \dfrac{2^3x^3(y^2)^3}{(x^{-2})^3(y^3)^3} = \dfrac{8x^3y^{\square}}{x^{\square}y^{\square}} = 8 \cdot \dfrac{x^3}{x^{\square}} \cdot \dfrac{y^{\square}}{y^{\square}}$

$\qquad = 8 \cdot x^{3-\square} \cdot y^{\square-\square} = 8x^{\square}y^{\square} = 8x^{\square} \cdot \square = \square$

(d) $\left(\dfrac{u^{-1}v^2}{uv^2}\right)^0 = \square$ (Hint: What is any nonzero number to the zero power?)

EXAMPLE 3 ■ Converting from Decimal Notation to Scientific Notation

Write each of the following real numbers in scientific notation.

(a) 289,700,000

(b) 0.00000721

Solution

eight places

(a) $289,700,000 = 2.897 \times 10^8$

Note that the exponent on 10 corresponds to the number of places the decimal point was moved.

six places

(b) $0.00000721 = 7.21 \times 10^{-6}$

Note that the exponent is negative, since the decimal point was moved to the right.

■

> **■ Starter Exercise 3** *Fill in the blanks.*
>
> Write each of the following real numbers in scientific notation.
>
> (a) $0.0002713 = 2.713 \times 10^{\square}$
>
> (b) $99,963,000 = \square \times 10^7$

EXAMPLE 4 ■ Converting from Scientific Notation to Decimal Notation

Convert each of the following numbers from scientific notation to decimal notation.

(a) 5.76×10^6

(b) 3.071×10^{-5}

Solution

six places

(a) $5.76 \times 10^6 = 5,760,000$

Note that the decimal point was moved six places since the exponent of 10 was six.

five places

(b) $3.071 \times 10^{-5} = 0.00003071$

Note that the decimal point was moved to the left since the exponent of 10 was negative.

■

Starter Exercise 4 | *Fill in the blanks.*

Convert each of the following numbers from scientific notation to decimal notation.

(a) $4.23 \times 10^{-4} = 0.\boxed{}423$

(b) $8.73 \times 10^8 = 873\boxed{}$

EXAMPLE 5 ■ Using Scientific Notation

Use scientific notation to evaluate the following expressions.

(a) $74{,}800{,}000 \times 0.000098$

(b) $239{,}400{,}000 \div 5{,}630{,}000$

Solution

(a) $74{,}800{,}000 \times 0.000098 = (7.48 \times 10^7)(9.8 \times 10^{-5})$

$= (7.48)(9.8) \times 10^{7+(-5)} = 73.304 \cdot 10^2 = 7330.4$

(b) $224{,}000{,}000 \div 5{,}600{,}000 = (2.24 \times 10^8) \div (5.6 \times 10^6)$

$= (2.24) \div (5.6) \times 10^{8-6} = 0.4 \times 10^2 = 40$

Starter Exercise 5 | *Fill in the blanks.*

Use scientific notation to evaluate the following expressions.

(a) $24{,}000{,}000 \times 0.00002 = \left(2.4 \times 10^{\boxed{}}\right)\left(2 \times 10^{\boxed{}}\right)$

$= (2.4)(2) \times 10^{\boxed{}+\boxed{}} = \boxed{} \times 10^{\boxed{}} = \boxed{}$

(b) $0.0008732 \div 0.00037 = \left(8.732 \times 10^{\boxed{}}\right) \div \left(3.7 \times 10^{\boxed{}}\right)$

$= (8.732) \div (3.7) \times 10^{\boxed{}-\boxed{}} = \boxed{} \times 10^{\boxed{}} = \boxed{}$

▦ EXAMPLE 6 ■ Using Scientific Notation

Find $278{,}000{,}000 \times 0.00004$ using a scientific calculator.

Solution

Since $278{,}000{,}000 = 2.78 \times 10^8$ and $0.00004 = 4 \times 10^{-5}$, we can multiply the two numbers using the following calculator steps.

Scientific: 2.78 [EE] 8 [×] 4 [EE] 5 [+/−] [=]

Graphing: 2.78 [EE] 8 [×] 4 [EE] [(−)] 5 [ENTER]

Both should yield 11,120.

■ **Solutions to Starter Exercises** ■

1. (a) $x^{-3} = \dfrac{1}{\boxed{x^3}}$

(b) $x^2 y^{-3} = x^2 \cdot \dfrac{1}{\boxed{y^3}} = \dfrac{x^2}{\boxed{y^3 \cdot}}$

(c) $\dfrac{1}{x^{-2}} = \dfrac{1}{1/x^2} = 1 \cdot \dfrac{x^2}{\boxed{1}} = \boxed{x^2}$

2. (a) $= 4x^{\boxed{4}}(3xy^2) = 4 \cdot \boxed{3} \cdot x^{\boxed{4}} \cdot xy^2$

$= \boxed{12}\, x^{\boxed{4} + \boxed{1}} y^2 = \boxed{12x^5 y^2}$

(b) $= x^{\boxed{-2} - \boxed{3}} y^{\boxed{4} - \boxed{1}} = \dfrac{1}{x^{\boxed{5}}} y^{\boxed{3}} = \boxed{\dfrac{y^3}{x^5}}$

(c) $= \dfrac{8x^3 y^{\boxed{6}}}{x^{\boxed{-6}} y^{\boxed{9}}} = 8 \cdot \dfrac{x^3}{x^{\boxed{-6}}} \cdot \dfrac{y^{\boxed{6}}}{y^{\boxed{9}}}$

$= 8x^{3 - \boxed{(-6)}} \cdot y^{\boxed{6} - \boxed{9}} = 8x^{\boxed{9}} y^{\boxed{-3}}$

$= 8x^{\boxed{9}} \cdot \dfrac{1}{\boxed{y^3}} = \boxed{\dfrac{8x^9}{y^3}}$

(d) $\left(\dfrac{u^{-1} v^2}{u v^2} \right)^0 = \boxed{1}$

3. (a) $2.713 \times 10^{\boxed{-4}}$

(b) $\boxed{9.9963} \times 10^7$

4. (a) $0.\boxed{000}423$

(b) $873\boxed{,000,000}$

5. (a) $= \left(2.4 \times 10^{\boxed{7}} \right)\left(2 \times 10^{\boxed{-5}} \right)$

$= (2.4)(2) \times 10^{\boxed{7} + \boxed{(-5)}} = \boxed{4.8} \times 10^{\boxed{2}} = \boxed{480}$

(b) $= \left(8.732 \times 10^{\boxed{-4}} \right) \div \left(3.7 \times 10^{\boxed{-4}} \right)$

$= (8.732) \div (3.7) \times 10^{\boxed{-4} - \boxed{(-4)}} = \boxed{2.36} \times 10^{\boxed{0}} = \boxed{2.36}$

5.4 EXERCISES

In Exercises 1–4, rewrite the quantity so that it has no negative exponents and then evaluate.

1. 2^{-3}

2. 5^{-2}

3. $\dfrac{5}{3^{-3}}$

4. $\left(\dfrac{3}{4} \right)^{-3}$

In Exercises 5–20, use the rules of exponents to write the expression without negative exponents and simplify. (Assume that each variable is not zero.)

5. $\dfrac{x^{-2}}{x^3}$

6. $\dfrac{z^4}{z^{-2}}$

7. $\dfrac{x^{-1}}{x^{-3}}$

8. $(a^3)^{-1}$

9. $(b^{-3})^1$

10. $(2x^{-3})^{-2}$

11. $(5x^2y^{-3}z^6)^0$

12. $\dfrac{x^2 \cdot x^{-3}}{x^{-4}}$

13. $\dfrac{3z^4}{z^{-2}}$

14. $\dfrac{8x^{-3}z^4}{6x^{-4}z^2}$

15. $(2x^2y^{-3}z)^{-2}$

16. $(-2a^3b^{-2})^{-2}(4ab^3)^2$

17. $(x^2y^2z^{-3})^{-1}(xy^{-3}z^6)^2$

18. $\left(\dfrac{xy^2}{x^{-2}}\right)^{-1}$

19. $\left(\dfrac{a^{-2}b}{a^{-3}b^2}\right)^{-2}\left(\dfrac{a}{a^{-1}b^2}\right)^{-2}$

20. $(2x+7y)^{10}(2x+7y)^{-10}$

In Exercises 21–36, write the given number in decimal form.

21. 5.7×10^5

22. 3.69×10^{-4}

23. 9.3721×10^{-3}

24. 4.79×10^7

25. 2×10^{-3}

26. 1.5×10^0

In Exercises 27–32, write the given number in scientific notation.

27. 105,000

28. 794,060,000

29. 0.083

30. 2.7

31. 0.00000437

32. 429

In Exercises 33–38, use scientific notation to evaluate the given quantity. (State your answer in scientific notation.)

33. $400,000 \times 6,000,000$

34. $26,000,000 \times 0.0000013$

35. $3,200,000 \div 0.000016$

36. $0.00000625 \div 0.000025$

37. $12,000,000^2$

38. $150,000,000^2$

Cumulative Practice Test for Chapters P–5

In Exercises 1–10, state whether the following statements are true or false. Assume that a, b and c represent real numbers.

1. $a \cdot (b \cdot c) = (a \cdot b) \cdot (a \cdot c)$

2. $a^0 = 0$

_____

_____

3. $a^{-1} = -a$

4. $a^m \cdot a^n = a^{m+n}$

_____

_____

5. $(a^m)^n = a^{mn}$

6. $a + 0 = 0$

_____

_____

7. $(a \cdot b)^n = a^n \cdot b^n$

_____

8. $a \cdot (b + c) = a \cdot b + a \cdot c$

_____

9. If $a \neq 0$, then $a \cdot (1/a) = 1$.

_____

10. $\dfrac{a}{0} = 0$

_____

In Exercises 11–20, perform the indicated operation and simplify. Your answer should involve no negative exponents.

11. $x(x + 2)$

12. $\dfrac{x^2}{x^3}$

13. $(2x^2 + 3x - 6) - (4x^2 + 1)$

14. $(2x + y)(2x - y)$

15. $(2x^{-3}y^2)(x^2y)^{-2}$

16. $\left(\dfrac{2ab}{a^{-2}}\right)^{-1}$

17. $(x + 6)^2$

18. $(x^2 + 2x + 3)(x^2 - x + 1)$

19. $(2a - b)(a + b) - ab$

20. $(2u^2v^3)^0$

In Exercises 21–26, solve the given equation.

21. $2x + 1 = 5$

22. $5x + 3 = 4x - 6$

23. $3(x + 1) = 3x + 1$

24. $2(2x + 1) = 2(x + 2)$

25. $(2x - 1) - (x + 7) = 5$

26. $\dfrac{1}{2}t - \dfrac{1}{3} = \dfrac{2}{3}t + \dfrac{1}{3}$

In Exercises 27–30, divide the polynomials.

27. $\dfrac{2x^2 + 4x + 8}{2x}$

28. $(5x^3 - 6x^2 - 7x) \div (3x)$

29. $(4x^2 + 3x - 2) \div (x + 1)$

30. $(3x^3 + 4x^2 + 1) \div (x - 2)$

31. What percent is 12 of 80?

32. 22% of what number is 88?

33. A bookstore sells a book for $26.80. If the book costs $20, what is the markup rate?

34. A pair of shoes are on sale for $20, which is 20% off the regular price. Find the regular price.

35. The sum of three consecutive even integers is 72. Find the three integers.

36. The sum of two integers is 24. Three less than three times the smaller integer is four more than twice the larger. Find the integers.

37. How many pounds of candy that costs $3 per pound must be mixed with 10 pounds of candy that costs $2 per pound to make a mixture that costs $2.25 per pound?

38. A total of $5000 is invested in two accounts. One account earns 7% and the other 8.5% simple annual interest. If the total interest earned on both accounts is $395, how much was invested in each account?

39. At 7:00 A.M., a jet plane left an airport traveling at 600 mph. At the same time a single engine plane left the same airport, in the opposite direction, at 120 mph. At what time are the two planes 1080 miles apart?

40. How much pure water must be added to 20 ounces of a 20% acid solution to make an 18% acid solution?

5.1	**Answers to Exercises**

	Standard Form	Degree	Leading Coefficient
1.	$2x - 1$	1	2
2.	$-3x^2 + 1$	2	-3
3.	$2x^2 + x - 3$	2	2
4.	$x^5 - 2x^3 + x$	5	1

5. binomial   **6.** monomial   **7.** trinomial   **8.** binomial

9. Cannot have the absolute value of a variable   **10.** Cannot have division by variables

11. Cannot have negative exponents   **12.** Cannot have division by variables

13. $9x^2 + 4x - 14$ **14.** $11x^3 - 2x^2 + 4x$ **15.** $x^3 + 3x - 5$

16. $3x^2 - x + 3$ **17.** $3x^2 - 7x - 4$ **18.** $x^3 + 6x^2 - x + 1$

19. $y^5 + y^2 + 2y$ **20.** $10t + 5$ **21.** $-3u - 7$

22. $x^2 - 4x + 8$ **23.** -2 **24.** $-x^3 - 2x^2 - 2x + 7$

25. $y + 8$ **26.** $-3x^2 + 5x - 4$ **27.** $x^3 - x^2 - x - 2$

28. $2y^3 - 2y^2 + 3y + 10$ **29.** $-12x - 12$ **30.** $2x^2 + 5x - 4$

31. $-2x^2 - 5x - 2$ **32.** $-y^3 + 6y^2 - 7y + 6$ **33.** $8x^2 + x - 3$

34. $-t^4 - 2t + 17$ **35.** $-4u + 10$ **36.** $-4x^2 - 8x + 3$

37. $-11u - 53$ **38.** $5x^6 - 2x - 6$ **39.** $2x^7 + 16x^5 + 8$

40. $-4y^2 + 8y - 7$

5.2 Answers to Exercises

1. $-3z^3$

2. $12x^2$

3. $64z^6$

4. $4x^3$

5. $5x^2$

6. $9x^4$

7. $6y^3 - 30y$

8. $20x^3 + 30x^2$

9. $x^2 + 2x$

10. $6x^2 - 2x$

11. $-3x^3 + 18x$

12. $10t^3 - 35t$

13. $3x^3 - 2x^2 + x$

14. $-2x^3 + 8x^2 - 6x$

15. $x^2 + 3x + 2$

16. $x^2 + 2x - 3$

17. $2x^2 - 11x - 6$

18. $6z^2 - 11z + 4$

19. $2x^2 + xy - y^2$

20. $4u^2 - uv - 14v^2$

21. $-2t^2 - 9t + 35$

22. $-4x^2 - 2x + 12$

23. $-2x^2 - 5x - 3$

24. $-2x^2 - 17x + 6$

25. $z^3 - z^2 - 5z - 3$

26. $8x^3 - 10x^2 - 15x + 18$

27. $2x^3 - 13x^2 - 9x - 1$

28. $2x^4 - 3x^3 + 13x^2 - 18x + 6$

29. $x^4 + 2x^3 - 7x^2 + 22x - 12$

30. $2x^4 - 4x^3 - 5x^2 - 11x - 3$

31. $x^3 + 6x^2 + 11x + 6$

32. $6x^3 - 31x^2 + 30x - 8$

33. $x^2 + 4x + 4$

34. $x^2 - 6x + 9$

35. $4x^2 - 16x + 16$

36. $x^2 - y^2$

37. $16x^2 - 9$

38. $-9x^2 + 4$

39. $4x$

40. $x^2 + 4x + 4 + 2xy + 4y + y^2$

41. $4x^2 - 4x + 1$

42. $3x^2 + 13x + 14$

43.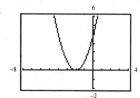

44.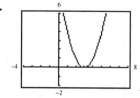

45.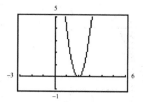

5.3 Answers to Exercises

1. $2x$

2. $-\dfrac{2}{x}$

3. $\dfrac{4z^3}{3}$

4. $\dfrac{7}{3x^2}$

5. $\dfrac{8}{3x^3}$

6. $\dfrac{a^4}{9b}$

7. $2x + 9$

8. $2x + 1$

9. $2x + 3$

10. $5x + 17$

11. $6x + 4$

12. $-5x + \frac{5}{3}$

13. $n^2 + 2n + 4$

14. $z^2 - 3z + 4$

15. $\frac{x}{2} - \frac{3}{2} + \frac{7}{2x}$

16. $2x^3 - 2x^2 + 3x$

17. $x + 6$

18. $y - 12$

19. $2x - 3 + \frac{4}{x + 1}$

20. $5y - 14 + \frac{25}{y + 2}$

21. $2z - 2 + \frac{15}{2z + 3}$

22. $5t + \frac{1}{3} + \frac{4/9}{3t - 1}$

23. $2x - 3$

24. $z + 5$

25. $x^2 + 4x + 16$

26. $y^2 - 3y + 9$

27. $z^2 - 2z + 6 + \frac{-11}{z + 2}$

28. $t^2 - t - \frac{1}{2} + \frac{11/2}{2t - 1}$

29. $x^2 + x + 2 + \frac{-3}{x + 1}$

30. $z^2 + 6 + \frac{13}{z - 3}$

31. $2x^2 + 3x + \frac{5}{2} + \frac{11}{2x - 2}$

32. $t^2 - t - \frac{5}{2} + \frac{5/2}{2t - 1}$

33. $1 + \frac{5}{x + 2}$

34. $3 + \frac{-6}{2t + 1}$

35. $2x^2$

36. $5z - 1 + \frac{1}{3z + 1}$

37. $x^3 + x^2 + x + 1$

38. $6x^2 + 6x + 6 + \frac{6}{x - 1}$

39. $-19x^2y^2$

40. $2x$

41.

42.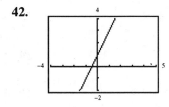

43.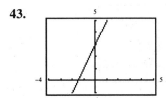

44.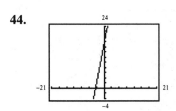

5.4 | Answers to Exercises

1. $\frac{1}{8}$

2. $\frac{1}{25}$

3. 135

4. $\frac{64}{27}$

5. $\frac{1}{x^5}$

6. z^6

7. x^2

8. $\frac{1}{a^3}$

9. $\frac{1}{b^3}$

10. $\frac{x^6}{4}$

11. 1   **12.** x^3 **13.** $3z^6$ **14.** $\dfrac{4xz^2}{3}$ **15.** $\dfrac{y^6}{4x^4z^2}$

16. $\dfrac{4b^{10}}{a^4}$ **17.** $\dfrac{z^{15}}{y^8}$ **18.** $\dfrac{1}{x^3y^2}$ **19.** $\dfrac{b^6}{a^6}$ **20.** 1

21. 570,000   **22.** 0.000369   **23.** 0.0093721   **24.** 47,900,000

25. 0.002   **26.** 1.5   **27.** 1.05×10^5 **28.** 7.9406×10^8

29. 8.3×10^{-2} **30.** 2.7×10^0 **31.** 4.37×10^{-6} **32.** 4.29×10^2

33. 2.4×10^{12} **34.** 3.38×10^1 **35.** 2×10^{11} **36.** 2.5×10^{-1}

37. 1.44×10^{14} **38.** 2.25×10^{16}

Answers to Cumulative Practice Test P–5

1. False   **2.** False   **3.** False   **4.** True   **5.** True

6. False   **7.** True   **8.** True   **9.** True   **10.** False

11. $x^2 + 2x$ **12.** $\dfrac{1}{x}$ **13.** $-2x^2 + 3x - 7$ **14.** $4x^2 - y^2$

15. $\dfrac{2}{x^7}$ **16.** $\dfrac{1}{2a^3b}$ **17.** $x^2 + 12x + 36$ **18.** $x^4 + x^3 + 2x^2 - x + 3$

19. $2a^2 - b^2$ **20.** 1   **21.** $\{2\}$ **22.** $\{-9\}$ **23.** $\emptyset$, no solution

24. $\{1\}$ **25.** $\{13\}$ **26.** $\{-4\}$ **27.** $x + 2 + \dfrac{4}{x}$ **28.** $\dfrac{5}{3}x^2 - 2x - \dfrac{7}{3}$

29. $4x - 1 + \dfrac{-1}{x+1}$ **30.** $3x^2 + 10x + 20 + \dfrac{41}{x-2}$ **31.** 15%   **32.** 400

33. 34%   **34.** \$25   **35.** 22, 24, 26   **36.** 11 and 13

37. $3\frac{1}{3}$ lbs   **38.** \$2000 at 7%   **39.** 8:30 A.M.   **40.** $2\frac{2}{9}$ oz
   \$3000 at 8.5%

C H A P T E R   S I X
Factoring and Solving Equations

6.1 | Factoring Polynomials with Common Factors

Section Highlights

1. The process of **factoring polynomials** is the *reverse* of the process of multiplying polynomials. The key to this process is the **Distributive Property.**

2. The **greatest common factor** of a polynomial is the product of all factors that are factors of every term in the polynomial.

EXAMPLE 1 ■ Finding the Greatest Common Factor

Find the greatest common factor of each of the following groups.

(a) 36 and 54

(b) $6x^2y^3$ and $8xy^2$

(c) $2x^2$, $10x$, and $12x^4$

Solution

(a) $36 = 2^2 \cdot 3^2 = (2 \cdot 3^2) \cdot 2$

$54 = 2 \cdot 3^3 = (2 \cdot 3^2) \cdot 3$

Thus, the greatest common factor is $2 \cdot 3^2 = 18$.

(b) $6x^2y^3 = 2 \cdot 3 \cdot x^2 \cdot y^3 = (2xy^2) \cdot 3xy$

$8xy^2 = 2^3 \cdot x \cdot y^2 = (2xy^2) \cdot 2^2$

Thus, the greatest common factor is $2xy^2$.

(c) $2x^2 = 2x^2 = (2x) \cdot x$

$10x = 2 \cdot 5 \cdot x = (2x) \cdot 5$

$12x^4 = 2^2 \cdot 3 \cdot x^4 = (2x) \cdot 2 \cdot 3 \cdot x^3$

Thus, the greatest common factor is $2x$.

Starter Exercise 1 | *Fill in the blanks.*

Find the greatest common factor of each of the following groups.

(a) 60 and 72

$60 = 2^2 \cdot 3 \cdot 5 = (2^{\square} \cdot 3) \cdot \square$

$72 = 2^{\square} \cdot 3^{\square} = (2^{\square} \cdot 3) \cdot \square$

The greatest common factor is $\square$.

(b) $12a^2b$ and $20a^2b^3$

$12a^2b = 2^{\boxed{2}} \cdot 3 \cdot a^2b = (2^{\boxed{2}}a^{\boxed{2}}b) \cdot 3$

$20a^2b^3 = 2^{\boxed{2}} \cdot 5 \cdot a^2b^3$

$= (2^{\boxed{2}}a^{\boxed{2}}b) \cdot 5b^{\boxed{3}}$

The greatest common factor is $\boxed{4}$.

(c) $4x^3$, $6y^2$, and $8z^4$

$4x^3 = 2^2x^3 = \square \cdot \square$

$6y^2 = 2 \cdot 3 \cdot y^2 = \square \cdot \square$

$8z^4 = 2^3z^4 = \square \cdot \square$

The greatest common factor is $\square$.

EXAMPLE 2 ■ Factoring Out the Greatest Common Monomial Factor

Factor out the greatest common monomial factor.

(a) $4x - 6$ (b) $6x^3 - 9x^2$ (c) $x^3 - 6x^2 + 12x$

Solution

(a) The greatest common factor of each term is 2. Thus,

$$4x - 6 = (2)(2x) - (2)(3) = 2(2x - 3).$$

(b) The greatest common factor of each term is $3x^2$. Thus,

$$6x^3 - 9x^2 = (3x^2)(2x) - (3x^2)(3) = 3x^2(2x - 3).$$

(c) The greatest common factor of each term is x. Thus,

$$x^3 - 6x^2 + 12x = x(x^2) - x(6x) + x(12) = x(x^2 - 6x + 12).$$

■

Starter Exercise 2   *Fill in the blanks.*

Factor out the greatest common monomial factor.

(a) $3x^2 - 2x = \boxed{}(3x) - \boxed{} \cdot (2) = \boxed{}(3x - 2)$

(b) $15x^3 + 9x^2 = (3x^2)\boxed{} + (3x^2)\boxed{} = 3x^2\left(\boxed{}\right)$

(c) $16a^2b + 12ab - 4b = \boxed{4}(4a^2) + \boxed{4}(3b) - \boxed{}(1) = \boxed{}\left(\boxed{}\right)$

EXAMPLE 3 ■ A Negative Common Monomial Factor

Factor the polynomial $-3x^2 + 6$ in two ways: first, by factoring out a 3, and then by factoring out a -3.

Solution

$$-3x^2 + 6 = 3(-x^2) + 3(2) = 3(-x^2 + 2)$$

$$-3x^2 + 6 = -3(x^2) + (-3)(-2) = -3(x^2 - 2)$$

■

Starter Exercise 3   *Fill in the blanks.*

Factor $-16x^3 + 12x^2 - 4x$ in two ways: first, by factoring out a $4x$, and then by factoring out a $-4x$.

$$-16x^3 + 12x^2 - 4x = \boxed{}(-4x^2) + \boxed{}(3x) - \boxed{}(1)$$

$$= \boxed{}\left(\boxed{} + \boxed{} - \boxed{}\right)$$

$$-16x^3 + 12x^2 - 4x = (-4x)\left(\boxed{}\right) + (-4x)\left(\boxed{}\right) - (-4x)\boxed{}$$

$$= -4x\left(\boxed{}\right)$$

EXAMPLE 4 ■ Common Binomial Factors

Factor (a) $5x(x+1)+7(x+1)$ and (b) $x^2(3x-1)+6(3x-1)$.

Solution

(a) The greatest common factor is the binomial $x+1$. Thus,

$$5x(x+1)+7(x+1)=(5x+7)(x+1).$$

(b) The greatest common factor is the binomial $3x-1$.

$$x^2(3x-1)+6(3x-1)=(x^2+6)(3x-1)$$

■ Starter Exercise 4 | *Fill in the blanks.*

Factor each of the following.

(a) $2x(3x+5)-3(3x+5)=\left(\boxed{}-\boxed{}\right)(3x+5)$

(b) $5x^2(x^2+1)-4(x^2+1)=\left(\boxed{}-4\right)\left(\boxed{}\right)$

EXAMPLE 5 ■ Factoring by Grouping

Factor the following polynomials.

(a) $6x^2-2x+9x-3$ 　　　　　(b) $5x^3-7x^2+30x-42$

Solution

(a) By grouping the first two terms together and the third and fourth terms together, we obtain the following.

$$6x^2-2x+9x-3=(6x^2-2x)+(9x-3)$$
$$=2x(3x-1)+3(3x-1)$$
$$=(2x+3)(3x-1)$$

(b) By grouping the first two terms together and the third and fourth terms together, we obtain the following.

$$5x^3-7x+30x-42=(5x^3-7x^2)+(30x-42)$$
$$=x^2(5x-7)+6(5x-7)$$
$$=(x^2+6)(5x-7)$$

> **Starter Exercise 5** *Fill in the blanks.*

Factor the following polynomials.

(a) $12x^2 - 20x - 3x + 5 = (12x^2 - 20x) - \left(\boxed{} - \boxed{}\right)$

$$= 4x\left(3x - \boxed{}\right) - \boxed{}\left(\boxed{} - \boxed{}\right)$$

$$= \left(4x - \boxed{}\right)\left(3x - \boxed{}\right)$$

(b) $2x^3 - x^2y + 6x - 3y = (2x^3 - x^2y) + (6x - 3y)$

$$= \boxed{}\left(2x - \boxed{}\right) + 3\left(\boxed{} - \boxed{}\right)$$

$$= \boxed{}$$

■ **Solutions to Starter Exercises** ■

1. (a) $60 = \left(2^{\boxed{2}} \cdot 3\right) \cdot \boxed{5}$

$72 = 2^{\boxed{3}} \cdot 3^{\boxed{2}} = \left(2^{\boxed{2}} \cdot 3\right) \cdot \boxed{6}$

GCF is $\boxed{12}$.

(b) $12a^2b = 2^{\boxed{2}} \cdot 3 \cdot a^2b = \left(2^{\boxed{2}} a^{\boxed{2}} b\right) \cdot 3$

$20a^2b^3 = 2^{\boxed{2}} \cdot 5 \cdot a^2b^3 = \left(2^{\boxed{2}} a^{\boxed{2}} b\right) \cdot 5b^{\boxed{2}}$

GCF is $\boxed{4a^2b}$.

(c) $4x^3 = 2^2x^3 = \boxed{2} \cdot \boxed{(2x^3)}$

$6y^2 = 2 \cdot 3y^2 = \boxed{2} \cdot \boxed{(3y^2)}$

$8z^4 = 2^3z^4 = \boxed{2} \cdot \boxed{(2^2z^4)}$

GCF is $\boxed{2}$.

2. (a) $3x^2 - 2x = \boxed{(x)} (3x) - \boxed{(x)} \cdot (2) = \boxed{x} (3x - 2)$

(b) $15x^3 + 9x^2 = (3x^2)\boxed{(5x)} + (3x^2)\boxed{(3)} = 3x^2\left(\boxed{5x + 3}\right)$

(c) $16a^2b + 12ab - 4b = \boxed{(4b)} (4a^2) + \boxed{(4b)} (3a) - \boxed{(4b)} (1)$

$$= \boxed{4b}\left(\boxed{4a^2 + 3a - 1}\right)$$

■   **Solutions to Starter Exercises**   ■

3. $-16x^3 + 12x^2 - 4x = \boxed{(4x)}(-4x^2) + \boxed{(4x)}(3x) - \boxed{(4x)}(1)$

$\qquad = \boxed{4x}\left(\boxed{-4x^2 + 3x - 1}\right)$

$-16x^3 + 12x^2 - 4x = (-4x)\left(\boxed{4x^2}\right) + (-4x)\left(\boxed{-3x}\right) - (-4x)\left(\boxed{-1}\right)$

$\qquad = -4x\left(\boxed{4x^2 - 3x + 1}\right)$

4. (a) $2x(3x + 5) - 3(3x + 5) = \left(\boxed{2x} - \boxed{3}\right)(3x + 5)$

(b) $5x^2(x^2 + 1) - 4(x^2 + 1) = \left(\boxed{5x^2} - 4\right)\left(\boxed{x^2 + 1}\right)$

5. (a) $(12x^2 - 20x) - \left(\boxed{3x} - \boxed{5}\right) = 4x\left(3x - \boxed{5}\right) - \boxed{1}\left(\boxed{3x} - \boxed{5}\right)$

$\qquad = \left(4x - \boxed{1}\right)\left(3x - \boxed{5}\right)$

(b) $(2x^3 - x^2 y) + (6x - 3y) = \boxed{x^2}\left(2x - \boxed{y}\right) + 3\left(\boxed{2x} - \boxed{y}\right)$

$\qquad = \boxed{(x^2 + 3)(2x - y)}$

6.1   EXERCISES

In Exercises 1–8, find the greatest common factor of the given expressions.

1. 28, 42

2. 56, 140, 168

3. $2x^3$, $4x$

4. $3x^4$, $5x^6$

5. $12xy^2$, $8x^2 y$

6. $36a^3 b^5$, $42a^2 b^7$

7. $14x^2$, $28x$, $7x^3$

8. $9x^2 y^2$, $3x^2 y$, $15x^3 y^6$

In Exercises 9–30, factor the given polynomial.

9. $3x - 6$

10. $10y + 15$

11. $-12x^2 - 24$

12. $4x^3 - 8$

13. $x^2 + x$

14. $2x^3 - x^2$

15. $14x^3 - 21x$

16. $16t^3 - 32t^2$

17. $3u^2 v + 7uv^2$

18. $16a^3 b^6 - 6a^4 b^2$

19. $11x^2 y^4 - 22x^4 y^2$

20. $42u^4 v^7 + 36u^4 v^8$

21. $x^3 - 14x^2 - 2x$

22. $2m^4 - 6m^2 + 4m$

23. $5a^2 b^3 + 10a^2 b^2 + 15a^2 b$

24. $35x^4 y^3 - 28x^3 y^4 - 21x^2 y^2$

25. $x(x - 1) + 7(x - 1)$

26. $2x(x + 6) + 9(x + 6)$

27. $x^2(2x - 3) + 5(2x - 3)$

28. $(x + y)(x - y) - (x - y)^2$

29. $x^3(x - y)^2 + x^2(x - y)$

30. $(2x + 1)(x - 2) + (2x + 1)(x - 2)^2$

In Exercises 31–36, factor the given expression by grouping.

31. $x^2 + 7x + 6x + 42$

32. $2x^2 + 4x - 3x - 6$

33. $15x^2 - 12x + 10x - 8$

34. $x^3 + x^2 + 4x + 4$

35. $10y^3 - 15y^2 - 6y + 9$

36. $2x^4 - 3x^2y^2 + 2x^2y^2 - 3y^4$

In Exercises 37 and 38, factor a negative number from the given polynomial.

37. $-2x + 6$

38. $-33x^2 - 9x + 27$

In Exercises 39 and 40, rewrite the given polynomial by factoring out the indicated fraction.

39. $\frac{2}{3}x + 1 = \frac{1}{3}\left(\boxed{}\right)$

40. $\frac{1}{2}t - \frac{5}{6} = \frac{1}{6}\left(\boxed{}\right)$

6.2   Factoring Trinomials

Section Highlights

1. **Strategy:** To factor $x^2 + bx + c$, find two integers m and n such that $m \cdot n = c$ and $m + n = b$. Then $x^2 + bx + c = (x + m)(x + n)$.
2. The strategy for factoring $x^2 + bxy + cy^2$ is the same. We still need two factors of c whose sum is b.
3. The first step in factoring a trinomial is to factor out the greatest common factor of all three terms, if there is one.

EXAMPLE 1 ■   Factoring Trinomials

Factor each of the following into a product of two binomials.

(a) $x^2 + 6x + 8$ (b) $x^2 - 5x + 6$ (c) $x^2 - 2x - 15$ (d) $x^2 + x - 12$

Solution

(a) We need two integers whose product is 8 and whose sum is 6. 2 and 4 are the desired numbers.

$$x^2 + 6x + 8 = (x + 2)(x + 4)$$

Note that you may check this by multiplying

$$(x + 2)(x + 4) = x^2 + 4x + 2x + 8 = x^2 + 6x + 8.$$

(b) We need two factors of 6 whose sum is -5. Note that the numbers must be negative. The desired numbers are -2 and -3.

$$x^2 - 5x + 6 = (x - 2)(x - 3)$$

(c) We need two factors of -15 whose sum is -2. Note that one number is going to be positive and one will be negative. The desired numbers are -5 and 3.

$$x^2 - 2x - 15 = (x - 5)(x + 3)$$

(d) We need two factors of -12 whose sum is 1. Again, one number will be positive and the other will be negative. The desired numbers are 4 and -3.

$$x^2 + x - 12 = (x + 4)(x - 3)$$

Starter Exercise 1 | *Fill in the blanks.*

Factor each of the following into a product of two binomials.

(a) $x^2 + 11x + 24$; Find two factors of 24 whose sum is 11.

$$x^2 + 11x + 24 = \left(x + \boxed{}\right)\left(x + \boxed{}\right)$$

(b) $x^2 + 5x - 24$; Find two factors of -24 whose sum is 5.

$$x^2 + 5x - 24 = \left(x + \boxed{}\right)\left(x - \boxed{}\right)$$

(c) $x^2 - 10x + 24 = \left(x + \boxed{}\right)(x - 12)$

(d) $x^2 - 8x - 9 = (x + 1)\left(x - \boxed{}\right)$

EXAMPLE 2 ■ **Factoring a Trinomial in Two Variables**

Factor the following trinomials.

(a) $x^2 - 9xy + 18y^2$

(b) $y^2 + 5yx - 36x^2$

Solution

(a) We need two factors of 18 whose sum is -9. Note that both numbers will be negative. The desired numbers are -6 and -3.

$$x^2 - 9xy + 18y^2 = (x - 6y)(x - 3y)$$

(b) We need two factors of -36 whose sum is 5. Note that one number will be positive and the other will be negative. The desired numbers are 9 and -4.

$$y^2 + 5yx - 36x^2 = (y + 9x)(y - 4x)$$

■ **Starter Exercise 2** *Fill in the blanks.*

(a) Factor $x^2 + 7xy + 12y^2$. Find two factors of 12 whose sum is 7.

$$x^2 + 7xy + 12y^2 = \left(x + \boxed{}\, y\right)\left(x + \boxed{}\, y\right)$$

(b) Factor $x^2 - 3xy - 28y^2$. Find two factors of -28 whose sum is -3.

$$x^2 - 3xy - 28y^2 = \left(x + \boxed{}\right)\left(\boxed{}\right)$$

EXAMPLE 3 ■ Factoring Completely

Factor (a) $5x^2 - 15x - 50$ and (b) $x^3 - 10x^2 + 16x$ completely.

Solution

(a) Notice the common factor of 5.

$$5x^2 - 15x - 50 = 5(x^2 - 3x - 10) = 5(x - 5)(x + 2)$$

Note that -5 and 2 are factors of -10 whose sum is -3.

(b) $x^3 - 10x^2 + 16x = x(x^2 - 10x + 16)$ Factor out the common factor of x.

$\qquad\qquad\qquad\;\; = x(x - 8)(x - 2)$ Factors of 16 whose sum is -10

■

■ **Starter Exercise 3** *Fill in the blanks.*

Factor the following trinomials completely.

(a) $2x^3 + 22x^2 + 36x = \boxed{}\,(x^2 + 11x + 18) = \boxed{}\left(x + \boxed{}\right)\left(x + \boxed{}\right)$

(b) $x^4 + 13x^3 - 48x^2 = \boxed{}\left(\boxed{} + \boxed{} - \boxed{}\right)$

$\qquad\qquad\qquad\; = \boxed{}\left(x + \boxed{}\right)\left(x - \boxed{}\right)$

■ **Solutions to Starter Exercises** ■

1. (a) $= \left(x + \boxed{8}\right)\left(x + \boxed{3}\right)$ (b) $= \left(x + \boxed{8}\right)\left(x - \boxed{3}\right)$

(c) $= \left(x + \boxed{2}\right)\left(x - \boxed{12}\right)$ (d) $= \left(x + \boxed{1}\right)\left(x - \boxed{9}\right)$

2. (a) $= \left(x + \boxed{3}\, y\right)\left(x + \boxed{4}\, y\right)$ (b) $= \left(x + \boxed{4y}\right)\left(\boxed{x - 7y}\right)$

3. (a) $= \boxed{2x}\,(x^2 + 11x + 18) = \boxed{2x}\left(x + \boxed{2}\right)\left(x + \boxed{9}\right)$

(b) $= \boxed{x^2}\left(\boxed{x^2} + \boxed{13x} - \boxed{48}\right) = \boxed{x^2}\left(x + \boxed{16}\right)\left(x - \boxed{3}\right)$

6.2 | EXERCISES

In Exercises 1–4, find the missing factor. Then check your answer by multiplying the two factors.

1. $x^2 + 6x + 5 = (x + 1)\left(\boxed{}\right)$

2. $x^2 - 13x + 22 = (x - 2)\left(\boxed{}\right)$

3. $x^2 - 23x - 24 = (x - 24)\left(\boxed{}\right)$

4. $x^2 + 2x - 35 = (x + 7)\left(\boxed{}\right)$

In Exercises 5–20, factor the given trinomial.

5. $x^2 + 12x + 11$

6. $x^2 + 8x + 15$

7. $x^2 - 9x + 20$

8. $x^2 - 12x + 27$

9. $x^2 + 9x - 36$

10. $x^2 + 14x - 32$

11. $x^2 - x - 42$

12. $x^2 - 25x - 54$

13. $x^2 - 20x + 19$

14. $x^2 - 17x - 38$

15. $x^2 - 7xy - 60y^2$

16. $x^2 + 34xz - 72z^2$

17. $a^2 + 4ab + 3b^2$

18. $u^2 - 4uv - 45v^2$

19. $x^2 + 22xy + 40y^2$

20. $a^2 - 2ab - 63b^2$

In Exercises 21–26, factor the given trinomial completely.

21. $3x^2 + 24x + 45$

22. $x^3 - 16x^2 + 39x$

23. $2x^3 + 44x^2 - 46x$

24. $5x^4 - 75x^3 - 170x^2$

25. $2x^2 + 2xy - 12y^2$

26. $x^4y^2 - 2x^3y^3 - 8x^2y^4$

In Exercises 27–30, find all positive integers b such that the trinomial can be factored.

27. $x^2 + bx + 6$

28. $x^2 + bx + 35$

29. $x^2 + bx - 15$

30. $x^2 + bx - 10$

In Exercises 31–34, find all positive integers c such that the trinomial can be factored.

31. $x^2 + 4x + c$

32. $x^2 + 5x + c$

33. $x^2 - 3x + c$

34. $x^2 - 6x + c$

6.3 | More About Factoring Trinomials

Section Highlights

1. **Trial and Error**

$$ax^2 + bx + c = \left(\boxed{} \ x + \boxed{} \right)\left(\boxed{} \ x + \boxed{} \right)$$

Factors of a (over the first and third boxes)

Factors of c (under the second and fourth boxes)

   i) Find all integer factors of a and all integer factors of c.
   ii) Form all binomial products, as above, using all combinations of factors of a in the first spots and factors of c in the second spots.
   iii) Pick the combination so that the outer and inner products add up to the middle term bx.

2. **Factoring by Grouping**
   i) Find integer factors, m and n, of the product ac, such that $m + n = b$.
   ii) Replace bx by $mx + nx$.
   iii) Factor by grouping as in Section 6.1.

3. If the trinomial has a common monomial factor, factor it out first.

EXAMPLE 1 ■   Factoring a Trinomial of the Form $ax^2 + bx + c$

Factor the following trinomials.

(a) $2x^3 + 9x + 9$

(b) $3x^2 - 4x - 15$

Solution

(a) The factors of $a = 2$ are $(1)(2)$. The factors of $c = 9$ are $(1)(9)$, $(-1)(-9)$, $(3)(3)$, and $(-3)(-3)$.

$$(x + 1)(2x + 9) = 2x^2 + 11x + 9$$
$$(x - 1)(2x - 9) = 2x^2 - 11x + 9$$
$$(x + 9)(2x + 1) = 2x^2 + 19x + 9$$
$$(x - 9)(2x - 1) = 2x^2 - 19x + 9$$
$$(x + 3)(2x + 3) = 2x^2 + 9x + 9 \qquad \text{Correct factorization}$$
$$(x - 3)(2x - 3) = 2x^2 - 9x + 9$$

Thus, we conclude that the correct factorization is

$$2x^2 + 9x + 9 = (x + 3)(2x + 3).$$

(b) The factors of $a = 3$ are $(1)(3)$. The factors of $c = -15$ are $(1)(-15)$, $(-1)(15)$, $(3)(-5)$, and $(-3)(5)$.

$$(x + 1)(3x - 15) = 3x^2 - 12x - 15$$
$$(x - 1)(3x + 15) = 3x^2 + 12x - 15$$
$$(x - 15)(3x + 1) = 3x^2 - 44x - 15$$
$$(x + 15)(3x - 1) = 3x^2 + 44x - 15$$
$$(x + 3)(3x - 5) = 3x^2 + 4x - 15$$
$$(x - 3)(3x + 5) = 3x^2 - 4x - 15 \qquad \text{Correct factorization}$$

$$\left.\begin{array}{l} (x - 5)(3x + 3) \\[6pt] (x + 5)(3x - 3) \end{array}\right\} \qquad \text{These are not valid possibilities since the second} \\ \text{factor has a common factor of 3.}$$

Thus, we can conclude that the correct factorization is

$$3x^2 - 4x - 15 = (x - 3)(3x + 5).$$

■

Starter Exercise 1   *Fill in the blanks.*

(a) Factor the trinomial $15x^2 + 2x - 1$. The factors of $a = 15$ are $(1)(15)$ and $(3)(5)$. The factors of $c = -1$ are $(1)(-1)$.

$$(x + 1)(15x - 1) = \boxed{} x^2 + \boxed{} x - \boxed{}$$
$$(15x + 1)(x - 1) = 15x^2 - \boxed{} x - \boxed{}$$
$$(3x + 1)\left(\boxed{} x - 1\right) = \boxed{} x^2 + \boxed{} x - \boxed{}$$
$$\left(\boxed{} x + 1\right)(3x - 1) = \boxed{} x^2 + \boxed{} x - \boxed{}$$

The correct factorization is $\boxed{}$.

(b) Factor the trinomial $7x^2 - x - 6$. The factors of $a = 7$ are $\left(\boxed{}\right)\left(\boxed{}\right)$. The factors of $c = -6$ are $(1)(-6)$, $(-1)(6)$, $(2)(-3)$, and $\left(\boxed{}\right)\left(\boxed{}\right)$.

$$(7x + 1)(x - 6) = \boxed{}$$
$$(7x - 1)\left(x + \boxed{}\right) = \boxed{}$$
$$(7x - 6)\left(x + \boxed{}\right) = \boxed{}$$
$$\left(7x + \boxed{}\right)(x - 1) = \boxed{}$$
$$(7x + 2)\left(x - \boxed{}\right) = \boxed{}$$
$$(7x - 2)(x + 3) = \boxed{}$$
$$(7x + 3)\left(x - \boxed{}\right) = \boxed{}$$
$$\left(7x - \boxed{}\right)(x + 2) = \boxed{}$$

The correct factorization is $\boxed{}$.

EXAMPLE 2 ■ Factoring a Trinomial by Grouping

Use factoring by grouping to factor the following trinomials.

(a) $6x^2 + 7x + 2$ (b) $10x^2 + 11x - 6$ (c) $6x^2 - 23x - 4$ (d) $-4x^2 + 13x - 3$

Solution

(a) Since $a = 6$ and $c = 2$, $ac = 12$. The factors of 12 whose sum is 7 are (4)(3).

$$6x^2 + 7x + 2 = 6x^2 + 4x + 3x + 2 \qquad \text{Replace } 7x \text{ by } 4x + 3x.$$
$$= 2x(3x + 2) + 1(3x + 2) \qquad \text{Factor by grouping.}$$
$$= (2x + 1)(3x + 2)$$

(b) $ac = -60$. Factors of -60 whose sum is 11 are $(-4)(15)$.

$$10x^2 + 11x - 6 = 10x^2 - 4x + 15x - 6 \qquad \text{Replace } 11x \text{ by } -4x + 15x.$$
$$= 2x(5x - 2) + 3(5x - 2) \qquad \text{Factor by grouping.}$$
$$= (2x + 3)(5x - 2)$$

(c) $ac = -24$. Factors of -24 whose sum is -23 are $(-24)(1)$.

$$6x^2 - 23x - 4 = 6x^2 - 24x + x - 4 \qquad \text{Replace } -23x \text{ by } -24x + x.$$
$$= 6x(x - 4) + 1(x - 4) \qquad \text{Factor by grouping.}$$
$$= (6x + 1)(x - 4)$$

(d) $ac = 12$. Factors of 12 whose sum is 13 are (1)(12).

$$-4x^2 + 13x - 3 = -4x^2 + x + 12x - 3 \qquad \text{Replace } 13x \text{ by } x + 12x.$$
$$= -x(4x - 1) + 3(4x - 1) \qquad \text{Factor by grouping.}$$
$$= (-x + 3)(4x - 1)$$

■

Starter Exercise 2 | *Fill in the blanks.*

(a) Use factoring by grouping to factor the trinomial $6x^2 + 13x + 6$. $ac = 36$; the factors of 36 whose sum is 13 are (4)(9).

$$6x^2 + 13x + 6 = 6x^2 + 4x + 9x + 6$$
$$= 2x\left(\boxed{} + \boxed{}\right) + \boxed{}(3x + 2) = \left(2x + \boxed{}\right)(3x + 2)$$

(b) Use factoring by grouping to factor the trinomial $4x^2 - 13x + 10$. $ac = \boxed{}$; the factors of 40 whose sum is -13 are $\left(\boxed{}\right)\left(\boxed{}\right)$.

$$4x^2 - 13x + 10 = 4x^2 - \boxed{} - \boxed{} + 10$$
$$= \boxed{}\left(\boxed{} - \boxed{}\right) - \boxed{}\left(\boxed{} - \boxed{}\right)$$
$$= \left(\boxed{}\right)\left(\boxed{}\right)$$

(c) Use factoring by grouping to factor the trinomial $21 - 4z - z^2$. $ac = -21$; the factors of -21 whose sum is -4 are $(-7)(3)$.

$$21 - 4z - z^2 = 21 - 7z + 3z - z^2$$
$$= \boxed{}(3 - z) + \boxed{}(3 - z) = \left(\boxed{}\right)\left(\boxed{}\right)$$

EXAMPLE 3 ■ Factoring Completely

Factor (a) $15x^2 - 24x - 12$ and (b) $4x^3 - 2x^2 - 56x$ completely.

Solution

(a) Note that the trinomial has a common factor of 3, so we factor it out first.

$$15x^2 - 24x - 12 = 3[5x^2 - 8x - 4]$$

Now we need to factor the trinomial $5x^2 - 8x - 4$. $ac = -20$; factors of -20 whose sum is -8 are $(-10)(2)$.

$$15x^2 - 24x - 12 = 3[5x^2 - 8x - 4] = 3[5x^2 - 10x + 2x - 4]$$
$$= 3[5x(x - 2) + 2(x - 2)] = 3(5x + 2)(x - 2)$$

(b) Note that the trinomial has a common factor of $2x$, so we factor it out first.

$$4x^3 - 2x^2 - 56x = 2x[2x^2 - x - 28]$$

Now we factor the trinomial $2x^2 - x - 28$. $ac = -56$; factors of -56 whose sum is -1 are -8 and 7.

$$4x^3 - 2x^2 - 56x = 2x[2x^2 - x - 28] = 2x[2x^2 - 8x + 7x - 28]$$
$$= 2x[2x(x - 4) + 7(x - 4)] = 2x(2x + 7)(x - 4)$$

■

Starter Exercise 3   *Fill in the blanks.*

Factor the following trinomials completely.

(a) $6x^3 + 5x^2 + x = \boxed{}[6x^2 + 5x + 1] = \boxed{}[6x^2 + 2x + 3x + 1]$

$$= \boxed{}\left[\boxed{}(3x + 1) + 1(3x + 1)\right]$$

$$= \boxed{}\left(\boxed{} + 1\right)(3x + 1)$$

(b) $16x^2y^2 - 52xy^2 + 30y^2 = 2y^2\left[8x^2 - 26x + \boxed{}\right]$

$$= 2y^2\left[8x^2 - 6x - 20x + \boxed{}\right]$$

$$= 2y^2\left[\boxed{}(4x - 3) - \boxed{}(4x - 3)\right]$$

$$= \boxed{}$$

■ **Solutions to Starter Exercises** ■

1. (a) $(x+1)(15x-1) = \boxed{15}\, x^2 + \boxed{14}\, x - \boxed{1}$

$(15x+1)(x-1) = 15x^2 - \boxed{14}\, x - \boxed{1}$

$(3x+1)\left(\boxed{5}\, x - 1\right) = \boxed{15}\, x^2 + \boxed{2}\, x - \boxed{1}$

$\left(\boxed{5}\, x + 1\right)(3x-1) = \boxed{15}\, x^2 - \boxed{2}\, x - \boxed{1}$

The correct factorization is $\boxed{(3x+1)(5x-1)}$.

(b) The factors of $a = 7$ are $\left(\boxed{1}\right)\left(\boxed{7}\right)$. The factors of $c = -6$ are

$(1)(-6),\ (-1)(6),\ (2)(-3),$ and $\left(\boxed{-2}\right)\left(\boxed{3}\right)$.

$(7x+1)(x-6) = \boxed{7x^2 - 41x - 6}$

$(7x-1)\left(x + \boxed{6}\right) = \boxed{7x^2 + 41x - 6}$

$(7x-6)\left(x + \boxed{1}\right) = \boxed{7x^2 + x - 6}$

$\left(7x + \boxed{6}\right)(x-1) = \boxed{7x^2 - x - 6}$

$(7x+2)\left(x - \boxed{3}\right) = \boxed{7x^2 - 19x - 6}$

$(7x-2)(x+3) = \boxed{7x^2 + 19x - 6}$

$(7x+3)\left(x - \boxed{2}\right) = \boxed{7x^2 - 11x - 6}$

$\left(7x - \boxed{3}\right)(x+2) = \boxed{7x^2 + 11x - 6}$

The correct factorization is $\boxed{(2x+6)(x-1)}$.

2. (a) $= 2x\left(\boxed{3x} + \boxed{2}\right) + \boxed{3}\,(3x+2) = \left(2x + \boxed{3}\right)(3x+2)$

(b) $ac = \boxed{40}$, $\left(\boxed{-5}\right)\left(\boxed{-8}\right)$

$= 4x^2 - \boxed{5x} - \boxed{8x} + 10$

$= \boxed{x}\left(\boxed{4x} - \boxed{5}\right) - \boxed{2}\left(\boxed{4x} - \boxed{5}\right)$

$= \left(\boxed{x-2}\right)\left(\boxed{4x-5}\right)$

(c) $= \boxed{7}\,(3-z) + \boxed{z}\,(3-z) = \left(\boxed{7+z}\right)\left(\boxed{3-z}\right)$

■ **Solutions to Starter Exercises** ■

3. (a) $= \boxed{x}\,[6x^2 + 5x + 1] = \boxed{x}\,[6x^2 + 2x + 3x + 1]$

$= \boxed{x}\,\Big[\boxed{2x}\,(3x+1) + 1(3x+1)\Big]$

$= \boxed{x}\,\Big(\boxed{2x}+1\Big)(3x+1)$

(b) $= 2y^2\Big[8x^2 - 26x + \boxed{15}\Big]$

$= 2y^2\Big[8x^2 - 6x - 20x + \boxed{15}\Big]$

$= 2y^2\Big[\boxed{2x}\,(4x-3) - \boxed{5}\,(4x-3)\Big]$

$= \boxed{2y^2(2x-5)(4x-3)}$

6.3 | EXERCISES

In Exercises 1–4, find the missing factors.

1. $2x^2 + 3x + 1 = (2x+1)\Big(\boxed{}\Big)$ **2.** $6x^2 + 7x - 5 = (2x-1)\Big(\boxed{}\Big)$

3. $3z^2 - z - 2 = (z-1)\Big(\boxed{}\Big)$ **4.** $15y^2 + 16y + 4 = (5y+2)\Big(\boxed{}\Big)$

In Exercises 5–24, factor the given trinomial. (*Note:* Some of the trinomials cannot be factored using integer coefficients.)

5. $2x^2 - 3x + 1$ **6.** $3x^2 + 8x - 3$ **7.** $6x^2 - 13x + 6$

8. $5x^2 - 9x - 2$ **9.** $7z^2 - 27z - 4$ **10.** $8x^2 + 26x + 15$

11. $10x^2 - 7x - 3$ **12.** $3x^2 + 9x - 1$ **13.** $10y^2 - 3y - 4$

14. $12x^2 + 17x + 5$ **15.** $49a^2 - 14a + 1$ **16.** $40x^2 + 39x + 9$

17. $2x^2 - 23x + 45$ **18.** $11x^2 - 29x - 12$ **19.** $18x^2 - 31x + 6$

20. $6x^2 - 31x + 18$ **21.** $2x^2 - 13x + 11$ **22.** $8x^2 - 14x - 15$

23. $12x^2 - 44x + 7$ **24.** $10x^2 + 19x - 15$

In Exercises 25–34, factor the polynomial completely. (*Note:* Some of the polynomials cannot be factored using integer coefficients.)

25. $12x^2 + 14x - 10$ **26.** $4x^3 + 35x^2 - 9x$ **27.** $6x^3 - 3x^2 - 3x$

28. $5x^3 - 9x^2 - 2x$ **29.** $5x^4 + x^3 + 7x^2$ **30.** $5x^4 - 26x^3 + 5x^2$

31. $8x^4 + 20x^3 - 168x^2$ **32.** $10 - 3z - z^2$ **33.** $-3x^2 + 31x - 56$

34. $10x^3y - 14y^3 + 11x^2$

In Exercises 35 and 36, find all integers b such that the trinomial can be factored.

35. $3x^2 + bx + 5$ **36.** $2x^2 + bx - 15$

In Exercises 37–40, use a graphing utility to check your answers to Exercises 5–8.

6.4   Factoring Polynomials with Special Forms

Section Highlights

1. Let a and b be real numbers, variables, or algebraic expressions.

   i) **Difference of Two Squares**

   $$a^2 - b^2 = (a + b)(a - b)$$

   ii) **Perfect Square Trinomials**

   $$a^2 + 2ab + b^2 = (a + b)^2$$
   $$a^2 - 2ab + b^2 = (a - b)^2$$

   iii) **Sum and Difference of Cubes**

   $$a^3 + b^3 = (a + b)(a^2 - ab + b)^2$$
   $$a^3 - b^3 = (a - b)(a^2 + ab + b)^2$$

2. Remember that with all factoring techniques we should first factor out any common factors.

EXAMPLE 1 ■   Factoring the Difference of Two Squares

Factor the following polynomials.

(a) $x^2 - 16$ (b) $4x^2 - 25$

Solution

(a) Since x^2 and 16 are both perfect squares, we recognize this polynomial as the difference of two squares. Therefore, the polynomial factors as follows.

$$x^2 - 16 = x^2 - 4^2 \qquad \text{Write as difference of two squares.}$$
$$= (x + 4)(x - 4) \qquad \text{Factored form}$$

(b) Since $4x^2$ and 25 are both perfect squares, we recognize this polynomial as the difference of two squares. Therefore, the polynomial factors as follows.

$$4x^2 - 25 = (2x)^2 - 5^2 \qquad \text{Write as difference of two squares.}$$
$$= (2x + 5)(2x - 5) \qquad \text{Factored form}$$

Starter Exercise 1 *Fill in the blanks.*

Factor the following polynomials.

(a) $4 - y^2 = \boxed{}^2 - y^2 = \left(\boxed{} + y\right)\left(\boxed{} - y\right)$

(b) $x^2 - 9y^2 = x^2 - \boxed{}^2 = \left(\boxed{} + \boxed{}\right)\left(\boxed{} - \boxed{}\right)$

EXAMPLE 2 ■ Factoring the Difference of Two Squares

Factor the expression $(x + 5)^2 - 1$.

Solution

Since $(x + 5)^2$ and 1 are both perfect squares, we recognize this expression as the difference of two squares. Therefore, the expression factors as follows.

$(x + 5)^2 - 1 = (x + 5)^2 - 1^2$ Write as difference of two squares.

$= [(x + 5) + 1][(x + 5) - 1]$ Factored form

$= (x + 6)(x + 4)$ Simplify.

Starter Exercise 2 *Fill in the blanks.*

Factor the expression $49 - (x - 1)^2$.

$$49 - (x - 1)^2 = \boxed{}^2 - (x - 1)^2 = \left[\boxed{} + (x - 1)\right]\left[\boxed{} - \left(\boxed{}\right)\right]$$

$$= \left(\boxed{}\right)\left(\boxed{}\right)$$

EXAMPLE 3 ■ Removing a Common Monomial Factor First

Factor the following polynomials.

(a) $2x^2 - 50$

(b) $4x - x^3$

Solution

(a) This polynomial has a common factor of 2. After factoring out 2, we are left with the difference of two squares.

$2x^2 - 50 = 2(x^2 - 25)$ Factor out 2.

$= 2(x^2 - 5^2)$ Write as difference of two squares.

$= 2(x + 5)(x - 5)$ Factored form

(b) This polynomial has a common factor of x. After factoring out x, we are left with the difference of two squares.

$4x - x^3 = x(4 - x^2)$ Factor out x.

$= x(2^2 - x^2)$ Write as difference of two squares.

$= x(2 + x)(2 - x)$ Factored form

Starter Exercise 3   *Fill in the blanks.*

Factor the following polynomials.

(a) $x^3 - 64x = \boxed{}(x^2 - 64) = \boxed{}\left(x^2 - \boxed{}^2\right)$

$= \boxed{}\left(x + \boxed{}\right)\left(x - \boxed{}\right)$

(b) $45x^2 - 5x^4 = \boxed{}(9 - x^2) = \boxed{}\left(\boxed{}^2 - x^2\right)$

$= \boxed{}\left(\boxed{}\right)\left(\boxed{}\right)$

EXAMPLE 4 ■   Factoring Completely

Factor the $x^4 - 81$ completely.

Solution

$x^4 - 81 = (x^2)^2 - 9^2 = (x^2 + 9)(x^2 - 9)$

Note that $x^2 - 9$ is the difference of two squares.

$x^4 - 81 = (x^2)^2 - 9^2 = (x^2 + 9)(x^2 - 9) = (x^2 + 9)(x + 3)(x - 3)$ ■

Starter Exercise 4   *Fill in the blanks.*

Factor the $16x^4 - 1$ completely.

$16x^4 - 1 = (4x^2 + 1)\left(\boxed{} - 1\right) = (4x^2 + 1)\left(\boxed{} + 1\right)\left(\boxed{} - 1\right)$

EXAMPLE 5 ■   Factoring Perfect Square Trinomials

Factor the following trinomials.

(a) $x^2 + 2x + 1$ 　　　(b) $4x^2 - 12x + 9$ 　　　(c) $x^2 + 6xy + 9y^2$

Solution

(a) $x^2 + 2x + 1 = x^2 + 2(x)(1) + 1^2 = (x + 1)^2$

(b) $4x^2 - 12x + 9 = (2x)^2 - 2(2x)(3) + 3^2 = (2x + 3)^2$

(c) $x^2 + 6xy + 9y^2 = x^2 + 2(x)(3y) + (3y)^2 = (x + 3y)^2$ ■

Starter Exercise 5   *Fill in the blanks.*

Factor the following trinomials.

(a) $x^2 - 2x + 1 = x^2 - 2(x)(1) + 1^2 = \left(\boxed{} - \boxed{}\right)^2$

(b) $25x^2 + 40x + 16 = \boxed{}^2 + 2\left(\boxed{}\right)(4) + 4^2 = \left(\boxed{}\right)^2$

(c) $4x^2 - 20xy + 25y^2 = \boxed{}^2 - 2\left(\boxed{}\right)\left(\boxed{}\right) + \boxed{}^2 = \left(\boxed{}\right)^2$

EXAMPLE 6 ■ Factoring Sums and Differences of Cubes

Factor the polynomials (a) $x^3 + 8$ and (b) $27 - x^3$.

Solution

(a) Since $x^3 + 8 = x^3 + 2^3$, this polynomial is the sum of two cubes. Therefore, we can factor the polynomial as follows.

$$x^3 + 8 = x^3 + 2^3 = (x + 2)(x^2 - 2x + 4)$$

(b) Since $27 - x^3 = 3^3 - x^3$, this polynomial is the difference of two cubes. Therefore, we can factor the polynomial as follows.

$$27 - x^3 = 3^3 - x^3 = (3 - x)(9 + 3x + x^2)$$

■

Starter Exercise 6 | *Fill in the blanks.*

Factor the following polynomials.

(a) $x^3 + 1 = x^3 + 1^3 = \left(x + \boxed{}\right)\left(x^2 - \boxed{} + 1\right)$

(b) $64x^3 - y^3 = \boxed{}^3 - y^3 = \left(\boxed{} - y\right)\left(\boxed{}\right)$

EXAMPLE 7 ■ Removing a Common Monomial Factor First

Factor the polynomials (a) $2x^2 + 12x + 18$ and (b) $x^4 - 64x$.

Solution

(a) $2x^2 + 12x + 18 = 2(x^2 + 6x + 9)$ Factor out common factor of 2.

$\qquad\qquad\qquad\quad = 2(x + 3)^2$ Factor as perfect square trinomial.

(b) $x^4 - 64x = x(x^3 - 64)$ Factor out common factor of x.

$\qquad\qquad\quad = x(x - 4)(x^2 + 4x + 16)$ Factor as difference of cubes.

■

Starter Exercise 7 | *Fill in the blanks.*

Factor the following polynomials.

(a) $3x^3 - 12x^2 + 12x = \boxed{}(x^2 - 4x + 4) = \boxed{}\left(\boxed{} - 2\right)^2$

(b) $2x^5 - 16x^2 = \boxed{}\left(x^3 - \boxed{}\right) = \boxed{}\left(x - \boxed{}\right)\left(x^2 + \boxed{} + \boxed{}\right)$

204   CHAPTER 6   Factoring and Solving Equations

■ Solutions to Starter Exercises ■

1. (a) $4 - y^2 = \boxed{2}^2 - y^2 = \left(\boxed{2} + y\right)\left(\boxed{2} - y\right)$

   (b) $x^2 - 9y^2 = x^2 - \boxed{(3y)}^2 = \left(\boxed{x} + \boxed{3y}\right)\left(\boxed{x} - \boxed{3y}\right)$

2. $49 - (x-1)^2 = \boxed{7}^2 - (x-1)^2 = \left[\boxed{7} + (x-1)\right]\left[\boxed{7} - \boxed{(x-1)}\right]$

   $= \left(\boxed{6+x}\right)\left(\boxed{8-x}\right)$

3. (a) $x^3 - 64x = \boxed{x}(x^2 - 64) = \boxed{x}\left(x^2 - \boxed{8}^2\right)$

   $= \boxed{x}\left(x + \boxed{8}\right)\left(x - \boxed{8}\right)$

   (b) $45x^2 - 5x^4 = \boxed{5x^2}(9 - x^2) = \boxed{5x^2}\left(\boxed{3}^2 - x^2\right)$

   $= \boxed{5x^2}\left(\boxed{3+x}\right)\left(\boxed{3-x}\right)$

4. $16x^4 - 1 = (4x^2 + 1)\left(\boxed{4x^2} - 1\right) = (4x^2 + 1)\left(\boxed{2x} + 1\right)\left(\boxed{2x} - 1\right)$

5. (a) $x^2 - 2x + 1 = x^2 - 2(x)(1) + 1^2 = \left(\boxed{x} - \boxed{1}\right)^2$

   (b) $25x^2 + 40x + 16 = \boxed{(5x)}^2 + 2\left(\boxed{5x}\right)(4) + 4^2 = \left(\boxed{5x+4}\right)^2$

   (c) $4x^2 - 20xy + 25y^2 = \boxed{(2x)}^2 - 2\left(\boxed{2x}\right)\left(\boxed{5y}\right) + \boxed{(5y)}^2$

   $= \left(\boxed{2x-5y}\right)^2$

6. (a) $x^3 + 1 = x^3 + 1^3 = \left(x + \boxed{1}\right)\left(x^2 - \boxed{x} + 1\right)$

   (b) $64x^3 - y^3 = \boxed{(4x)}^3 - y^3 = \left(\boxed{4x} - y\right)\left(\boxed{16x^2 + 4xy + y^2}\right)$

7. (a) $3x^3 - 12x^2 + 12x = \boxed{3x}(x^2 - 4x + 4) = \boxed{3x}\left(\boxed{x} - 2\right)^2$

   (b) $2x^5 - 16x^2 = \boxed{2x^2}\left(x^3 - \boxed{8}\right)$

   $= \boxed{2x^2}\left(x - \boxed{2}\right)\left(x^2 + \boxed{2x} + \boxed{4}\right)$

6.4 | EXERCISES

In Exercises 1–6, factor the given difference of two squares.

1. $x^2 - 4$

2. $4x^2 - 25$

3. $100 - z^2$

4. $49x^2 - 36$

5. $1 - 64y^2$

6. $(x+3)^2 - 9$

In Exercises 7–14, factor the perfect square trinomial.

7. $x^2 + 4x + 4$ **8.** $x^2 - 6x + 9$ **9.** $x^2 - 2xy + y^2$

10. $x^2 + 8xy + 16y^2$ **11.** $9x^2 + 42x + 49$ **12.** $25x^2 - 20x + 4$

13. $4x^2 - 12xy + 9y^2$ **14.** $9x^2 + 48xy + 64y^2$

In Exercises 15–18, factor the given sum or difference of cubes.

15. $z^3 - 1$ **16.** $x^3 + 8$ **17.** $8x^3 + 27$ **18.** $64x^3 - 1$

In Exercises 19–34, factor the polynomial completely.

19. $2x^2 - 4x + 2$ **20.** $x^3 - 9x$ **21.** $x^4 + x$

22. $25x^3 + 10x^2 + x$ **23.** $3x^2 - 10x - 8$ **24.** $16x^4 - 54x$

25. $3x^5 + 3x^2$ **26.** $x^4 - 16$ **27.** $3(2x - 1) + x(2x - 1)$

28. $x^3 + 6x^2 + 8x$ **29.** $27x^2 - x^5$ **30.** $4x^3y - 28x^2y^2 + 49xy^3$

31. $x^4 - 256$ **32.** $98x^3 + 140x^2 + 50x$

33. $x^6 - 64$ **34.** $(x - 1)^2 - 25$

In Exercises 35–38, find all integers b so that the algebraic expression is a perfect square.

35. $x^2 + bx + 9$ **36.** $x^2 - bx + 25$

37. $4x^2 + bx + 81$ **38.** $9x^2 - bxy + y^2$

In Exercises 39 and 40, find c so that the algebraic expression is a perfect square.

39. $x^2 + 4x + c$ **40.** $x^2 - 16x + c$

6.5 | Solving Equations and Problem Solving

Section Highlights

1. **The Zero-Factor Property**
   Let a and b be real numbers, variables, or algebraic expressions. If a and b are factors such that $a \cdot b = 0$, then $a = 0$ or $b = 0$. This means if the product of two factors equals zero, then one factor or the other must be zero.

2. In order to apply the **Zero-Factor Property**, the polynomial equation must first be in standard form, $(ax^2 + bx + c = 0)$.

3. The **Zero-Factor Property** can be applied ONLY when one side of the equation is zero.

EXAMPLE 1 ■ Using Factoring to Solve an Equation

Solve the equation $x^2 + x - 20 = 0$.

Solution

The strategy for this equation is to first check to see that one side of the equation is zero. Next, we factor the polynomial. Finally, we apply the Zero-Factor Property to find the solutions.

$x^2 + x - 20 = 0$	Given equation
$(x - 4)(x + 5) = 0$	Factor.
$x - 4 = 0$ or $x + 5 = 0$	Set both factors equal to zero.
$x = 4$ or $x = -5$	Solve each equation.

Check first solution:

$x^2 + x - 20 = 0$	Given equation
$(-5)^2 + (-5) - 20 \stackrel{?}{=} 0$	Replace x by -5.
$25 - 5 - 20 \stackrel{?}{=} 0$	
$0 = 0$	Solution checks.

Check second solution:

$x^2 + x - 20 = 0$	Given equation
$(4)^2 + 4 - 20 \stackrel{?}{=} 0$	Replace x by 4.
$16 + 4 - 20 \stackrel{?}{=} 0$	
$0 = 0$	Solution checks.

Starter Exercise 1 | *Fill in the blanks.*

Use factoring to solve $2x^2 + x - 1 = 0$. The strategy for this equation is to first check to see if one side of the equation is zero. Next, we factor the polynomial. Finally, we apply the Zero-Factor Property to find the solutions. Check the solutions in the original equation.

$$2x^2 + x - 1 = 0$$

$$\left(\boxed{} - 1 \right)(x + 1) = 0$$

$$\boxed{} - 1 = 0 \quad \text{or} \quad x + 1 = 0$$

$$x = \boxed{} \quad \text{or} \quad x = \boxed{}$$

EXAMPLE 2 ■ An Equation with a Repeated Solution

Solve the equation $x^2 + 6x + 9 = 0$.

Solution

$$x^2 + 6x + 9 = 0 \qquad \text{Given equation}$$
$$(x + 3)^2 = 0 \qquad \text{Factor.}$$
$$x + 3 = 0 \qquad \text{Set factor equal to zero.}$$
$$x = -3$$

Check this in the original equation.

■

| Starter Exercise 2 | *Fill in the blanks.*

Solve the equation $4x^2 + 12x + 9 = 0$.

$$4x^2 + 12x + 9 = 0$$
$$\left(\boxed{} + 3 \right)^2 = 0$$
$$\boxed{} + 3 = 0$$
$$x = \boxed{}$$

EXAMPLE 3 ■ Solving a Polynomial Equation

Solve the following equations.

(a) $x^2 + 7x = -10$ (b) $x(x - 8) = -12$

Solution

(a)
$$x^2 + 7x = -10 \qquad \text{Given equation}$$
$$x^2 + 7x + 10 = 0 \qquad \text{Standard form}$$
$$(x + 2)(x + 5) = 0 \qquad \text{Factor.}$$
$$x + 2 = 0 \quad \text{or} \quad x + 5 = 0 \qquad \text{Set both factors equal to zero.}$$
$$x = -2 \quad \text{or} \quad x = -5$$

(b)
$$x(x - 8) = -12 \qquad \text{Given equation}$$
$$x^2 - 8x = -12 \qquad \text{Multiply factors.}$$
$$x^2 - 8x + 12 = 0 \qquad \text{Standard form}$$
$$(x - 2)(x - 6) = 0 \qquad \text{Factor.}$$
$$x - 2 = 0 \quad \text{or} \quad x - 6 = 0 \qquad \text{Set both factors equal to zero.}$$
$$x = 2 \quad \text{or} \quad x = 6$$

■

■ **Starter Exercise 3** *Fill in the blanks.*

Solve the following equations.

(a)
$$6x^2 = 3 - 7x$$
$$6x^2 + 7x - 3 = 0$$
$$\left(\boxed{} - 1\right)\left(\boxed{} + 3\right) = 0$$
$$\boxed{} - 1 = 0 \quad \text{or} \quad \boxed{} + 3 = 0$$
$$x = \boxed{} \quad \text{or} \quad x = \boxed{}$$

(b)
$$(x + 5)(x - 4) = -8$$
$$x^2 + x - 20 = -8$$
$$x^2 + x - \boxed{} = 0$$
$$\left(x \boxed{}\right)\left(x \boxed{}\right) = 0$$
$$x \boxed{} = 0 \quad \text{or} \quad x \boxed{} = 0$$
$$x = \boxed{} \quad \text{or} \quad x = \boxed{}$$

EXAMPLE 4 ■ Solving a Polynomial Equation with Three Factors

Solve the equation $2x^3 + 16x = 12x^2$.

Solution

$$
\begin{aligned}
2x^3 + 16x &= 12x^2 && \text{Given equation} \\
2x^3 - 12x^2 + 16x &= 0 && \text{Standard form} \\
2x(x^2 - 6x + 8) &= 0 && \text{Factor out common factor.} \\
2x(x - 2)(x - 4) &= 0 && \text{Factor completely.}
\end{aligned}
$$

$2x = 0 \quad \text{or} \quad x - 2 = 0 \quad \text{or} \quad x - 4 = 0$ Set all factors equal to zero

$x = 0 \quad \text{or} \quad x = 2 \quad \text{or} \quad x = 4$

■

■ **Starter Exercise 4** *Fill in the blanks.*

Solve the following equation.

$$6x^3 = x^2 + x$$
$$6x^3 - x^2 - x = 0$$
$$\boxed{}(6x^2 - x - 1) = 0$$
$$\boxed{}\left(\boxed{} + 1\right)\left(\boxed{} - 1\right) = 0$$
$$\boxed{} = 0 \quad \text{or} \quad \boxed{} + 1 = 0 \quad \text{or} \quad \boxed{} - 1 = 0$$
$$x = \boxed{} \qquad x = \boxed{} \qquad x = \boxed{}$$

EXAMPLE 5 ■ An Application

A corral is to have a length that is six feet more than twice its width. It is to enclose 360 square feet. Find the dimensions of the corral.

Solution

Verbal model: | Length | · | Width | = | Area |

Labels: Width $= w$
                   Length $= 2w + 6$
                   Area $= 360$

Equation:
$$w(2w + 6) = 360$$
$$2w^2 + 6w = 360$$
$$2w^2 + 6w - 360 = 0$$
$$2(w^2 + 3w - 180) = 0$$
$$2(w + 15)(w - 12) = 0$$
$$2(w + 15) = 0 \quad \text{or} \quad w - 12 = 0$$
$$w = -15 \quad \text{or} \quad w = 12$$

$w = -15$ is not a valid solution since it is negative. Thus, the width is 12 and the length is $2w + 6 = 30$. ■

■ Starter Exercise 5   *Fill in the blanks.*

Find two consecutive integers whose product is 506.

Verbal model: | Integer | · | Next integer | = 506

Labels: First integer $= n$
                Second integer $= \boxed{}$

Equation:
$$n \left(\boxed{} \right) = 506$$
$$\boxed{} = 506$$
$$\boxed{} - 506 = 0$$
$$\left(\boxed{} \right)\left(\boxed{} \right) = 0$$
$$\boxed{} = 0 \quad \text{or} \quad \boxed{} = 0$$
$$n = \boxed{} \quad \text{or} \quad n = \boxed{}$$

■ Solutions to Starter Exercises ■

1. $\left(\boxed{2x} - 1 \right)(x + 1) = 0$

$\boxed{2x} - 1 = 0 \qquad$ or $x + 1 = 0$

$x = \boxed{\frac{1}{2}} \qquad x = \boxed{-1}$

2. $\left(\boxed{2x} + 3 \right)^2 = 0$

$\boxed{2x} + 3 = 0$

$x = \boxed{-\frac{3}{2}}$

3. (a) $\left(\boxed{3x} - 1 \right)\left(\boxed{2x} + 3 \right) = 0$

$\boxed{3x} - 1 = 0 \qquad$ or $\boxed{2x} + 3 = 0$

$x = \boxed{\frac{1}{3}} \qquad\qquad x = \boxed{-\frac{3}{2}}$

(b) $\qquad x^2 + x - \boxed{12} = 0$

$\left(x \boxed{+4} \right)\left(x \boxed{-3} \right) = 0$

$x \boxed{+4} = 0 \qquad$ or $x \boxed{-3} = 0$

$x = \boxed{-4} \qquad\qquad x = \boxed{3}$

4. $\qquad\qquad \boxed{x}\,(6x^2 - x - 1) = 0$

$\boxed{x}\left(\boxed{3x} + 1 \right)\left(\boxed{2x} - 1 \right) = 0$

$\boxed{x} = 0 \qquad$ or $\boxed{3x} + 1 = 0 \qquad$ or $\boxed{2x} - 1 = 0$

$x = \boxed{0} \qquad\qquad x = \boxed{-\frac{1}{3}} \qquad\qquad x = \boxed{\frac{1}{2}}$

5. Second integer $= \boxed{n + 1}$

$n\left(\boxed{n + 1} \right) = 506$

$\boxed{n^2 + n} = 506$

$\boxed{n^2 + n} - 506 = 0$

$\left(\boxed{n - 22} \right)\left(\boxed{n + 23} \right) = 0$

$\boxed{n - 22} = 0 \qquad$ or $\boxed{n + 23} = 0$

$n = \boxed{22} \qquad\qquad n = \boxed{-23}$

Thus, the two integers are 22 and 23 or −23 and −22.

6.5 | EXERCISES

In Exercises 1–26, solve the given equation.

1. $x(x - 1) = 0$

2. $z(2z + 3) = 0$

3. $(x + 1)(x - 3) = 0$

4. $(x - 7)(x + 8) = 0$

5. $(5x + 3)(2x - 3) = 0$

6. $(z - 1)(z + 2)(z - 3) = 0$

7. $x^2 - 7x + 10 = 0$

8. $z^2 + 10z + 21 = 0$

9. $x^2 - 9 = 0$

10. $y^2 - 25 = 0$

11. $144 - z^2 = 0$

12. $(x + 1)^2 - 1 = 0$

13. $4x^2 - x = 0$

14. $x^2 - 8x = 0$

15. $z(z - 1) + 6(z - 1) = 0$

16. $2x(x + 6) + 4(x + 6) = 0$

17. $6x^2 + 5x + 1 = 0$

18. $5x^2 - 16x + 3 = 0$

19. $14x^2 + x - 3 = 0$

20. $x^2 - 3x = 28$

21. $(x + 3)(x - 5) = -7$

22. $(x + 4)(x - 8) = 13$

23. $x^3 + 4x^2 + 4x = 0$

24. $5x^3 - 50x^2 = -125x$

25. $y^2(y + 3) - 4(y + 3) = 0$

26. $z^3 + 2z^2 - 9z - 18 = 0$

27. *Number Problem*   Find two consecutive positive even integers whose product is 224.

28. *Number Problem*   Two positive numbers have a sum of 33. Their product is 216. Find the two numbers.

29. *Dimensions of a Rectangle*   The length of a room is 7 feet more than its width. The area of the floor of this room is 228 square feet. Find the dimensions of the room.

30. *Height of an Object*   A ball is dropped from the top of a 64-foot building. The height (in feet) of the ball is given by

Height $= -16t^2 + 64$

where t is measured in seconds. How long will it take the ball to hit the ground?

31. *Dimensions of a Box*   An open box with a square base is constructed from 240 square inches of cardboard. The height of the box is 2 inches. Find the dimensions of the base of the box. (*Hint:* The surface area is given by $s = x^2 + 4xh$.)

In Exercises 32–40, use a graphing utility to check your answers to Exercises 1–9 by determining whether your solutions are the x-intercepts.

Cumulative Practice Test for Chapters P–6

In Exercises 1–5, match the property from the list below to the statement it best justifies.

(a) Commutative Property of Addition

(b) Distributive Property

(c) Inverse Property of Multiplication

(d) Associative Property of Multiplication

1. $2x + (1 + 2x) = 2x + (2x + 1)$

2. $5y + 3y = (5 + 3)y$

_____

_____

3. $3(17z) = (3 \cdot 17)z$

4. $\left(\frac{1}{2} \cdot 2\right)x = 1x$

_____

_____

5. $(6x - 1) + 8x = 8x + (6x - 1)$

_____

In Exercises 6–10, factor the polynomial completely.

6. $x^2 - 1$

7. $x^3 - x^2 - 6x$

8. $6x^2 - 7x + 1$

9. $x^4 - 16$

10. $4x^3 - 12x^2y + 9xy^2$

In Exercises 11–20, perform the indicated operations and simplify. Your answer should not involve negative exponents.

11. $x^2(2x^3 - x + 1)$

12. $(x^3 - x) - (2x^2 + x)$

13. $-|-3| - \dfrac{2^2 - 5}{3^2 - 2^3}$

14. $\dfrac{1}{2} + \dfrac{1}{3} - \dfrac{1}{4}$

15. $2[2x - (3x - 7y)] - y$

16. $x^2(xy^3) - x^3y^3$

17. $(x^3 - 4x^2 + 3x - 1) \div (x + 1)$

18. $(2x^2 + 4x + 7) \div (x)$

19. $(x^2 + 2x + 3)(x^2 - x + 7)$

20. $\left(\dfrac{x^2y^{-2}}{x^3}\right)^{-1}$

In Exercises 21–30, solve the given equation or inequality.

21. $\dfrac{3x}{2} - \dfrac{1}{3} = \dfrac{x}{4} + 2$

22. $\dfrac{1}{2}y = 6$

23. $-2x < 4$

24. $2x + 1 = 7$

25. $x^2 + 6x = -9$

26. $x - [4x - (2x - 1)] = x + 1$

27. $2(x+3) = 2x+1$

28. $6x^3 - 7x^2 = -x$

29. $z^2(2z+1) = 2z+1$

30. $2x - (1+5x) \geq 2x - 3$

31. *Percent*   13.2 is 22% of what number?

32. *Area of a Rectangle*   The length of a rectangle is one more than twice the width. If the area of the rectangle is 136 square inches, then find its dimensions.

33. *Integers*   The sum of three consecutive even integers is 102. Find the integers.

34. *Mixture*   How many ounces of pure water must be added to 60 ounces of 15% acid solution to make a 10% acid solution?

35. *Sale*   A coat is on sale for $40. This is $\frac{1}{3}$ off the regular price. Find the regular price.

36. *Commission Rate*   Determine the commission rate for an employee who earned $450 in commission on sales of $5000.

37. *Coins*   A person has 40 coins in dimes and quarters with a combined value of $7.30. Determine the number of coins of each type.

38.   Two joggers start out at the same time from the same point and jog in the same direction. If one jogger is jogging at 8 mph and the other at 10 mph, how long before they are 2 miles apart?

6.1   Answers to Exercises

1. 14

2. 28

3. $2x$

4. x^4

5. $4xy$

6. $6a^2b^5$

7. $7x$

8. $3x^2y$

9. $3(x-2)$

10. $5(2y+3)$

11. $12(-x^2-2)$

12. $4(x^3-2)$

13. $x(x+1)$

14. $x^2(2x-1)$

15. $7x(2x^2-3)$

16. $16t^2(t-2)$

17. $uv(3u+7v)$

18. $2a^3b^2(8b^4-3a)$

19. $11x^2y^2(y^2-2x^2)$

20. $6u^4v^7(7+v)$

21. $x(x^2-14x-2)$

22. $2m(m^3-3m+2)$

23. $5a^2b(b^2+2b+3)$

24. $7x^2y^2(5x^2y-4xy^2-3)$

25. $(x+7)(x-1)$

26. $(2x+9)(x+6)$

27. $(x^2+5)(2x-3)$

28. $2y(x-y)$

29. $x^2(x-y)(x^2-xy+1)$

30. $(2x+1)(x-2)(x-1)$

31. $(x+6)(x+7)$

32. $(2x-3)(x+2)$

33. $(3x+2)(5x-4)$

34. $(x^2+4)(x+1)$

35. $(5y^2-3)(2y-3)$

36. $(x^2 + y^2)(2x^2 - 3y^2)$

37. $-2(x - 3)$

38. $-3(11x^2 + 3x - 9)$

39. $\frac{1}{3}(2x + 3)$

40. $\frac{1}{6}(3t - 5)$

6.2 Answers to Exercises

1. $(x + 5)$ **2.** $(x - 11)$ **3.** $(x + 1)$ **4.** $(x - 5)$

5. $(x + 1)(x + 11)$ **6.** $(x + 3)(x + 5)$ **7.** $(x - 4)(x - 5)$

8. $(x - 3)(x - 9)$ **9.** $(x + 12)(x - 3)$ **10.** $(x + 16)(x - 2)$

11. $(x + 6)(x - 7)$ **12.** $(x - 27)(x + 2)$ **13.** $(x - 1)(x - 19)$

14. $(x - 19)(x + 2)$ **15.** $(x - 12y)(x + 5y)$ **16.** $(x + 36z)(x - 2z)$

17. $(a + b)(a + 3b)$ **18.** $(u - 9v)(u + 5v)$ **19.** $(x + 2y)(x + 20y)$

20. $(a - 9b)(a + 7b)$ **21.** $3(x + 3)(x + 5)$ **22.** $x(x - 13)(x - 3)$

23. $2x(x + 23)(x - 1)$ **24.** $5x^2(x - 17)(x + 2)$ **25.** $2(x + 3y)(x - 2y)$

26. $x^2y^2(x - 4y)(x + 2y)$ **27.** $7, -7, 5, -5$ **28.** $36, -36, 12, -12$

29. $14, -14, 2, -2$ **30.** $9, -9, 3, -3$ **31.** $3, 4$

32. $4, 6$ **33.** 2 **34.** $5, 8, 9$

6.3 Answers to Exercises

1. $x + 1$ **2.** $3x + 5$ **3.** $3z + 2$ **4.** $3y + 2$

5. $(2x - 1)(x - 1)$ **6.** $(x + 3)(3x - 1)$ **7.** $(2x - 3)(3x - 2)$

8. $(5x + 1)(x - 2)$ **9.** $(7z + 1)(z - 4)$ **10.** $(2x + 5)(4x + 3)$

11. $(10x + 3)(x - 1)$ **12.** Cannot be factored   **13.** $(2y + 1)(5y - 4)$

14. $(12x + 5)(x + 1)$ **15.** $(7a - 1)(7a - 1)$ **16.** $(8x + 3)(5x + 3)$

17. $(2x - 5)(x - 9)$ **18.** $(11x + 4)(x - 3)$ **19.** $(9x - 2)(2x - 3)$

20. $(3x - 2)(2x - 9)$ **21.** $(2x - 11)(x - 1)$ **22.** $(4x + 3)(2x - 5)$

23. $(2x - 7)(6x - 1)$ **24.** $(5x - 3)(2x + 5)$ **25.** $2(3x + 5)(2x - 1)$

26. $x(x + 9)(4x - 1)$ **27.** $3x(2x + 1)(x - 1)$ **28.** $x(5x + 1)(x - 2)$

29. $x^2(5x^2 + x + 7)$ **30.** $x^2(5x - 1)(x - 5)$ **31.** $4x^2(2x - 7)(x + 6)$

32. $(5 + z)(2 - z)$ **33.** $(-3x + 7)(x - 8)$ **34.** Cannot be factored

35. $16, -16, 8, -8$ **36.** $-29, 29, -13, 13, -7, 7, -1, 1$

37.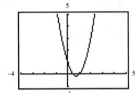

38.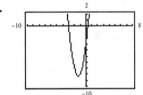

39.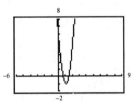

40.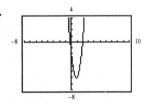

6.4   **Answers to Exercises**

1. $(x + 2)(x - 2)$ **2.** $(2x + 5)(2x - 5)$ **3.** $(10 + z)(10 - z)$ **4.** $(7x + 6)(7x - 6)$

5. $(1 + 8y)(1 - 8y)$ **6.** $x(x + 6)$ **7.** $(x + 2)^2$ **8.** $(x - 3)^2$

9. $(x - y)^2$ **10.** $(x + 4y)^2$ **11.** $(3x + 7)^2$ **12.** $(5x - 2)^2$

13. $(2x - 3y)^2$ **14.** $(3x + 8y)^2$ **15.** $(z - 1)(z^2 + z + 1)$

16. $(x + 2)(x^2 - 2x + 4)$ **17.** $(2x + 3)(4x^2 + 6x + 9)$ **18.** $(4x - 1)(16x^2 + 4x + 1)$

19. $2(x - 1)^2$ **20.** $x(x + 3)(x - 3)$ **21.** $x(x + 1)(x^2 - x + 1)$

22. $x(5x + 1)^2$ **23.** $(3x + 2)(x - 4)$ **24.** $2x(2x - 3)(4x^2 + 6x + 9)$

25. $3x^2(x + 1)(x^2 - x + 1)$ **26.** $(x^2 + 4)(x + 2)(x - 2)$ **27.** $(3 + x)(2x - 1)$

28. $x(x + 2)(x + 4)$ **29.** $x^2(3 - x)(9 + 3x + x^2)$ **30.** $xy(2x - 7y)^2$

31. $(x^2 + 16)(x + 4)(x - 4)$ **32.** $2x(7x + 5)^2$

33. $(x-2)(x+2)(x^2+2x+4)(x^2-2x+4)$ **34.** $(x+4)(x-6)$

35. 6, −6   **36.** 10, −10   **37.** 36, −36   **38.** 6, −6   **39.** 4   **40.** 64

6.5 | Answers to Exercises

1. $x = 0, 1$ **2.** $z \geq -\frac{3}{2}, 0$ **3.** $x \geq -1, 3$ **4.** $x = -8, 7$

5. $x = -\frac{3}{5}, \frac{3}{2}$ **6.** $z = -2, 1, 3$ **7.** $x = 2, 5$ **8.** $x = -7, -3$

9. $x = -3, 3$ **10.** $y = -5, 5$ **11.** $z = -12, 12$ **12.** $x = 0, -2$

13. $x = 0, \frac{1}{4}$ **14.** $x = 0, 8$ **15.** $z = -6, 1$ **16.** $x = -6, -2$

17. $x = -\frac{1}{2}, -\frac{1}{3}$ **18.** $x = \frac{1}{5}, 3$ **19.** $x = -\frac{1}{2}, \frac{3}{7}$ **20.** $x = -4, 7$

21. $x = -2, 4$ **22.** $x = -5, 9$ **23.** $x = -2, 0$ **24.** $x = 0, 5$

25. $y = -3, -2, 2$ **26.** $z = -3, -2, 3$ **27.** 14 and 16   **28.** 9 and 24

29. 12 feet by 19 feet   **30.** 2 seconds   **31.** 12 feet by 12 feet

---

Answers to Cumulative Practice Test P–6

1. (a)   **2.** (b)   **3.** (d)   **4.** (c)   **5.** (a)

6. $(x+1)(x-1)$ **7.** $x(x-3)(x+2)$ **8.** $(6x-1)(x-1)$

9. $(x^2+4)(x+2)(x-2)$ **10.** $x(2x-3y)^2$ **11.** $2x^5 - x^3 + x^2$

12. $x^3 - 2x^2 - 2x$ **13.** −2   **14.** $\frac{7}{12}$ **15.** $-2x + 13y$ **16.** 0

17. $x^2 - 5x + 8 + \dfrac{-9}{x+1}$ **18.** $2x + 4 + \dfrac{7}{x}$ **19.** $x^4 + x^3 + 8x^2 + 11x + 21$

20. xy^2 **21.** $x = \frac{28}{15}$ **22.** $y = 12$ **23.** $x > -2$ **24.** $x = 3$

25. $x = -3$ **26.** $x = -2$ **27.** No solution   **28.** $x = 0, \frac{1}{6}, 1$ **29.** $x = -1, -\frac{1}{2}, 1$

30. $x \leq \frac{2}{5}$ **31.** 60   **32.** 8 in. by 17 in.   **33.** 32, 34, 36   **34.** 30 oz

35. $60   **36.** 9%   **37.** 18 dimes, 22 quarters   **38.** 1 hr

CHAPTER SEVEN
Functions and Their Graphs

7.1 Relations, Functions, and Function Notation

Section Highlights

1. Any set of ordered pairs is a *relation*.
2. The set of all first components of a relation is the *domain* of the relation, and the set of all second components is the *range*.
3. A *function* is a relation in which each domain element corresponds to one and only one range element.
4. An equation in two variables x and y represents y as a function of x if all values of x determine only one value of y. Note that $y = f(x)$.
5. **Vertical Line Test** If every vertical line meets the graph of an equation in two variables in at most one point, then the equation represents y as a function of x.
6. If an equation in x and y can be uniquely solved for y, then the equation represents y as a function of x.

EXAMPLE 1 ■ Analyzing a Relation

Determine the domain and range of the relation $\{(-1, 2), (5, 1), (7, -4)\}$.

Solution

The domain and range are $D = \{-1, 5, 7\}$ and $R = \{2, 1, -4\}$.

Starter Exercise 1   *Fill in the blanks*

Determine the domain and range of the relation $\{(-2, 0), (4, 1), (7, -6), (5, 0)\}$.

$D = \left\{ -2, 4, \boxed{}, \boxed{} \right\}$

$R = \left\{ \boxed{}, \boxed{}, \boxed{} \right\}$

EXAMPLE 2 ■ Testing Whether a Relation is a Function

Which of the following relations are functions?

(a) $\{(1, 2), (2, 3), (3, 4)\}$ (b) $\{(1, 2), (1, 3), (1, -1)\}$ (c) $\{(2, 1), (3, 1), (-1, 1)\}$

Solution

(a) Each domain element corresponds one and only range elements, therefore the relation is a function.

(b) The domain element 1 corresponds to three different range elements, therefore the relation is not a function.

(c) Though there is only one range element, each domain element corresponds to one and only one range element, therefore the relation is a function.

> **Starter Exercise 2**

Which of the following relations are functions?

(a) $\{(2, 1), (3, -6), (5, 1)\}$ (b) $\{(-1, 1), (7, 5), (-1, 2)\}$

Solution

(a) Each domain element corresponds to one and only range element. Conclusion: ☐

(b) -1 corresponds to 1 and 2. Conclusion: ☐

EXAMPLE 3 ■ Using the Vertical Line Test

Which of the following represent y as a function of x?

(a) $x - y = 5$ (b) $x = |y|$ (c) $x^2 + y = 4$

Solution

(a) Notice that no vertical lines intersect the graph in more than one point. Hence, the equation *does* represent y as a function of x.

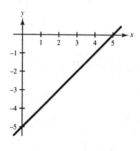

(b) From the graph we can see that many vertical lines meet the graph in more than one point. Hence, the equation *does not* represent y as a function of x. Also notice that $(2, -2)$ and $(2, 2)$ are in the solution set. This means an x-value of 2 determines two different y-values. This also indicates the equation does not represent y as a function of x.

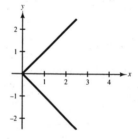

(c) We can see from the graph that no vertical lines meet the graph in more than one point. Hence, the equation represents y as a function of x.

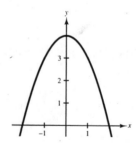

Starter Exercise 3   *Fill in the blanks.*

Graph the following equations to determine which represent y as a function of x.

(a) $y = 2$

x	-2	-1	0	1	2
$y = 2$	2	2	2	$\square$	$\square$
Solution	$(-2,\ 2)$	$(-1,\ 2)$	$(\square,\ \square)$	$(\square,\ \square)$	$(\square,\ \square)$

Conclusion: $\boxed{}$

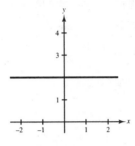

(b) $x = y^2 - 3$

y	-2	-1	0	1	2
$x = y^2 - 3$	1	$\square$	$\square$	$\square$	$\square$
Solution	$(1,\ -2)$	$(-2,\ -1)$	$(\square,\ \square)$	$(\square,\ \square)$	$(\square,\ \square)$

Conclusion: $\boxed{}$

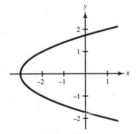

(c) $y = |x - 1|$

x	-1	0	1	2	3
$y = |x - 1|$	2	1	0	$\square$	$\square$
Solution	$(-1,\ 2)$	$(\square,\ \square)$	$(\square,\ \square)$	$(\square,\ \square)$	$(\square,\ \square)$

Conclusion: $\boxed{}$

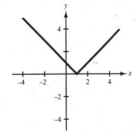

EXAMPLE 4 ■   Testing Whether an Equation is a Function

Which of the following equations represent a function of x?

(a) $2x - 4y = 12$ (b) $x^2 + y^2 = 9$

Solution

(a) Solve for y.

$2x - 4y = 12$ Original equation

$y = \frac{1}{2}x - 3$ Solved for y

In this form, we can see that each value of x would produce a unique value of y. Hence, y is a function of x.

(b) In solving $x^2 + y^2 = 9$ for y we see that $y^2 = 9 - x^2$. Note that if $x = 0$, then $y = 3$ and $y = -3$. Hence y is not a function of x. ■

Starter Exercise 4 | *Fill in the blanks*

Which of the following equations represent a function of x?

(a) $|y| = 2 + x$ (b) $x^2 - y = 4$

Solution

(a) If $x = 1$, then $y = 3$ and $y = \boxed{}$.
   Conclusion: $\boxed{}$

(b) Solve $x^2 - y = 4$ for y obtaining $y = \boxed{}$.
   Conclusion: $\boxed{}$

EXAMPLE 5 ■   Evaluating a Function

Given $f(x) = 2x^2 - 6$, find the following.

(a) $f(0)$ (b) $f(a)$

Solution

(a) $f(x) = 2x^2 - 6$ Given function

$f(0) = 2(0)^2 - 6$ Replace x by 0.

$= 0 - 6$ Simplify.

$= -6$ Simplify.

(b) $f(x) = 2x^2 - 6$ Given function

$f(a) = 2a^2 - 6$ Replace x by a. ■

Starter Exercise 5

Given $g(x) = 12x - 7$, find $g(3)$.

$g(x) = 12x - 7$

$g(3) = 12\boxed{} - 7$

$= \boxed{} - 7$

$= \boxed{}$

Solutions to Starter Exercises ■

1. $D = \left\{ -2, 4, \boxed{7}, \boxed{5} \right\}$

    $R = \left\{ \boxed{0}, \boxed{1}, \boxed{-6} \right\}$

2. (a) The relation is a function.
    (b) The relation is not a function.

3. (a)

x	-2	-1	0	1	2
$y = 2$	2	2	2	2	2
Solution	$(-2, 2)$	$(-1, 2)$	$(0, 2)$	$(1, 2)$	$(2, 2)$

Conclusion: The equation *does* represent y as a function of x.

(b)

y	-2	-1	0	1	2
$x = y^2 - 3$	1	-2	-3	-2	1
Solution	$(1, -2)$	$(-2, -1)$	$(-3, 0)$	$(-2, 1)$	$(1, 2)$

Conclusion: The equation *does not* represent y as a function of x.

(c)

x	-1	0	1	2	3
$y = |x - 1|$	2	1	0	1	2
Solution	$(-1, 2)$	$(0, 1)$	$(1, 0)$	$(2, 1)$	$(3, 2)$

Conclusion: The equation *does* represent y as a function of x.

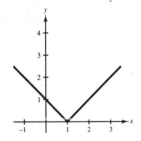

4. (a) $y = \boxed{-3}$

    Conclusion: The equation does not represent a function of x.

    (b) $y = \boxed{x^2 - 4}$

    Conclusion: The equation does not represent a function of x.

5. $g(3) = 12\boxed{3} - 7$

    $= \boxed{36} - 7$

    $= \boxed{29}$

| 7.1 | **EXERCISES** |

In Exercises 1–8, is the relation a function?

1.
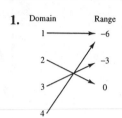

2.
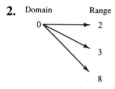

3.
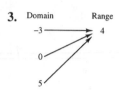

4.
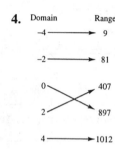

5. $\{(5, 0), (6, -1), (9, 18), (16, 7)\}$

6. $\{(0, 4), (1, 4), (2, 4), (3, 4)\}$

7. $\{(-1, 6), (-1, 8), (-1, 10)\}$

8. $\{(2, -3), (5, 21), (9, 73), (2, 3), (25, 624)\}$

In Exercises 9–16, use the Vertical Line Test to determine whether y is a function of x.

9. $x - y = 2$

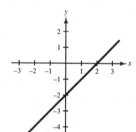

10. $x = 2$

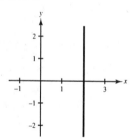

11. $y = x^2 - 2x + 1$

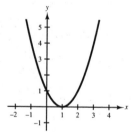

12. $x = y^2 - 2y + 1$

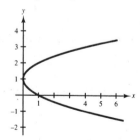

13. $x^2 + y^2 = 16$

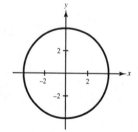

14. $4x^2 + 9y^2 = 36$

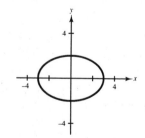

15. $y = |x| + 2$

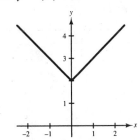

16. $|y| + 2$

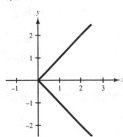

 In Exercises 17–22, use a graphing utility to graph the function and estimate the intercepts.

17. $f(x) = x - 1$ **18.** $g(x) = 5 - 3x$ **19.** $h(x) = |x| - 2$

20. $f(x) = x^2 + 4x + 4$ **21.** $g(x) = x^2 - x - 6$ **22.** $h(x) = x^3 + 3x^2 - 10x$

In Exercises 23–30, evaluate the function

23. $f(x) = 2x + 1$
   (a) $f(0)$
   (b) $f(2)$
   (c) $f(-1)$
   (d) $f\left(\frac{1}{5}\right)$

24. $g(x) = x - 7$
   (a) $g(1)$
   (b) $g(-3)$
   (c) $g(9)$
   (d) $g(0.24)$

25. $h(x) = |x| - 3$
   (a) $h(4)$
   (b) $h(-3)$
   (c) $h(10)$
   (d) $h\left(-\frac{1}{2}\right)$

26. $f(t) = |t - 6|$
   (a) $f(-5)$
   (b) $f(-1)$
   (c) $f(-5)$
   (d) $f\left(\frac{1}{6}\right)$

27. $g(u) = \frac{1}{2}u^3$
   (a) $g(0)$
   (b) $g(z)$
   (c) $g(-1)$
   (d) $g\left(\frac{1}{2}\right)$

28. $h(r) = r^2 + 1$
   (a) $h(3)$
   (b) $h(-5)$
   (c) $h(9)$
   (d) $h(0.63)$

29. $f(v) = v^2 - v - 12$
   (a) $f(4)$
   (b) $f(-3)$
   (c) $f(0)$
   (d) $f(-0.33)$

30. $g(x) = |2x^2 - 3|$
   (a) $g(0)$
   (b) $g(4)$
   (c) $g\left(\frac{1}{2}\right)$
   (d) $g(0.75)$

In Exercises 31–35, determine the range R of the function for the specified domain D.

31. $f(x) = 2x - 1$, $D = \{-2, -1, 0\}$ **32.** $g(x) = -x^2 + 1$, $D = \{-2, -1, 0, 1, 2\}$

33. $h(x) = |x| - 2$, $D = \{-2, -1, 0, 1, 2\}$ **34.** $f(t) = |x - 2|$, $D = \{0, 1, 2, 3, 4\}$

35. $g(t) = x^2 - 4x + 4$, $D = \{0, 1, 2, 3, 4\}$

36. The function $d(t) = 55t$ gives the distance (in miles) that a car will travel in t hours at an average speed of 55 mph. Find the distance traveled for (a) $t = 1$, (b) $t = 3$, (c) $t = 10$.

37. The function $h(t) = -16t^2 + 100$ gives the height (in feet) of an object after falling for t seconds from a height of 100 feet. Find the height of the object for (a) $t = 1$, and (b) $t = 2$.

38. The function $f(p) = 25 - 0.4p$ gives the demand for a product that sells for p dollars per unit. Find the demand for (a) $p = \$2.50$ and (b) $p = \$5$.

39. Find a function that gives the area a of a square in terms of the length of a side s.

40. Find a function that gives the circumference of a circle in terms of its radius r.

7.2 | Slope and Graphs of Linear Functions

Section Highlights

1. The **slope** m of the nonvertical line passing through the points (x_1, y_1) and (x_2, y_2) is

$$m = \frac{y_2 - y_1}{x_2 - x_1} = \frac{\text{Change in } y}{\text{Change in } x}.$$

   where $x_1 \neq x_2$. Vertical lines do not have slope.
2. The graph of the equation $y = mx + b$ is a line whose slope is m and whose y-intercept is $(0, b)$.
3. We only need to know two points on a line to graph that line.
4. Two distinct nonvertical lines are parallel if and only if they have the same slope.
5. Two nonvertical lines with slopes m_1 and m_2 are perpendicular if and only if $m_1 = -1/m_2$.

EXAMPLE 1 ■   Finding the Slope of a Line Passing Through Two Points

Find the slopes of the lines passing through the following pairs of points.

(a) $(2, 4)$, $(-1, 1)$ (b) $(-1, 3)$, $(3, -4)$ (c) $(2, 1)$, $(4, 1)$

Solution

(a) $m = \dfrac{y_2 - y_1}{x_2 - x_1} = \dfrac{1 - 4}{-1 - 2} = \dfrac{-3}{-3} = 1$ (b) $m = \dfrac{y_2 - y_1}{x_2 - x_1} = \dfrac{-4 - 3}{3 - (-1)} = \dfrac{-7}{4} = -\dfrac{7}{4}$

(c) $m = \dfrac{y_2 - y_1}{x_2 - x_1} = \dfrac{1 - 1}{4 - 2} = \dfrac{0}{2} = 0$

Starter Exercise 1 | *Fill in the blanks.*

Find the slopes of the lines passing through the following points.

(a) Points: $(3, 3)$, $(-1, 2)$

$$m = \frac{y_2 - y_1}{x_2 - x_1} = \frac{2 - \Box}{-1 - \Box} = \frac{-1}{-4} = \frac{\Box}{\Box}$$

(b) Points: $(2, 4)$, $(-3, 4)$

$$m = \frac{y_2 - y_1}{x_2 - x_1} = \frac{\Box - \Box}{\Box - \Box} = \frac{0}{-5} = \Box$$

EXAMPLE 2 ■   Using Slope to Describe Lines

Use slope to determine whether the line through each of the following pairs of points rises, falls, is horizontal, or is vertical.

(a) $(1, 1)$, $(2, 4)$ (b) $(1, 2)$, $(1, 4)$

(c) $(-1, -1)$, $(1, -6)$ (d) $(-4, 3)$, $(1, 3)$

Solution

(a) Since

$$m = \frac{4-1}{2-1} = \frac{3}{1} = 3 > 0$$

the line rises from left to right.

(c) Since

$$m = \frac{-6-(-1)}{1-(-1)} = \frac{-5}{2} < 0$$

the line falls from left to right.

(b) Since

$$m = \frac{4-2}{1-1} = \frac{2}{0}$$

which is undefined, the line is vertical.

(d) Since

$$m = \frac{3-3}{1-(-4)} = \frac{0}{-5} = 0$$

the line is horizontal.

Starter Exercise 2   *Fill in the blanks.*

Use the slope to determine whether the line through each of the following pairs of points rises, falls, is horizontal, or vertical.

(a) Points: $(1,\ 2),\ (-1,\ 2)$

$$m = \frac{\Box - \Box}{\Box - \Box} = \frac{0}{\Box} = 0$$

Horizontal

(b) Points: $(2, 4),\ (4, 2)$

$$m = \frac{2-4}{4-2} = \frac{\Box}{\Box} = -1 < 0$$

$\boxed{}$

(c) Points: $(-3,\ -1),\ (0,\ 2)$

$$m = \frac{2-\Box}{0-\Box} = \frac{\Box}{\Box} = \Box > 0$$

$\boxed{}$

(d) Points: $(-2,\ 2),\ (-2,\ -2)$

$$m = \frac{\Box - \Box}{\Box - \Box} = \frac{\Box}{\Box} \quad \text{Undefined}$$

$\boxed{}$

EXAMPLE 3 ■   Finding the Slope of a Line

Find the slope and y-intercept of the line given by $3x + 2y = 6$.

Solution

We start by solving the equation for y.

$$3x + 2y = 6$$
$$2y = -3x + 6$$
$$y = -\tfrac{3}{2}x + 3$$

Now we compare this equation to $y = mx + b$ (slope-intercept form).

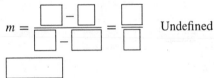

We can see that $m = -\tfrac{3}{2}$ and $b = 3$. Thus, the slope is $-\tfrac{3}{2}$ and the y-intercept is $(0, 3)$.

■ **Starter Exercise 3** | *Fill in the blanks.*

Find the slope and y-intercept of the line given by $4x - 2y = 2$. First solve the equation for y.

$$4x - 2y = 2$$
$$-2y = \boxed{}\,x + 2$$
$$y = \boxed{}\,x - \boxed{}, \quad m = \boxed{}, \quad b = \boxed{}$$

Slope is $\boxed{}$ and y-intercept is $\left(\boxed{}, \boxed{}\right)$.

EXAMPLE 4 ■ Using the Slope and y-intercept to Sketch a Line

Use the slope and y-intercept to sketch the graph of $5x - 3y = 9$.

Solution

First we find the slope-intercept of the equation.

$$5x - 3y = 9$$
$$-3y = -5x + 9$$
$$y = \tfrac{5}{3}x - 3$$

Now we note the slope is $\tfrac{5}{3}$ and the y-intercept is $(0, -3)$. To sketch the graph we:

i) Plot $(0, -3)$.

ii) Use the slope to locate a second point on the line by moving from $(0, -3)$ up 5 units and to the right 3 units to the point $(3, 2)$.

iii) Draw the line through these two points.

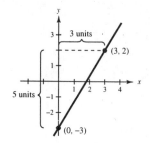

■ **Starter Exercise 4** | *Fill in the blanks.*

Use the slope and y-intercept to sketch the graph of $y = -\tfrac{4}{3}x + 1$.

The slope is $\boxed{}$.

The y-intercept is $\left(\boxed{}, \boxed{}\right)$.

Plot $(0, 1)$ and from there go down 4 units and to the right 3 units. Note that we went *down* 4 units since the slope is negative.

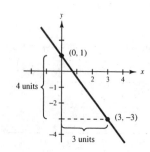

EXAMPLE 5 ■   Determining Whether Lines are Parallel or Perpendicular

Determine whether the following pairs of lines are parallel, perpendicular, or neither.

(a) Line 1: $y = 2x$

   Line 2: $y = 2x + 4$

(b) Line 1: $y = \frac{1}{3}x + 1$

   Line 2: $y = -3x - 3$

Solution

(a) Line 1 has slope 2 and Line 2 also has slope 2. Therefore, the two lines are parallel since they have the same slope.

(b) Line 1 has slope $\frac{1}{3}$ and Line 2 has slope -3. Since these two slopes are negative reciprocals of each other, they are perpendicular.   ■

■ **Starter Exercise 5** | *Fill in the blanks.*

Determine whether the following pairs of lines are parallel, perpendicular, or neither.

(a) Line 1: $y = \frac{3}{4}x + 3$

   Line 2: $y = -\frac{4}{3}x - 1$

   Slope of Line 1 is ☐ .

   Slope of Line 2 is ☐ .

   The two lines are ☐ .

(b) Line 1: $y = 2x + 3$

   Line 2: $x - 2y = 4$

   Slope of Line 1 is ☐ .

   Slope of Line 2 is ☐ .

   The two lines are ☐ .

■   **Solutions to Starter Exercises**   ■

1. (a) $m = \dfrac{y_2 - y_1}{x_2 - x_1} = \dfrac{2 - \boxed{3}}{-1 - \boxed{3}} = \dfrac{-1}{-4} = \dfrac{\boxed{1}}{4}$

(b) $m = \dfrac{y_2 - y_1}{x_2 - x_1} = \dfrac{\boxed{4} - \boxed{4}}{\boxed{-3} - \boxed{2}} = \dfrac{0}{-5} = \boxed{0}$

2. (a) $m = \dfrac{\boxed{2} - \boxed{2}}{\boxed{-1} - \boxed{1}} = \dfrac{0}{\boxed{-2}} = 0$

   Horizontal

(b) $m = \dfrac{2 - 4}{4 - 2} = \dfrac{\boxed{-2}}{\boxed{2}} = -1 < 0$

   $\boxed{\text{Falls}}$

(c) $m = \dfrac{2 - \boxed{(-1)}}{0 - \boxed{(-3)}} = \dfrac{\boxed{3}}{3} = \boxed{1} > 0$

   $\boxed{\text{Rises}}$

(d) $m = \dfrac{\boxed{-2} - \boxed{2}}{\boxed{-2} - \boxed{(-2)}} = \dfrac{\boxed{-4}}{0}$ Undefined

   $\boxed{\text{Vertical}}$

3. $-2y = \boxed{-4}\, x + 2$

   $y = \boxed{2}\, x - \boxed{1}$, $m = \boxed{2}$, $b = \boxed{-1}$

   Slope is $\boxed{2}$ and y-intercept is $\left(\boxed{0} , \boxed{-1} \right)$

■ **Solutions to Starter Exercises** ■

4. The slope is $\boxed{-\frac{4}{3}}$, the y-intercept is $\left(\boxed{0}, \boxed{1}\right)$.

5. (a) Slope of Line 1 is $\boxed{\frac{3}{4}}$.

Slope of Line 2 is $\boxed{-\frac{4}{3}}$.

The two lines are $\boxed{\text{perpendicular}}$.

(b) Slope of Line 1 is $\boxed{2}$.

Slope of Line 2 is $\boxed{\frac{1}{2}}$.

The two lines are $\boxed{\text{neither parallel nor perpendicular}}$.

7.2 | EXERCISES

In Exercises 1–4, estimate the slope of the given line from the graph.

1.

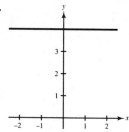

2.

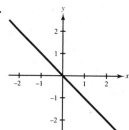

3.

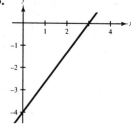

4.

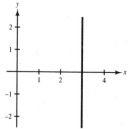

In Exercises 5–14, find the slope of the line that passes through the given points.

5. $(0, 0)$, $(2, 4)$

6. $(3, -6)$, $(9, 1)$

7. $(-1, -1)$, $(-2, -3)$

8. $(5, 7)$, $(7, 7)$

9. $(0, 1)$, $\left(\frac{1}{2}, \frac{3}{4}\right)$

10. $(0, 2)$, $(3, 0)$

11. $(-3, 2)$, $(-3, 7)$

12. $\left(\frac{1}{4}, -\frac{1}{2}\right)$, $\left(-\frac{1}{3}, \frac{5}{6}\right)$

13. $(-3, -2)$, (a, b), $b \neq -2$

14. (a, b), (c, d), $b \neq d$

In Exercises 15–18, find the unknown coordinate so that the line through the two points will have the given slope.

Points	Slope		Points	Slope

15. $(1, 1)$, $(x, 2)$ $m = 1$ **16.** $(-2, y)$, $(4, 1)$ $m = -2$

17. $(3, 2)$, $(0, y)$ $m = -\frac{2}{3}$ **18.** $(x, 8)$, $(-4, 6)$ $m = -\frac{1}{4}$

In Exercises 19–22, sketch the graph of the line through the point $(0, 3)$ having the given slope.

19. $m = \frac{2}{3}$ **20.** m is undefined.   **21.** $m = 0$ **22.** $m = -2$

In Exercises 23–28, write the equation in slope-intercept form. Use the slope and y-intercept to sketch the graph of the line.

23. $5x - 3y + 9 = 0$ **24.** $2x + 2y - 6 = 0$ **25.** $3x + y + 5 = 0$

26. $x + y = 0$ **27.** $y + 2 = 0$ **28.** $x - 2y + 4 = 0$

In Exercises 29 and 30, determine whether the lines L_1 and L_2, passing through the given pairs of points, are parallel, perpendicular, or neither.

29. L_1: $(2, 1)$, $(4, 2)$
   L_2: $(-1, -3)$, $(1, -2)$

30. L_1: $(1, -2)$, $(5, 6)$
   L_2: $(-1, -1)$, $(3, 1)$

In Exercises 31–32, determine whether the lines given by the following equations are parallel, perpendicular, or neither. Use a graphing utility to graph both equations to help verify your response. Note: you should use the "square" setting.

31. L_1: $y = 2x + 3$
   L_2: $y = -\frac{1}{2}x - 3$

32. L_1: $y = 4x - 3$
   L_2: $y = 4x + 7$

7.3 | Equations of Lines

Section Highlights

1. The **point-slope** form of the equation of the line with slope m passing through the point (x_1, y_1) is $y - y_1 = m(x - x_1)$.
2. The **general form** of a linear equation is $ax + by + c = 0$ where a and b are not both zero.
3. An equation of a vertical line through the point $(a, 0)$ is $x = a$.
4. An equation of a horizontal line through the point $(0, b)$ is $y = b$.

EXAMPLE 1 ■ The Point-Slope Form of the Equation of a Line

Find an equation of the line with slope -2 and passing through the point $(-1, 2)$.

Solution

Using the point-slope form with $x_1 = -1$, $y_1 = 2$, and $m = -2$, we have the following.

$$y - y_1 = m(x - x_1) \qquad \text{Point-slope form}$$
$$y - 2 = -2(x - (-1)) \qquad \text{Substitute } y_1 = 2, \ x_1 = -1 \text{ and } m = -2.$$
$$y - 2 = -2x - 2$$
$$y = -2x \qquad \text{Equation of line}$$

■

Starter Exercise 1 *Fill in the blanks.*

Find an equation of a line with slope $\frac{1}{2}$ and passing through the point $(2, 3)$.

$$x_1 = \boxed{}, \quad y_1 = \boxed{}, \quad m = \boxed{}$$
$$y - \boxed{} = \boxed{}\left(x - \boxed{}\right)$$
$$y - \boxed{} = \boxed{}x - \boxed{}$$
$$y = \boxed{}x + \boxed{}$$

EXAMPLE 2 ■ Finding an Equation of a Line Passing Through Two Points

Find an equation of the line that passes through the points $(-2, 1)$ and $(3, 6)$.

Solution

We first find the slope

$$m = \frac{6 - 1}{3 - (-2)} = \frac{5}{5} = 1.$$

Now, using point-slope form, we find the equation of the line.

$$y - y_1 = m(x - x_1)$$
$$y - 1 = 1(x - (-2))$$
$$y - 1 = x + 2$$
$$y = x + 3$$

■

Starter Exercise 2 *Fill in the blanks.*

Find an equation of the line that passes through the points $(-1, 4)$ and $(2, -5)$.

$$m = \frac{-5 - \boxed{}}{2 - \boxed{}} = \frac{\boxed{}}{\boxed{}} = \boxed{}$$
$$y - y_1 = m(x - x_1)$$
$$y - \boxed{} = \boxed{}\left(x - \boxed{}\right)$$
$$y - \boxed{} = \boxed{}x - \boxed{}$$
$$y = \boxed{}x + \boxed{}$$

EXAMPLE 3 ■ Writing Equations of Horizontal and Vertical Lines

Write an equation for each of the following lines.

(a) Line passes through $(1, 2)$ and $(1, 4)$

(b) Line passes through $(2, -3)$ and $(0, -3)$

Solution

(a) The line that passes through $(1, 2)$ and $(1, 4)$ is vertical. Thus, its equation is $x = 1$.

(b) The line that passes through $(2, -3)$ and $(0, -3)$ is horizontal. Thus, its equation is $y = -3$. ■

■ Starter Exercise 3 | *Fill in the blanks.*

Write an equation for each of the following lines.

(a) Line passes through $(-1, 2)$ and $(2, 2)$. This line is [＿＿＿] (vertical or horizontal).

$y = \boxed{}$

(b) Line passes through $(-1, 2)$ and $(-1, 4)$. This line is [＿＿＿].

$\boxed{} = \boxed{}$

EXAMPLE 4 ■ Equations of Parallel Lines

Find an equation of the line that passes through the point $(-1, 3)$ and is parallel to the line $3x - 4y + 1 = 0$.

Solution

We start by writing the given equation in slope-intercept form.

$$3x - 4y + 1 = 0 \qquad \text{Given equation}$$
$$3x - 4y = -1$$
$$-4y = -3x - 1$$
$$y = \tfrac{3}{4}x + \tfrac{1}{4}$$

Therefore, the given equation has slope $m = \tfrac{3}{4}$. Since any line parallel to the given line also has slope $m = \tfrac{3}{4}$, the required line through $(-1, 3)$ has the following equation.

$$y - y_1 = m(x - x_1)$$
$$y - 3 = \tfrac{3}{4}(x - (-1))$$
$$y - 3 = \tfrac{3}{4}x + \tfrac{3}{4}$$
$$y = \tfrac{3}{4}x + \tfrac{15}{4}$$

■

Starter Exercise 4 *Fill in the blanks.*

Find an equation of the line that passes through $(0, -1)$ and is parallel to the line $2x - y = 1$.

$$2x - y = 1$$
$$-y = \boxed{} + 1$$
$$y = \boxed{} - 1$$
$$m = \boxed{}$$
$$y - y_1 = m(x - x_1)$$
$$y - \boxed{} = \boxed{}(x - 0)$$
$$y + \boxed{} = \boxed{}x$$
$$y = \boxed{}x - \boxed{}$$

EXAMPLE 5 ■ Equations of Perpendicular Lines

Find an equation of the line that passes through the point $(-1, 1)$ and is perpendicular to the line $3x + 5y = 2$.

Solution

We start by finding the slope of the given line.

$$3x + 5y = 2$$
$$5y = -3x + 2$$
$$y = -\tfrac{3}{5}x + \tfrac{2}{5}$$

Thus, the slope of the given line is $-\tfrac{3}{5}$. Hence, any line perpendicular to this line must have slope $\tfrac{5}{3}$. Therefore, the equation of the line through $(-1, 1)$ and perpendicular to $3x + 5y = 2$ is found as follows.

$$y - y_1 = m(x - x_1)$$
$$y - 1 = \tfrac{5}{3}(x - (-1))$$
$$y - 1 = \tfrac{5}{3}x + \tfrac{5}{3}$$
$$y = \tfrac{5}{3}x + \tfrac{8}{3}$$

Starter Exercise 5 *Fill in the blanks.*

Find the equation of the line that passes through the point $(3, -1)$ and is perpendicular to the line $2x - y = 3$. The slope of the given line is $\boxed{}$. The slope of any line perpendicular to the given line is $\boxed{}$. The equation is

$$y - y_1 = m(x - x_1)$$
$$y - \boxed{} = \boxed{}\left(x - \boxed{}\right)$$
$$y + \boxed{} = \boxed{}x + \boxed{}$$
$$y = \boxed{}x + \boxed{}.$$

EXAMPLE 6 ■ An Application: Linear Interpolation

A business purchases a photocopier for $5000. It is estimated that after four years, its depreciated value will be $1000. Assuming straight-line depreciation, find the depreciated value of the photocopier after two years.

Solution

We will use a linear model, with y representing the value of the photocopier and t representing years. We start by finding the slope using the points $(0, 5000)$ and $(4, 1000)$.

$$m = \frac{1000 - 5000}{4 - 0} = -1000$$

Now, using point-slope form, we find the equation of the line as follows.

$$y - y_1 = m(t - t_1)$$
$$y - 5000 = -1000(t - 0)$$
$$y - 5000 = -1000t$$
$$y = -1000t + 5000$$

Finally, we estimate the value of the photocopier after two years ($t = 2$) to be $y = -1000(2) + 5000 = \$3000$. ■

Starter Exercise 6 | An Application—Fill in the blanks.

In a month in which Nancy's total sales were $20,000, her pay was $2400. In a month in which her total sales were $18,000, her pay was $2260. Assuming the relationship between sales and pay is linear, find Nancy's pay in a month if her sales were $22,000.

$$m = \frac{2400 - 2260}{20,000 - \boxed{}} = \boxed{}$$

$$y - y_1 = m(x - x_1)$$
$$y - 2400 = \boxed{}\left(x - \boxed{}\right)$$
$$y - 2400 = \boxed{} - \boxed{}$$
$$y = \boxed{} + \boxed{}$$

Now replace x by $\boxed{}$.

$$y = \boxed{} + \boxed{} = \boxed{}$$

■ **Solutions to Starter Exercises** ■

1. $x_1 = \boxed{2}$, $y_1 = \boxed{3}$, $m = \boxed{\frac{1}{2}}$

$$y - \boxed{3} = \boxed{\tfrac{1}{2}}\left(x - \boxed{2}\right)$$
$$y - \boxed{3} = \boxed{\tfrac{1}{2}}x - \boxed{1}$$
$$y = \boxed{\tfrac{1}{2}}x + \boxed{2}$$

■ Solutions to Starter Exercises ■

2.
$$m = \frac{-5 - \boxed{4}}{2 - \boxed{(-1)}} = \frac{\boxed{-9}}{\boxed{3}} = \boxed{-3}$$

$$y - \boxed{4} = \boxed{-3}\left(x - \boxed{(-1)}\right)$$

$$y - \boxed{4} = \boxed{-3}x - \boxed{3}$$

$$y = \boxed{-3}x + \boxed{1}$$

3. (a) This line is $\boxed{\text{horizontal}}$.

$$y = \boxed{2}$$

(b) This line is $\boxed{\text{vertical}}$.

$$\boxed{x} = \boxed{-1}$$

4.
$$-y = \boxed{-2x} + 1$$

$$y = \boxed{2x} - 1$$

$$m = \boxed{2}$$

$$y - \boxed{(-1)} = \boxed{2}(x - 0)$$

$$y + \boxed{1} = \boxed{2}x$$

$$y = \boxed{2}x - \boxed{1}$$

5. The slope of the given line is $\boxed{2}$. The slope of any line perpendicular to the given line is $\boxed{-\tfrac{1}{2}}$. The equation is

$$y - \boxed{(-1)} = \boxed{-\tfrac{1}{2}}\left(x - \boxed{3}\right)$$

$$y + \boxed{1} = \boxed{-\tfrac{1}{2}}x + \boxed{\tfrac{3}{2}}$$

$$y = \boxed{-\tfrac{1}{2}}x + \boxed{\tfrac{1}{2}}.$$

6.
$$m = \frac{2400 - 2260}{20,000 - \boxed{18,000}} = \boxed{0.07}$$

$$y - 2400 = \boxed{0.07}\left(x - \boxed{20,000}\right)$$

$$y - 2400 = \boxed{0.07x} - \boxed{1400}$$

$$y = \boxed{0.07x} + \boxed{1000}$$

Now replace x by $\boxed{22,000}$.

$$y = \boxed{0.07(22,000)} + \boxed{1000} = \boxed{\$2540}$$

7.3 EXERCISES

In Exercises 1–6, find the slope of the given line.

1. $y - 1 = 2(x - 3)$

2. $y = \frac{1}{2}x + \frac{7}{8}$

3. $y = 5$

4. $x + 3 = 0$

5. $5x - 2y + 1 = 0$

6. $3x + 4y - 6 = 0$

In Exercises 7–10, find the equation of the line that passes through the given point and has the specified slope. (Write your answer in slope-intercept form.)

	Point	Slope			Point	Slope
7.	$(0, 0)$	$m = 3$		**8.**	$(-1, 2)$	$m = -1$
9.	$\left(0, \frac{7}{8}\right)$	$m = -\frac{1}{2}$		**10.**	$\left(\frac{1}{2}, \frac{3}{4}\right)$	$m = -2$

In Exercises 11–20, find an equation of the line through the given points. (Write your answer in general form.)

11. $(2, 4)$, $(1, 3)$ **12.** $(-1, -1)$, $(0, -1)$ **13.** $(-2, 1)$, $(-1, 6)$

14. $(2, -3)$, $(3, -2)$ **15.** $(1, 4)$, $(1, -4)$ **16.** $(-1, 2)$, $(2, 5)$

17. $(5, 9)$, $(-1, -11)$ **18.** $(0, 0)$, $(2, 1)$ **19.** $\left(\frac{1}{2}, 1\right)$, $\left(2, \frac{3}{4}\right)$

20. $\left(-\frac{1}{4}, \frac{1}{2}\right)$, $\left(\frac{1}{3}, \frac{3}{4}\right)$

In Exercises 21–24, find an equation of the line passing through the given point with the specified slope. (Write your answer in general form.)

21. $(0, 2)$; $m = 3$ **22.** $(2, -1)$; $m = -2$

23. $(2, 4)$; $m = 0$ **24.** $(2, 4)$; m is undefined.

In Exercises 25–28, write an equation of the line through the indicated point (a) parallel to the given line, and (b) perpendicular to the given line.

	Point	Line			Point	Line
25.	$(0, 1)$	$y = 2x - 1$		**26.**	$(2, 3)$	$2x - 4y = 7$
27.	$(5, 1)$	$y - 4 = 0$		**28.**	$(2, -1)$	$x - 6 = 0$

29. The relationship between Fahrenheit and Celsius temperature scales is linear. If 32° Fahrenheit is 0° Celsius and 212° Fahrenheit is 100° Celsius, find an equation describing this relationship.

30. A sales representative receives a salary of $1000 per month plus 7% commission of the total monthly sales. Write a linear equation giving the wages, W, in terms of the sales, S.

31. In a 100-unit apartment complex, all 100 units are rented when the rent is $600. However, when the rent is $625, the number of units rented drops to 95. Assume the relationship between the monthly rent and the number of units rented is linear.

(a) Write an equation giving the number of units rented, y, in terms of the rent, x.

(b) Use this equation to predict the number of units rented if the rent is $650.

(c) Use this equation to predict the number of units rented if the rent is $610.

In Exercises 32 and 33, use a graphing calculator to sketch the graphs of the pair of equations. Set the zoom feature to the *square* setting. Then determine whether the lines are parallel, perpendicular, or neither.

32. $y = \dfrac{x - 1}{2}$

$y = \dfrac{x + 1}{2}$

33. $y = -0.2x + 0.07$

$y = 5x + 2$

7.4 Graphs of Linear Inequalities

Section Highlights

1. The graph of a linear inequality in two variables is a half-plane.
2. To graph a linear inequality in two variables:

   (a) Replace the inequality sign by an equal sign, and sketch the graph of the resulting equation. (Use a dashed line for < or > and a solid line for ≤ or ≥.)

   (b) In the original inequality, test one point from each half-plane formed by the graph in Step (a). If the point satisfies the inequality, then shade the entire half-plane to denote that every point in the region satisfies the inequality.

EXAMPLE 1 ■ Verifying Solutions of Linear Inequalities

Determine whether the following points are solutions of the linear inequality $2x + 3y < 1$.

(a) (1, 1) (b) (−1, −1) (c) (0, 0)

Solution

(a) Substitute (1, 1) into the given inequality.

$$2x + 3y < 1 \qquad \text{Given inequality}$$

$$2(1) + 3(1) \overset{?}{<} 1$$

$$5 \not< 1$$

Since the point (1, 1) does not satisfy the inequality, (1, 1) is not a solution of the inequality.

(b)
$$2x + 3y < 1$$
$$2(-1) + 3(-1) \overset{?}{<} 1$$
$$-5 < 1$$
(−1, −1) is a solution.

(c)
$$2x + 3y < 1$$
$$2(0) + 3(0) \overset{?}{<} 1$$
$$0 < 1$$
(0, 0) is a solution.

Starter Exercise 1 *Fill in the blanks.*

Determine whether the following points are solutions of the inequality $3x - y \geq 4$.

(a) Point: (1, −2)

$$3x - y \geq 4$$
$$3(1) - \boxed{} \overset{?}{\geq} 4$$
$$\boxed{} \geq 4$$

Conclusion: $\boxed{}$.

(b) Point: (1, −1)

$$3x - y \geq 4$$
$$3\boxed{} - \boxed{} \overset{?}{\geq} 4$$
$$\boxed{} \geq 4$$

Conclusion: $\boxed{}$.

(c) Point: (0, 0)

$$3x - y \geq 4$$
$$3\boxed{} - \boxed{} \overset{?}{\geq} 4$$
$$\boxed{} \geq 4$$

Conclusion: $\boxed{}$.

EXAMPLE 2 ■ Sketching the Graph of a Linear Inequality

Sketch the graphs of the following linear inequalities.

(a) $y \leq 2$

(b) $2x - y > 1$

Solution

(a) The graph of the corresponding equation, $y = 2$, is a horizontal line. Note that the point $(0, 0)$ satisfies the original inequality. Hence, the points that satisfy the inequality are those that lie on or below the line as shown at the right.

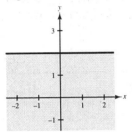

(b) First we graph the equation $2x - y = 1$ using a dashed line. Since $(0, 0)$ does not satisfy the inequality, but $(2, 0)$ does, we shade the half-plane below the line.

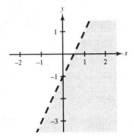

Starter Exercise 2 | *Fill in the blanks.*

Sketch the graphs of the following inequalities.

(a) $x > -2$

First sketch the line given by the equation [＿＿＿], using a [＿＿＿] line. Then shade [＿＿＿] the line.

(b) $3x + 2y \leq -4$

First sketch the line given by the equation [＿＿＿], using a [＿＿＿] line. Then shade [＿＿＿] the line.

▦ EXAMPLE 3 ■ A Graphing Calculator Problem

Use a graphing calculator to sketch the graph of $y \leq x + 3$.

Solution

First graph the equation $y = x + 3$. Since $(0, 0)$ satisfies the original inequality, we want to shade below the line. To shade this portion of the screen, press [DRAW], move cursor to **Shade (**, press [ENTER], and enter: Shade $(-10, x + 3)$ by pressing [(−)] [10] [ALPHA] [,] [X|T] [+] 3. Then press [ENTER].

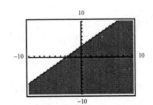

▦ **Starter Exercise 3** | *Fill in the blanks.*

Use a graphing calculator to sketch the graph of $y \geq -x + 1$. First graph the equation, then to shade [＿＿＿] the line, use these keystrokes: [DRAW], move cursor to **Shade (**, [ENTER] [(−)] [X|T] [＿] 1 [ALPHA] [＿] 10 [ENTER] .

■ Solutions to Starter Exercises ■

1. (a) $3(1) - \boxed{(-2)} \overset{?}{\geq} 4$

$\boxed{5} \geq 4$

Conclusion:

$\boxed{(1, -2) \text{ is a solution}}$.

(b) $3\boxed{(1)} - \boxed{(-1)} \overset{?}{\geq} 4$

$\boxed{4} \geq 4$

Conclusion:

$\boxed{(1, -1) \text{ is a solution}}$.

(c) $3\boxed{(0)} - \boxed{0} \overset{?}{\geq} 4$

$\boxed{0} \not\geq 4$

Conclusion:

$\boxed{(0, 0) \text{ is not a solution}}$.

2. (a) First sketch the line given by the equation $\boxed{x = -2}$, using a $\boxed{\text{dashed}}$ line. Then shade $\boxed{\text{right of}}$ the line.

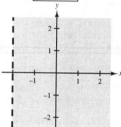

(b) First sketch the line given by the equation $\boxed{3x + 2y = -4}$, using a $\boxed{\text{solid}}$ line. Then shade $\boxed{\text{below}}$ the line.

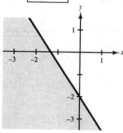

3. First graph the equation, then to shade $\boxed{\text{above}}$ the line, use these keystrokes: $\boxed{\text{DRAW}}$, move cursor to **Shade (,**

$\boxed{\text{ENTER}}$ $\boxed{(-)}$ $\boxed{\text{X|T}}$ $\boxed{+}$ 1 $\boxed{\text{ALPHA}}$ $\boxed{,}$ 10 $\boxed{\text{ENTER}}$.

7.4 | EXERCISES

In Exercises 1 and 2, determine whether the points are solutions of the inequality.

Inequality   *Points*

1. $x + y \leq 3$ (a) $(0, 0)$
           (b) $(2, -1)$
           (c) $(1, 2)$
           (d) $\left(3, \frac{1}{2}\right)$

Inequality   *Points*

2. $5x - 3y > 6$ (a) $(0, 0)$
            (b) $(2, -2)$
            (c) $(5, -1)$
            (d) $(2, 2)$

In Exercises 3 and 4, state whether the boundary of the graph of the given inequality should be dashed or solid.

3. $2x + 3y \geq 6$

4. $x - 7y < 5$

In Exercises 5–8, match the given inequality with its graph.

5. $x \geq 1$ **6.** $y < 4$ **7.** $2x + y \leq 2$ **8.** $x - y \geq 2$

(a)

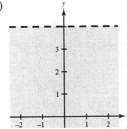

(b)

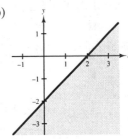

(c)

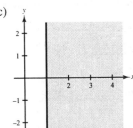

(d)

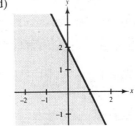

In Exercises 9–22, sketch the graph of the inequality.

9. $x \geq 2$ **10.** $y < 3$ **11.** $x - y < 1$ **12.** $x + y \geq -2$

13. $2x + 3y < 6$ **14.** $x - 2y \geq 4$ **15.** $y < -2x + 1$ **16.** $y \leq \frac{1}{2}x - 3$

17. $-x - 3y - 6 > 9$ **18.** $2x + y - 7 \leq -5$ **19.** $x \leq 2y + 4$

20. $6 + 2y > x$ **21.** $\dfrac{x}{2} + \dfrac{y}{3} < 2$ **22.** $\dfrac{x}{4} - \dfrac{y}{2} \geq 1$

In Exercises 23–26, find an inequality that represents the graph.

23.

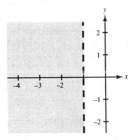

24.

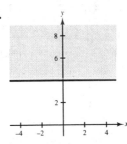

25.

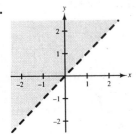

26.

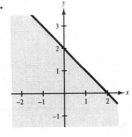

In Exercises 27 and 28, use a graphing calculator to sketch the graph of the inequality.

27. $y \leq -\frac{1}{2}x + 2$ **28.** $y \geq 2x - 1$

Cumulative Practice Test for Chapters P–7

In Exercises 1–6, determine whether the statement is true or false.

1. $\dfrac{x}{0} = 0$

2. $x^m \cdot x^n = x^{mn}$

3. $\dfrac{x+2}{x+3} = \dfrac{2}{3}$

4. $a \cdot (b \cdot c) = (a \cdot b) \cdot (a \cdot c)$

5. $2a + 2b = 2b + 2a$ is an application of the Commutative Property of Addition.

6. $3x + 6x = (3+6)x$ is an application of the Distributive Property.

In Exercises 7–16, simplify the given expression.

7. $(2x + 3y) + 6x$

8. $\left(\dfrac{x}{2} + \dfrac{1}{2}\right)^2$

9. $2[x - (2x + 1)]$

10. $-|-3|^2 + \dfrac{2^2 - 1}{2^3 - 5}$

11. $(x + 2)(x - 6)$

12. $x[x - (2x + 3)] - (x + 1)^2$

13. $\left(\dfrac{a^2 b^3}{a^{-4}}\right)^{-2}$

14. $(x - 3)(x + 3)$

15. $(2x^3 + 6x + 1) \div (x - 1)$

16. $\dfrac{2x^2 - 3x + 2}{x}$

In Exercises 17–22, factor the given polynomial.

17. $3x^2 + 4x$

18. $x^2 + 6x + 8$

19. $16x^2 - 25$

20. $8x^2 + 10x - 3$

21. $x^2 - 6x + 9$

22. $6x^3 + 22x^2 - 8x$

In Exercises 23–30, solve the given equation or inequality.

23. $2x + 7 = 5$

24. $3(x - 2) = 4(x + 1)$

25. $x^2 - 16 = 0$

26. $x^2 - x - 6 = 0$

27. $2(5 - x) \le 12$

28. $2x^2 + 8x + 8 = 0$

29. $(x - 2)^2 - 16 = 0$

30. $2(x - 6) = 3(x + 3)$

31. Company A charges $15 a day and $0.15 per mile for a rental car. Company B charges $45 a day and unlimited mileage for their rental cars. How many miles can you drive a car from Company A in one day and have it cost less than a Company B car?

32. The sum of three consecutive odd integers is 105. Find the integers.

33. The area of a rectangular patio is 280 square feet. The length of the patio is eight feet less than twice its width. Find the dimensions of the patio.

34. How much pure water must be added to 60 liters of an 8% acid solution to make a 6% acid solution?

In Exercises 35–37, which equations represent functions of x?

35. $2x - 3y = 6$ **36.** $x^2 + y^2 = 1$ **37.** $x^2 - y = 4$

In Exercises 38–40, use a graphing utility to estimate the intercepts of the following functions.

38. $f(x) = x + 1$ **39.** $g(x) = |x|$ **40.** $h(x) = x^2 + 8x + 12$

| 7.1 | **Answers to Exercises** |

1. Function   **2.** Not a function   **3.** Function   **4.** Function

5. Function   **6.** Function   **7.** Not a function   **8.** Not a function

9. Function   **10.** Not a function   **11.** Function   **12.** Not a function

13. Not a function   **14.** Not a function   **15.** Function   **16.** Not a function

17. $(0, -1), (1, 0)$ **18.** $(0, 5), \left(\frac{5}{3}, 0\right)$ **19.** $(0, -2), (2, 0), (-2, 0)$

20. $(0, 4), (-2, 0)$ **21.** $(0, -6), (3, 0), (-2, 0)$ **22.** $(0, 0), (-5, 0), (2, 0)$

23. (a) 1 (b) 5 (c) -1 (d) $\frac{7}{5}$ **24.** (a) -6 (b) -10 (c) 2 (d) -6.76 **25.** (a) 1 (b) 0 (c) 7 (d) $-\frac{5}{2}$ **26.** (a) 5 (b) 7 (c) 11 (d) $\frac{35}{6}$

27. (a) 0 (b) 4 (c) $-\frac{1}{2}$ (d) $\frac{1}{16}$ **28.** (a) 10 (b) 26 (c) 82 (d) 1.3969   **29.** (a) 0 (b) 0 (c) -12 (d) -11.5611 **30.** (a) 3 (b) 29 (c) $\frac{5}{2}$ (d) 1.875

31. $R = \{-5, -3, -1\}$ **32.** $R = \{-3, 0, 1, 0, -3\}$ **33.** $R = \{0, -1, -2, -1, 0\}$

34. $R = \{2, 1, 0, 1, 2\}$ **35.** $R = \{4, 1, 0, 1, 4\}$ **36.** (a) 55 miles (b) 165 miles (c) 550 miles

37. (a) 84 feet (b) 36 feet   **38.** (a) 24 units (b) 23 units   **39.** $A = s^2$ **40.** $C = 2\pi r$

| 7.2 | **Answers to Exercises** |

1. $m = 0$ **2.** $m = -1$ **3.** $\frac{4}{3}$ **4.** Undefined

5. $m = 2$ **6.** $m = \frac{7}{6}$ **7.** $m = 2$ **8.** $m = 0$

9. $m = -\frac{1}{2}$ **10.** $m = -\frac{2}{3}$ **11.** Undefined   **12.** $m = -\frac{16}{7}$

13. $m = \dfrac{b+2}{a+3}$ **14.** $m = \dfrac{d-b}{c-a}$ **15.** $x = 2$ **16.** $y = 13$

17. $y = 4$ **18.** $x = -12$

19.
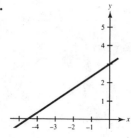

20.
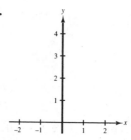

21.
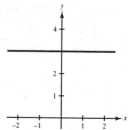

22.
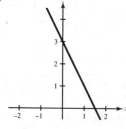

23. $y = \frac{5}{3}x + 3$
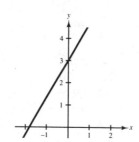

24. $y = -x + 3$
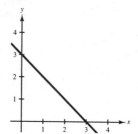

25. $y = -3x - 5$
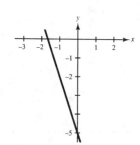

26. $y = -x$
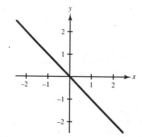

27. $y = -2$

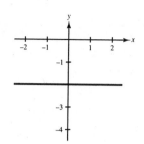

28. $y = \frac{1}{2}x + 2$

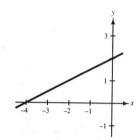

29. Parallel   **30.** Neither   **31.** Perpendicular   **32.** Parallel

---

7.3   Answers to Exercises

1. $m = 2$ **2.** $m = \frac{1}{2}$ **3.** $m = 0$ **4.** Undefined

5. $m = \frac{5}{2}$ **6.** $m = -\frac{3}{4}$ **7.** $y = 3x$ **8.** $y = -x + 1$

9. $y = -\frac{1}{2}x + \frac{7}{8}$ **10.** $y = -2x + \frac{7}{4}$ **11.** $x - y + 2 = 0$ **12.** $y + 1 = 0$

13. $5x - y + 11 = 0$ **14.** $x - y - 5 = 0$ **15.** $x - 1 = 0$ **16.** $x - y + 3 = 0$

17. $10x - 3y - 23 = 0$ **18.** $x - 2y = 0$ **19.** $2x + 12y - 13 = 0$

20. $12x - 28y + 17 = 0$ **21.** $3x - y + 2 = 0$ **22.** $2x + y - 3 = 0$

23. $y - 4 = 0$ **24.** $x - 2 = 0$ **25.** (a) $y = 2x + 1$ **26.** (a) $y = \frac{1}{2}x + 2$
                                                          (b) $y = -\frac{1}{2}x + 1$ (b) $y = -2x + 7$

27. (a) $y = 1$ **28.** (a) $x = 2$ **29.** $F = \frac{9}{5}C + 32$ **30.** $W = 0.07S + 1000$
       (b) $x = 5$ (b) $y = -1$

31. (a) $y = -\frac{1}{5}x + 220$ **32.** Parallel   **33.** Perpendicular
       (b) 90 units rented
       (c) 98 units rented

7.4   Answers to Exercises

1. (a) Solution   **2.** (a) Not a solution   **3.** Solid   **4.** Dashed
    (b) Solution   (b) Solution
    (c) Solution   (c) Solution
    (d) Not a solution   (d) Not a solution

5. (c)   **6.** (a)   **7.** (d)   **8.** (b)

9.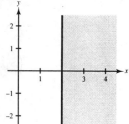

10.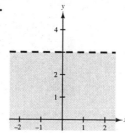

11.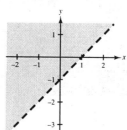

12.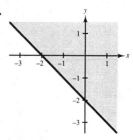

13.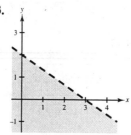

14.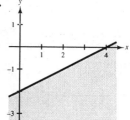

15.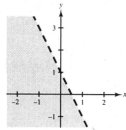

16.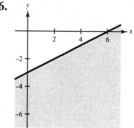

17.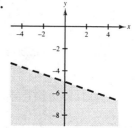

18.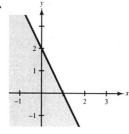

19.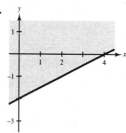

20.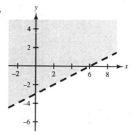

21.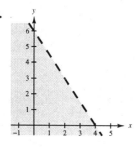

22.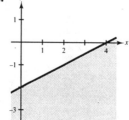

23. $x < -1$ **24.** $y \geq 4$ **25.** $y > x$ **26.** $y \leq -x + 2$

27. 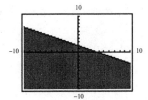   **28.**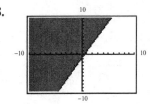

Answers to Cumulative Practice Test P–7

1. False   **2.** False   **3.** False   **4.** False

5. True   **6.** True   **7.** $8x + 3y$ **8.** $\dfrac{x^2}{4} + \dfrac{x}{2} + \dfrac{1}{4}$

9. $-2x - 2$ **10.** -8 **11.** $x^2 - 4x - 12$ **12.** $-2x^2 - 5x - 1$

13. $\dfrac{1}{a^{12}b^6}$ **14.** $x^2 - 9$ **15.** $2x^2 + 2x + 8 + \dfrac{9}{x - 1}$

16. $2x - 3 + \dfrac{2}{x}$ **17.** $x(3x + 4)$ **18.** $(x + 2)(x + 4)$

19. $(4x + 5)(4x - 5)$ **20.** $(2x + 3)(4x - 1)$ **21.** $(x - 3)^2$

22. $2x(x + 4)(3x - 1)$ **23.** $x = -1$ **24.** $x = -10$

25. $x = -4$ or $x = 4$ **26.** $x = -2$ or $x = 3$ **27.** $x \geq -1$

28. $x = -2$ **29.** $x = -2$ or $x = 6$ **30.** $x = -21$

31. Less than 200 miles   **32.** 33, 35, 37   **33.** 14 ft by 20 ft

34. 20 liters   **35.** Function   **36.** Not a function

37. Function   **38.** $(0, 1), (-1, 0)$ **39.** $(0, 0)$

40. $(0, 12), (-6, 0), (-2, 0)$

CHAPTER EIGHT
Systems of Linear Equations

| 8.1 | Solving Systems of Equations by Graphing |

Section Highlights

1. A **system of equations** is two or more equations with the same variables, considered simultaneously.
2. A **solution** of a system of equations in two variables is an ordered pair of real numbers that satisfies all the equations in the system.
3. Steps to solving by graphing:
   (a) Graph all equations in the system on the same rectangular coordinate system.
   (b) Estimate the coordinates of any points of intersection.
   (c) Check the ordered pairs from part (b) to see if they are solutions of the original system of equations.

EXAMPLE 1 ■ Checking Solutions of a System of Linear Equations

Determine whether the point $(1, 1)$ is a solution of the given system of linear equations.

$$3x + y = 4$$
$$2x + y = 3$$

Solution

Substitute $(1, 1)$ into both equations.

$$3x + y = 4 \qquad 2x + y = 3$$

$$3(1) + 1 \overset{?}{=} 4 \qquad 2(1) + 1 \overset{?}{=} 3$$

$$3 + 1 \overset{?}{=} 4 \qquad 2 + 1 \overset{?}{=} 3$$

$$4 = 4 \qquad 3 = 3$$

$(1, 1)$ is a solution of both equations. Therefore, $(1, 1)$ is a solution of the system of linear equations. ■

| **Starter Exercise 1** | *Fill in the blanks.* |

$$2x - y = -2$$
$$x + y = -1$$

Determine whether the point $(-2, \ -2)$ is a solution of the given system of linear equations.

$$2x - y = -2 \qquad\qquad x + y = -1$$

$$2(-2) - \boxed{} \overset{?}{=} -2 \qquad \boxed{} + (-2) \overset{?}{=} -1$$

$$\boxed{} - \boxed{} \overset{?}{=} -2 \qquad \boxed{}\boxed{} -1$$

$$\boxed{}\boxed{} -2$$

Conclusion: $\boxed{}$ is not a solution.

EXAMPLE 2 ■   Solving a System of Linear Equations by Graphing

Solve the following system by graphing.

$$3x + 2y = 2$$
$$x + 2y = -2$$

Solution

To graph each equation, we first write each in slope-intercept form.

$$y = -\tfrac{3}{2}x + 1$$
$$y = -\tfrac{1}{2}x - 1$$

Now, by using the slope and y-intercept, we graph each line. It appears that the two lines intersect at $(2, -2)$, so we check $(2, -2)$ in the equations.

$$3(2) + 2(-2) = 6 - 4 = 2$$
$$2 + 2(-2) = 2 - 4 = -2$$

Since $(2, -2)$ satisfies both equations, it is a solution of the system. From our graph we can see that $(2, -2)$ is the only solution.

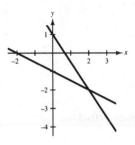

Starter Exercise 2   *Fill in the blanks.*

Solve the following system by graphing.

$$2x + y = -1$$
$$2x - y = -3$$

Start by writing the equations in slope-intercept form and graphing each.

$$y = -2x - 1$$

$$y = 2x + 3$$

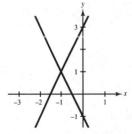

The solution appears to be $\left(\boxed{}, \boxed{}\right)$. Check this in the system.

EXAMPLE 3 ■   A System of Linear Equations with No Solution

Solve the following system of linear equations.

$$2x - y = 1$$
$$4x - 2y = 6$$

Solution

We start by writing each equation in slope-intercept form.

$$y = 2x - 1$$
$$y = 2x - 3$$

We see that the two lines are parallel (since they have the same slope), and hence, do not intersect. Therefore, there is no solution. ■

Starter Exercise 3   *Fill in the blanks.*

Solve the following system of linear equations.

$$4x - 2y = 6$$
$$6x - 3y = 6$$

Write each equation in slope-intercept form.

$y = \boxed{}$

$y = \boxed{}$

The two lines are $\boxed{}$.

Conclusion: $\boxed{}$

EXAMPLE 4 ■   A System of Linear Equations with Infinitely Many Solutions

Solve the following system of linear equations.

$$2x - 3y = 1$$
$$4x - 6y = 2$$

Solution

We start by writing each equation in slope-intercept form.

$$y = \tfrac{2}{3}x - \tfrac{1}{3}$$
$$y = \tfrac{2}{3}x - \tfrac{1}{3}$$

We can see that these two equations are equivalent. Hence, they have the same graph and solution set. We can describe the solution set by saying every point on the line $y = \tfrac{2}{3}x - \tfrac{1}{3}$ is a solution of the system. ■

Starter Exercise 4   *Fill in the blanks.*

Solve the following system of linear equations.

$$x - 2y = 2$$
$$3x - 6y = 6$$

Start by writing each equation in slope-intercept form.

$y = \tfrac{1}{2}x - 1$

$y = \boxed{}$

Conclusion: $\boxed{}$

▦ EXAMPLE 5 ■   A Graphing Calculator Problem

Use a graphing calculator to approximate the solution of the following system of linear equations.

$$2x - y = -1$$
$$x + y = 4$$

Solution

We start by writing each equation in slope-intercept form.

$$y = 2x + 1$$

$$y = -x + 4$$

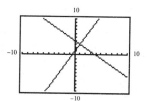

Next, graph these two equations on the same screen. The display should look like the figure on the right. Now press the [TRACE] key. Move the cursor until it is as close as possible to the point of intersection. Your calculator may display the coordinates as $x = .94736842$ and $y = 2.8947368$. We will conjecture that the solution is $(1, 3)$. We must check this point in the equations $2(1) + 1 = 3$ and $-1 + 4 = 3$. Since $(1, 3)$ satisfies both equations, we conclude that $(1, 3)$ is the solution of the system.

■

■   Solutions to Starter Exercises   ■

1. $2(-2) - \boxed{(-2)} \overset{?}{=} -2$ $\boxed{-2} + (-2) \overset{?}{=} -1$

$\boxed{-4} - \boxed{(-2)} \overset{?}{=} -2$ $\boxed{-4} \boxed{\neq} -1$

$\boxed{-2} \boxed{=} -2$

Conclusion: $\boxed{(-2, -2)}$ is not a solution.

2. $\left(\boxed{-1}, \boxed{1} \right)$ **3.** $y = \boxed{2x - 3}$

$y = \boxed{2x - 2}$

The two lines are $\boxed{\text{parallel}}$.

Conclusion: $\boxed{\text{No solution}}$

4. $y = \boxed{\frac{1}{2}x - 1}$

Conclusion: $\boxed{\text{All points on the line } y = \frac{1}{2}x - 1 \text{ are solutions.}}$

8.1   EXERCISES

In Exercises 1–4, determine which ordered pair is a solution of the system of equations.

	System	*Points*		*System*	*Points*
1.	$2x + 3y = 1$	(a) $(-1, 1)$	**2.**	$x + y = 2$	(a) $(3, 4)$
	$x - y = -2$	(b) $(2, 1)$		$2x - y = 1$	(b) $(1, 1)$
3.	$x + 2y = 4$	(a) $\left(1, \frac{1}{2}\right)$	**4.**	$3x - 4y = 11$	(a) $\left(\frac{21}{5}, \frac{2}{5}\right)$
	$x - 3y = -6$	(b) $(0, 2)$		$x - 3y = 3$	(b) $(0, 0)$

In Exercises 5–8, use the graph of the linear equations to determine the number of solutions of the given system.

5. $\begin{aligned} x + y &= 2 \\ -x + y &= -2 \end{aligned}$

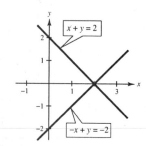

6. $\begin{aligned} x + 3y &= 9 \\ x + 3y &= 3 \end{aligned}$

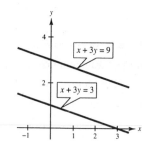

7. $\begin{aligned} x - y &= -2 \\ -2x + 2y &= 4 \end{aligned}$

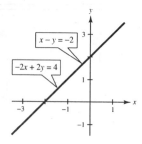

8. $\begin{aligned} 3x + 2y &= 12 \\ x + 2y &= 0 \end{aligned}$

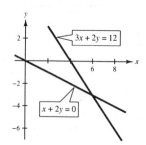

In Exercises 9–18, solve the system of linear equations by graphing.

9. $\begin{aligned} x - 2y &= 2 \\ x + y &= 2 \end{aligned}$

10. $\begin{aligned} x - y &= 0 \\ 2x + y &= 3 \end{aligned}$

11. $\begin{aligned} x + y &= 4 \\ x + 3y &= 6 \end{aligned}$

12. $\begin{aligned} x + 2y &= 6 \\ 4x - 2y &= 4 \end{aligned}$

13. $\begin{aligned} 2x - 3y &= -9 \\ 2x + 3y &= -3 \end{aligned}$

14. $\begin{aligned} 3x + 2y &= 4 \\ \tfrac{3}{2}x + y &= 2 \end{aligned}$

15. $\begin{aligned} -x + y &= 1 \\ x + y &= 1 \end{aligned}$

16. $\begin{aligned} x - y &= -3 \\ x + y &= -1 \end{aligned}$

17. $\begin{aligned} x + 2y &= 8 \\ x + 2y &= 2 \end{aligned}$

18. $\begin{aligned} 3x - 2y &= 0 \\ x &= -2 \end{aligned}$

In Exercises 19–22, write each equation of the given system in slope-intercept form and determine the number of solutions of the system. (Do not graph.)

19. $\begin{aligned} 2x - 3y &= 7 \\ x - 6y &= 4 \end{aligned}$

20. $\begin{aligned} x - 2y &= 4 \\ x - 2y &= 1 \end{aligned}$

21. $\begin{aligned} 2x - y &= 1 \\ 4x - 2y &= 2 \end{aligned}$

22. $\begin{aligned} 5x - 4y &= 7 \\ 3x + 3y &= 9 \end{aligned}$

In Exercises 23 and 24, use a graphing calculator to approximate the solution of the system of equations.

23. $\begin{aligned} x + y &= 4 \\ -2x + y &= 1 \end{aligned}$

24. $\begin{aligned} x + 2y &= 2 \\ x - y &= 2 \end{aligned}$

8.2 | Solving Systems of Equations by Substitution

Section Highlights

Steps for solving by the method of substitution:

1. Solve one of the equations for one variable in terms of the other.
2. Substitute the expression obtained in Step 1 into the other equation and solve the resulting equation.
3. Back-substitute the solution in Step 2 into the expression found in Step 1 to find the value of the other variable.
4. Check your answer to see that it satisfies both of the original equations.

EXAMPLE 1 ■ Method of Substitution: One-Solution Case

Solve the following system of linear equations.

$$x + y = 4$$
$$x + 3y = 6$$

Solution

(1) We will solve the first equation for x. (This is only one of three good options.)

$x + y = 4$ First equation

$x = -y + 4$ Revised first equation

(2) Now we substitute for x in the second equation and solve for y.

$x + 3y = 6$ Second equation

$(-y + 4) + 3y = 6$ Replace x by $-y + 4$.

$2y + 4 = 6$

$2y = 2$

$y = 1$ Solve for y.

(3) Back-substitute for y into the revised first equation.

$x = -y + 4$ Revised first equation

$x = -1 + 4$ Replace y by 1.

$x = 3$ Solve for x.

(4) The solution is (3, 1). Check (3, 1) in the original equations.

Starter Exercise 1 *Fill in the blanks.*

Solve the following system of linear equations.

$$2x + 3y = 3$$
$$2x - y = 7$$

(1) Solve the second equation for y.

$$2x - y = 7$$

$$-y = \boxed{} + 7$$

$$y = \boxed{}$$

(2) Substitute for y in the first equation.

$$2x + 3y = 3$$

$$2x + 3\boxed{} = 3$$

$$2x + \boxed{} - \boxed{} = 3$$

$$\boxed{}x - \boxed{} = 3$$

$$\boxed{}x = 24$$

$$x = \boxed{}$$

(3) Back-substitute for x in the revised second equation.

$$y = \boxed{}$$

$$y = \boxed{}$$

$$y = \boxed{}$$

$$y = \boxed{}$$

(4) The solution is $\boxed{}$. Check this in the system.

EXAMPLE 2 ■ Method of Substitution: No-Solution Case

Solve the following system of linear equations.

$$x - 2y = 1$$
$$-2x + 4y = 5$$

Solution

(1) Solve the first equation for x.

$$x - 2y = 1 \qquad \text{First equation}$$

$$x = 2y + 1$$

(2) Substitute for x in the second equation.

$$-2x + 4y = 5 \qquad \text{Second equation}$$
$$-2(2y + 1) + 4y = 5 \qquad \text{Replace } x \text{ by } 2y+1.$$
$$-4y - 2 + 4y = 5$$
$$-2 = 5 \qquad \text{False statement}$$

Since the substitution resulted in a false statement, we conclude that the given system has no solution.

Starter Exercise 2 | *Fill in the blanks.*

Solve the following system of linear equations.

$$2x + y = 1$$
$$6x + 3y = 2$$

(1) Solve the first equation for y.

$$2x + y = 1$$
$$y = \boxed{}$$

(2) Substitute for y in the second equation.

$$6x + 3y = 2$$
$$6x + 3\boxed{} = 2$$
$$\boxed{} = 2$$
$$\boxed{} = 2$$

Conclusion: $\boxed{}$

EXAMPLE 3 ■ Method of Substitution: Many-Solutions Case

Solve the following system of linear equations.

$$2x + 6y = 12$$
$$3x + 9y = 18$$

Solution

(1) Solve the first equation for x.

$$2x + 6y = 12$$
$$2x = -6y + 12$$
$$x = -3y + 6$$

(2) Substitute for x in the second equation.

$$3x + 9y = 18$$
$$3(-3y + 6) + 9y = 18$$
$$-9y + 18 + 9y = 18$$
$$18 = 18$$

This last equation is true for all values of x. This implies that any solution of the second equation is a solution of the first equation. The solutions consists of all (x, y) such that $2x + 6y = 12$.

| **Starter Exercise 3** | *Fill in the blanks.*

Solve the following system of linear equations.

$$4x - 2y = 6$$
$$2x - y = 3$$

(1) Solve the second equation for y.

$$2x - y = 3$$

$$-y = \boxed{}$$

$$y = \boxed{}$$

(2) Substitute for y in the first equation.

$$4x - 2y = 6$$

$$4x - 2\boxed{} = 6$$

$$\boxed{} = 6$$

$$\boxed{} = 6$$

Conclusion: $\boxed{}$

■   **Solutions to Starter Exercises**   ■

1. (1) $-y = \boxed{-2x} + 7$

$y = \boxed{2x - 7}$

(3) $y = \boxed{2x - 7}$

$y = \boxed{2(3) - 7}$

$y = \boxed{6 - 7}$

$y = \boxed{-1}$

(2) $2x + 3\boxed{(2x - 7)} = 3$

$2x + \boxed{6x} - \boxed{21} = 3$

$\boxed{8}\,x - \boxed{21} = 3$

$\boxed{8}\,x = 24$

$x = \boxed{3}$

(4) $\boxed{(3, -1)}$

2. (1) $y = \boxed{-2x + 1}$

(2) $6x + 3\boxed{(-2x + 1)} = 2$

$\boxed{6x - 6x + 3} = 2$

$\boxed{3} = 2$

Conclusion: $\boxed{\text{No solution}}$

3. (1) $-y = \boxed{-2x + 3}$

$y = \boxed{2x - 3}$

(2) $4x - 2\boxed{(2x - 3)} = 6$

$\boxed{4x - 4x + 6} = 6$

$\boxed{6} = 6$

Conclusion:

$\boxed{\text{Solution are all } (x, \ y) \text{ such that } 2x - y = 3.}$

8.2 | EXERCISES

In Exercises 1–18, use the method of substitution to solve the given system of linear equations.

1. $x + y = 1$
$x - y = 3$

2. $2x - y = 0$
$x - 2y = 3$

3. $3x - y = -1$
$x - y = -3$

4. $x - 2y = 2$
$x + y = 2$

5. $-16x - 3y = 4$
$3x + y = 1$

6. $5x + 2y = 4$
$x - 2y = 8$

7. $-x + 3y = 6$
$x + y = -2$

8. $2x - y = 5$
$3x + y = 10$

9. $2x - 4y = 6$
$2x + 4y = 6$

10. $3x - 2y = 4$
$2x + y = 5$

11. $6x - 4y = 8$
$-3x + 2y = 12$

12. $x + 5y = 3$
$-2x - 4y = 0$

13. $x - 2y = 5$
$-2x + 4y = -10$

14. $\frac{1}{2}x - \frac{1}{3}y = 1$
$3x - 2y = 18$

15. $4x + 5y = 9$
$y = 1$

16. $\frac{1}{4}x + \frac{1}{2}y = 4$
$\frac{1}{8}x + \frac{1}{4}y = 2$

17. $6x - 15y = 90$
$12x - 24y = 108$

18. $x = 2$
$2x + 5y = -1$

In Exercises 19 and 20, find a such that the given system is inconsistent.

19. $ax + 2y = 4$
$2x + 4y = 5$

20. $3x - 6y = 1$
$ax - 12y = \frac{1}{3}$

In Exercises 21 and 22, find b such that the given system has infinitely many solutions.

21. $5x + by = 15$
$x + 2y = 3$

22. $4x - 3y = 1$
$8x + by = 2$

8.3 | Solving Systems of Equations by Elimination

Section Highlights

Steps for solving by elimination:

1. Obtain coefficients for x (or y) that differ only in sign by multiplying all terms of one or both equations by suitably chosen constants.
2. Add the equations to eliminate one variable and solve the resulting equation.
3. Back-substitute the value obtained in Step 2 into either of the original equations and solve for the other variable.
4. Check your solution in both of the original equations.

EXAMPLE 1 ■   The Method of Elimination

Solve the following system of linear equations.

$$5x - 2y = 3$$
$$4x + 2y = 6$$

Solution

Note that the y-coefficients are opposites. Therefore, by adding the two equations we can eliminate y.

$$5x - 2y = 3$$
$$4x + 2y = 6$$
$$\overline{ }$$
$$9x = 9 \qquad \text{Add equations.}$$

From this equation, we can see $x = 1$. By back-substituting this value of x into the first equation, we can solve for y.

$$5x - 2y = 3 \qquad \text{First equation}$$
$$5(1) - 2y = 3 \qquad \text{Replace } x \text{ by 1.}$$
$$5 - 2y = 3$$
$$-2y = -2$$
$$y = 1 \qquad \text{Solve for } y.$$

Therefore, the solution is (1, 1). Check (1, 1) in both original equations. ∎

Starter Exercise 1   *Fill in the blanks.*

Solve the following system of equations.

$$\tfrac{1}{2}x + \tfrac{3}{4}y = \tfrac{1}{4}$$
$$\tfrac{2}{3}x + \tfrac{5}{6}y = 1$$

To get integer coefficients, multiply the first equation by 4, and the second by 6.

$$\tfrac{1}{2}x + \tfrac{3}{4}y = \tfrac{1}{4} \quad \Rightarrow \quad 2x + 3y = 1$$
$$\tfrac{2}{3}x + \tfrac{5}{6}y = 1 \quad \Rightarrow \quad 4x + 5y = 6$$

Now, get opposite coefficients on x by multiplying the first equation by $\boxed{}$.

$$2x + 3y = 1 \quad \Rightarrow \quad \boxed{} - \boxed{} = \boxed{}$$
$$4x + 5y = 6 \quad \Rightarrow \quad \underline{4x \ + \ 5y \ = \ 6}$$
$$\boxed{} = \boxed{}$$
$$y = \boxed{}$$

Back-substitute into the revised first equation and solve for x.

$$2x + 3y = 1$$
$$2x + 3\boxed{} = 1$$
$$2x - \boxed{} = 1$$
$$2x = \boxed{}$$
$$x = \boxed{}$$

The solution is $\left(\boxed{}, \boxed{} \right)$. Check your solution in the original equations.

EXAMPLE 2 ■   **The Method of Elimination: No Solution Case**

Solve the following system of linear equations.

$$4x - 8y = 2$$
$$6x - 12y = 5$$

Solution

To obtain coefficients of x that are opposites, multiply the first equation by 3, and the second by -2.

$$4x - 8y = 2 \quad \Rightarrow \quad 12x - 24y = 6$$
$$6x - 12y = 5 \quad \Rightarrow \quad \underline{-12x + 24y = -10}$$
$$0 = -4 \qquad \text{Add equations.}$$

Since there are no values of x and y for which $0 = -4$, we conclude that the system is inconsistent and has no solution.   ■

Starter Exercise 2   *Fill in the blanks.*

Solve the following system of linear equations.

$$3x - 5y = 1$$
$$6x - 10y = 4$$

Multiply the first equation by $\boxed{}$ to get coefficients of x that are opposites.

$$3x - 5y = 1 \quad \Rightarrow \quad \boxed{} + \boxed{} = \boxed{}$$
$$6x - 10y = 4 \quad \Rightarrow \quad \underline{6x - 10y = 4}$$
$$\boxed{} = \boxed{}$$

Conclusion: $\boxed{}$

EXAMPLE 3 ■   **The Method of Elimination: Many-Solutions Case**

Solve the following system of linear equations.

$$5x - 3y = 1$$
$$15x - 9y = 3$$

Solution

To obtain coefficients of x that are opposites, multiply the first equation by -3.

$$5x - 3y = 1 \quad \Rightarrow \quad -15x + 9 = -3$$
$$15x - 9y = 3 \quad \Rightarrow \quad \underline{15x - 9 = 3}$$
$$0 = 0 \qquad \text{Add equations.}$$

Since the two equations are equivalent, we conclude that the solutions are all (x, y) such that $5x - 3y = 1$.   ■

■ **Starter Exercise 3**   *Fill in the blanks.*

Solve the following system of linear equations.

$$15x + 5y = 10$$
$$6x + 2y = 4$$

To obtain coefficients that are opposites, multiply the first equation by ☐ , and the second by ☐ .

$$15x + 5y = 10 \quad \Rightarrow \quad 30x + \boxed{} = \boxed{}$$
$$6x + 2y = 4 \quad \Rightarrow \quad \boxed{} - \boxed{} = \boxed{}$$
$$\boxed{} = \boxed{}$$

Conclusion: $\boxed{}$

■   **Solutions to Starter Exercises**   ■

1. Multiply the first equation by $\boxed{-2}$.

$$\boxed{-4x} - \boxed{6y} = \boxed{-2}$$
$$\underline{4x + 5y = 6}$$
$$\boxed{-y} = \boxed{4}$$
$$y = \boxed{-4}$$

$$2x + 3\boxed{(-4)} = 1$$
$$2x - \boxed{12} = 1$$
$$2x = \boxed{13}$$
$$x = \boxed{\tfrac{13}{2}}$$

Solution: $\left(\boxed{\tfrac{13}{2}} , \boxed{-4} \right)$

2. Multiply the first equation by $\boxed{-2}$.

$$\boxed{-6x} + \boxed{10y} = \boxed{-2}$$
$$\underline{6x - 10y = 4}$$
$$\boxed{0} = \boxed{2}$$

Conclusion: $\boxed{\text{No solution}}$

3. Multiply the first equation by $\boxed{2}$, and the second by $\boxed{-5}$.

$$30x + \boxed{10y} = \boxed{20}$$
$$\boxed{-30x} - \boxed{10y} = \boxed{-20}$$
$$\boxed{0} = \boxed{0}$$

Conclusion: $\boxed{\text{The solutions are all } (x, y) \text{ such that } 6x = 2y = 4.}$

| 8.3 | **EXERCISES** |

In Exercises 1–18, use the method of elimination to solve the given system of equations.

1. $x + y = 1$
$x - y = 3$

2. $x - 2y = 4$
$-x + 3y = 1$

3. $5x - 2y = 3$
$3x + 2y = 5$

4. $2x - 4y = 1$
$3x + 4y = 9$

5. $3x + 2y = 1$
$2x + 2y = 6$

6. $4x - 3y = 3$
$4x + 5y = -5$

7. $2x - 3y = 7$
$x - 2y = 1$

8. $3x - 5y = 1$
$6x - 9y = 4$

9. $3x + 4y = 1$
$2x + 3y = 4$

10. $5x - 7y = 2$
$3x - 4y = 1$

11. $6x + 3y = 9$
$2x + y = 3$

12. $2x - 4y = 6$
$5x - 10y = 1$

13. $\frac{1}{2}x - \frac{1}{4}y = \frac{3}{4}$
$\frac{1}{6}x + \frac{1}{3}y = \frac{5}{6}$

14. $\frac{1}{3}x - \frac{2}{3}y = 2$
$\frac{2}{5}x + \frac{1}{10}y = \frac{3}{5}$

15. $-2x - 3y = 1$
$10x + 15y = 2$

16. $10x - 5y = 15$
$6x - 3y = 9$

17. $0.01x + 0.03y = 0.02$
$0.05x - 0.07y = -0.12$

18. $0.02x - 0.05y = 0.07$
$0.3x - 0.4y = 1.2$

In Exercises 19–22, use the more convenient method (substitution or elimination) to solve the given system.

19. $x + 2y = 1$
$5x + 3y = -2$

20. $2x - 4y = 1$
$3x + 4y = 4$

21. $5x - 3y = 4$
$5x - 2y = 2$

22. $3x + 2y = 5$
$2x + y = 1$

In Exercises 23 and 24, find a such that the system of linear equations is inconsistent.

23. $ax + 3y = 4$
$4x + 6y = 9$

24. $6x - 12y = 9$
$ax - 8y = 1$

In Exercises 25 and 26, find k such that the system of linear equations has infinitely many solutions.

25. $5x - 3y = 2$
$10x - 6y = k$

26. $5x + 25y = k$
$x + 5y = -3$

| 8.4 | Applications of Systems of Linear Equations

Section Highlights

In this section we solve some application problems involving systems of equations. Some of these problems are the same problems we did earlier. You may want to review the formulas in Chapter 4 before doing this section.

EXAMPLE 1 ■ A Number Problem

The sum of two numbers is 42. The difference between two times one number and three times the other is −6. Find the two numbers.

Solution

Verbal model: First number + Second number = 42

2 · First number − 3 · Second number = −6

Labels: x = first number

y = second number

System of equations: $x + y = 42$

$2x − 3y = −6$

We will solve by the substitution method. Solve the first equation for x and substitute for x in the second equation.

$$x + y = 42 \Rightarrow \quad x = -y + 42$$

$$2x - 3y = -6 \qquad \text{Second equation}$$

$$2(-y + 42) - 3y = -6 \qquad \text{Replace } x \text{ by } -y + 42.$$

$$-2y + 84 - 3y = -6$$

$$-5y + 84 = -6$$

$$-5y = -90$$

$$y = 18 \qquad \text{Solve for } y.$$

Now we back-substitute $y = 18$ into the first equation.

$$x + y = 42$$

$$x + 18 = 42$$

$$x = 24$$

Thus, the two numbers are 24 and 18. ∎

Starter Exercise 1   *Fill in the blanks.*

A board is 10 feet long. The board must be cut so that twice the length of one piece is three times the length of the other. Find the length of each piece.

Verbal model: First piece + Second piece = 10

2 · First piece = 3 · Second piece

Labels: x = length of first piece

y = length of second piece

System of equations: $x + \boxed{} = 10$

$\boxed{} = 3y$

We will solve this system by the method of substitution. From the first equation, $x = 10 - y$. Substitute this into the second equation and solve.

$$2x = 3y$$

$$2(\boxed{}) = 3y$$

$$\boxed{} - 2y = 3y$$

$$\boxed{} = 5y$$

$$\boxed{} = y$$

Since $y = 4$, $x = 10 - 4 = 6$. The two pieces must have lengths of 6 feet and 4 feet.

EXAMPLE 2 ■ Investment Problem

A total of $50,000 is to be invested in two accounts. One account pays 7% and the other 9% simple annual interest. How much should be invested in each account so that the total interest earned is $4200?

Solution

Verbal model:

| Amount invested at 7% | + | Amount invested at 9% | = | $50,000 |

| Interest earned at 7% | + | Interest earned at 9% | = | $4200 |

Labels: 7%: amount $= x$, interest $= 0.07x$
9%: amount $= y$, interest $= 0.09y$

System of equations:
$$x + y = 50,000$$
$$0.07x + 0.09y = 4200$$

We will use the method of elimination to eliminate the x. Multiply the first equation by -7 and the second equation by 100.

$$
\begin{array}{lcl}
x + y = 50,000 & \Rightarrow & -7x - 7y = -350,000 \\
0.07x + 0.09y = 4200 & \Rightarrow & \underline{7x + 9y = 420,000}
\end{array}
$$

$$2y = 70,000 \quad \text{Add equations.}$$
$$y = 35,000$$

Now, from the first equation, $x = 50,000 - 35,000 = 15,000$. Thus, $15,000 must be invested at 7% at $35,000 at 9%. ■

■ **Starter Exercise 2** | *Fill in the blanks.*

A total of $7200 is to be invested in two accounts paying 8% and 10% simple annual interest. How much should be invested in each so that the interest earned on each account is the same?

Verbal model:

| Amount invested at 8% | + | Amount invested at 10% | = | $7200 |

| Interest earned at 8% | = | Interest earned at 10% |

Labels: 8%: amount $= x$, interest $=$ ☐

10%: amount $= y$, interest $=$ ☐

System of equations: $x + y = 7200$

☐ $=$ ☐

We will solve by the substitution method. Solve the first equation for x.

$x =$ ☐

Substitute this into the second equation and solve for y.

$0.08($ ☐ $) =$ ☐

☐ $=$ ☐

☐ $=$ ☐

☐ $=$ ☐

Invest ☐ at 10% and ☐ at 8%.

EXAMPLE 3 ■ Coin Problem

A bank contains $28.50 in dimes and quarters. If there are 150 coins in the bank, how many of each coin are in the bank?

Solution

Verbal model:

$$\boxed{\text{Number of dimes}} + \boxed{\text{Number of quarters}} = \boxed{150}$$
$$\boxed{\text{Value of dimes}} + \boxed{\text{Value of quarters}} = \boxed{\$28.50}$$

Labels: Dimes: number $= x$, value $= 0.10x$
Quarters: number $= y$, value $= 0.25y$

System of equations:
$$x + y = 150$$
$$0.10x + 0.25y = 28.50$$

We will solve by the method of elimination.

$$
\begin{array}{llll}
x + y = 150 & \Rightarrow & -10x - 10y = -1500 & \text{Multiply by } -10. \\
0.10x + 0.25y = 28.50 & \Rightarrow & \underline{10x + 25y = 2850} & \text{Multiply by 100.} \\
& & 15y = 1350 & \text{Add equations.} \\
& & y = 90 &
\end{array}
$$

Since $y = 90$ from the first equation, $x = 60$. Thus, there are 60 dimes and 90 quarters in the bank. ■

Starter Exercise 3 | *Fill in the blanks.*

Sam bought some 29¢ and 23¢ stamps. He spent $24.20 on 100 stamps. How many of each type of stamp did he buy?

Verbal model:

$$\boxed{} + \boxed{} = 100$$
$$\boxed{} + \boxed{} = \$24.20$$

Labels: 29¢ stamps: number $= x$, value $= 0.29x$
23¢ stamps: number $= y$, value $= 0.23y$

System of equations:
$$x + y = 100$$
$$0.29x + 0.23y = 24.20$$

Solve by the elimination method.

$\boxed{}$ 29¢ stamps and $\boxed{}$ 23¢ stamps

EXAMPLE 4 ■ Mixture Problem

How many pounds of chocolates that cost $3.00 per pound must be mixed with caramels that cost $2.00 per pound to make a 10-pound mixture that costs $2.60 per pound?

Solution

Verbal model:

$$\boxed{\begin{array}{c}\text{Amount of}\\\text{chocolates}\end{array}} + \boxed{\begin{array}{c}\text{Amount of}\\\text{caramels}\end{array}} = \boxed{10}$$

$$\boxed{\begin{array}{c}\text{Value of}\\\text{chocolates}\end{array}} + \boxed{\begin{array}{c}\text{Value of}\\\text{caramels}\end{array}} = \boxed{\begin{array}{c}\text{Value of}\\\text{mixture}\end{array}}$$

Labels: Chocolates: amount $= x$, value $= 3x$
Caramels: amount $= y$, value $= 2x$
Mixture: amount $= 10$, value $= 2.60(10)$

System of equations:
$$x + y = 10$$
$$3x + 2y = 2.60(10)$$

We will solve by the substitution method. From the first equation, $x = 10 - y$. Substitute this into the second equation and solve.

$$3x + 2y = 26$$
$$3(10 - y) + 2y = 26$$
$$30 - 3y + 2y = 26$$
$$30 - y = 26$$
$$-y = -4$$
$$y = 4$$
$$x = 10 - y = 10 - 4 = 6$$

Thus, 6 pounds of chocolates must be used.

Starter Exercise 4   *Fill in the blanks.*

You need 10 liters of a $11\frac{1}{2}\%$ saline solution. You have 15% and 10% saline solutions. How many liters of each must be used to make the $11\frac{1}{2}\%$ saline solution?

Verbal model:

$$\boxed{\text{Amount of } 15\% \text{ solution}} + \boxed{\text{Amount of } 10\% \text{ solution}} = \boxed{10}$$

$$\boxed{\text{Amount of saline in } 15\% \text{ solution}} + \boxed{\text{Amount of saline in } 10\% \text{ solution}} = \boxed{\text{Amount of saline in } 11\frac{1}{2}\% \text{ solution}}$$

Labels:

15% solution: amount $= x$, amount of saline $= 0.15x$
10% solution: amount $= y$, amount of saline $= 0.10y$
$11\frac{1}{2}\%$ solution: amount $= 10$, amount of saline $= 0.115(10) = 1.15$

System of equations:

$$x \quad + \quad y \quad = \quad 10$$
$$\boxed{} + \boxed{} = 1.15$$

Solve by substitution.

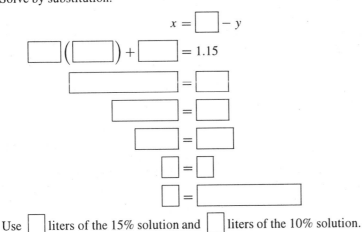

Use ☐ liters of the 15% solution and ☐ liters of the 10% solution.

EXAMPLE 5 ■ Two Rate Problem

With the wind, a plane could fly 390 miles in 3 hours. On the return trip, against the wind, the plane could only fly 330 miles in 3 hours. Find the rate of the wind and the rate of the plane in calm air.

Solution

Verbal model:

$$\boxed{\text{With the wind distance}} = \boxed{\text{With the wind rate}} \cdot \boxed{\text{With the wind time}}$$

$$\boxed{\text{Against the wind distance}} = \boxed{\text{Against the wind rate}} \cdot \boxed{\text{Against the wind time}}$$

Labels:
p = rate of the plane in calm air
w = rate of the wind
$p + w$ = with the wind rate
$p - w$ = against the wind rate

System of equations:
$390 = (p + w) \cdot 3 \Rightarrow 390 = 3p + 3w$
$330 = (p - w) \cdot 3 \Rightarrow 330 = 3p - 3w$

We will solve by the method of elimination.

$$
\begin{array}{ll}
390 = 3p + 3w & \\
330 = 3p - 3w & \\
\hline
720 = 6p & \text{Add equations.} \\
120 = p &
\end{array}
$$

From the first equation:

$390 = 3(120) + 3w$

$30 = 3w$

$10 = w$

Thus, the rate of the plane in calm air is 120 mph and the rate of the wind is 10 mph.

■

| **Starter Exercise 5** | *Fill in the blanks.* |

A tire is on sale for $42.50, which is 15% off the regular price. Find the regular price of the tire.

Verbal model:

$$\boxed{\text{Regular price}} - \boxed{\text{Discount}} = \boxed{\text{Sale price}}$$

$$\boxed{\text{Discount}} = 0.15 \cdot \boxed{\text{Regular price}}$$

Labels:
R = regular price
D = discount

System of equations:
$R - D = 42.50$
$D = 0.15R$

Solve by the method of substitution.

$R - D = 42.50$

$R - \boxed{} = 42.50$

$\boxed{} = \boxed{}$

$\boxed{} = \boxed{}$

$\boxed{}$ is the regular price of the tire.

■ **Solutions to Starter Exercises** ■

1. $x + \boxed{y} = 10$

$\boxed{2x} = 3y$

$2\left(\boxed{10-y}\right) = 3y$

$\boxed{20} - 2y = 3y$

$\boxed{20} = 5y$

$\boxed{4} = y$

2. 8%: interest $= \boxed{0.08x}$

10%: interest $= \boxed{0.10y}$

$\boxed{0.08x} = \boxed{0.10y}$

$x = \boxed{7200 - y}$

$0.08\left(\boxed{7200 - y}\right) = \boxed{0.10y}$

$\boxed{576 - 0.08y} = \boxed{0.10y}$

$\boxed{576} = \boxed{0.18y}$

$\boxed{3200} = \boxed{y}$

Invest $\boxed{\$3200}$ at 10% and $\boxed{\$4000}$ at 8%.

3. $\boxed{\text{Number of 29¢ stamps}} + \boxed{\text{Number of 23¢ stamps}} = 100$

$\boxed{\text{Value of 29¢ stamps}} + \boxed{\text{Value of 23¢ stamps}} = \$24.20$

$\boxed{x} + \boxed{y} = 100$

$\boxed{0.29x} + \boxed{0.23y} = \$24.20$

$$\begin{array}{r} -29x - 29y = -2900 \\ 29x + 2y = 2420 \\ \hline -6y = -480 \\ y = 80 \\ x = 90 \end{array}$$

$\boxed{\text{Twenty}}$ 29¢ stamps and $\boxed{80}$ 23¢ stamps

4. $x + y = 10$

$\boxed{0.15x} + \boxed{0.10y} = 1.15$

$x = \boxed{10} - y$

$\boxed{0.15}\left(\boxed{10-y}\right) + \boxed{0.10y} = 1.15$

$\boxed{1.5 - 0.15y + 0.10y} = \boxed{1.15}$

$\boxed{1.5 - 0.05y} = \boxed{1.15}$

$\boxed{-0.05y} = \boxed{-0.35}$

$\boxed{y} = \boxed{7}$

$\boxed{x} = \boxed{10 - y = 10 - 7 = 3}$

Use $\boxed{3}$ liters of the 15% solution and $\boxed{7}$ liters of the 10% solution.

■ Solutions to Starter Exercises ■

5. $R - \boxed{0.15R} = 42.50$

$\boxed{0.85R} = \boxed{42.50}$

$\boxed{R} = \boxed{\$50}$

$\boxed{\$50}$ is the regular price of the tire.

8.4 | EXERCISES

Number Problems   In Exercises 1–3, find two numbers that satisfy the given requirements.

1. The sum is 39, and their difference is 9.

2. The sum is 52, and the larger number is three times the smaller number.

3. The difference is 10, and the sum of the larger number and three times the smaller number is 86.

Coin Problems   In Exercises 4–6, determine the number of each type of coin.

	Number of coins	Types of coins	Value
4.	50	Nickels and dimes	$3.65
5.	75	Nickels and quarters	$8.75
6.	100	Dimes and quarters	$18.25

Dimensions of a Rectangle   In Exercises 7–9, find the dimensions of the rectangle that meet the specified conditions.

	Perimeter	Relationship between length and width
7.	24 inches	The length is twice the width.
8.	46 feet	The length is two more than the width.
9.	50 meters	The length is 150% of the width.

10. *Wholesale Cost*   The selling price of a book is $19.50. The markup rate is 30% of the wholesale cost. Find the wholesale cost.

11. *List Price*   The sale price of a stereo receiver is $280. The discount is 20% of the list price. Find the list price.

12. *Investment*   A total of $10,500 is invested in two accounts that pay $7\frac{1}{2}\%$ and $9\frac{1}{4}\%$ simple interest. If the annual interest earned is $883.75, how much is invested in each account?

13. *Ticket Sales*   A total of 300 tickets were sold to a movie. The receipts totaled $1,637.50. Adult tickets were $6.50 and children's tickets were $4.00. How many of each type of ticket were sold?

14. *Gasoline Mixture*   The total cost of 6 gallons of unleaded gasoline and 10 gallons of super unleaded gasoline was $19.64. Super unleaded gasoline is 14¢ per gallon more than unleaded gasoline. Find the price per gallon of each grade of gasoline.

15. *Seed Mixture*   How many pounds of seeds at 80¢ per pound must be purchased with seeds at 70¢ per pound to have 40 pounds of seeds with an average cost of 74¢ per pound?

16. *Mixture Problem*   How many liters of 20% alcohol solution must be mixed with 30% alcohol solution to make 20 liters of 24% alcohol solution?

17. *Average Speed*   A car travels for one hour at 50 mph. How much longer must the car travel at 40 mph so that the average speed for the total trip will be 45 mph?

18. *Boat Speed*   A boat can travel 36 miles downstream in two hours, but only 24 miles upstream in two hours. Find the rate of the boat in calm water and the rate of the current.

19. *Digit Problem*   The sum of the digits of a two-digit number is 9. If the digits are reversed, the number is increased by 9. Find the number.

20. *Best-Fitting Line*   The line $y = mx + b$ that best fits the three points (1, 2), (3, 3), and (5, 6) is given by the following system of linear equations.

$$35m + 9b = 41$$
$$9m + 3b = 11$$

Solve the system and find the best-fitting line.

Cumulative Practice Test for Chapters P–8

In Exercises 1–4, complete the statement by using the Distributive Property.

1. $2(x + 2) = $ _____   **2.** $2x + 3x = $ _____   **3.** $6ab - 5ab = $ _____   **4.** $(a+b)(x+y) = $ _____

In Exercises 5–9, match the equation or inequality with the graphs below.

5. $y = 2x + 1$ **6.** $y = x^2 + 1$ **7.** $y = |x|$ **8.** $x = 2$ **9.** $y > -\frac{1}{2}x + 3$

(a)
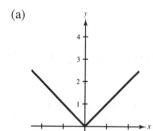

(b)
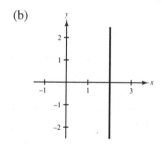

(c)
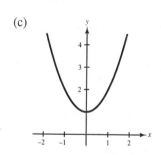

(d)
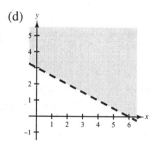

(e)
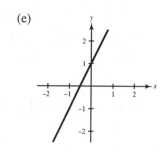

In Exercises 10–12, find an equation of the line through the given points. (Write your answer in general form.)

10. $(2, 1)$, $(-3, 3)$

11. $(-1, -1)$, $(-1, 5)$

12. $(3, -1)$, $(-2, 5)$

In Exercises 13–18, simplify the given expression.

13. $2x - (6x + 4)$

14. $(x + y)^2 - 2xy$

15. $(2x - 3)(2x + 3)$

16. $3[x - (2x - 3)] + 6$

17. $\dfrac{a^2 b^{-3}}{ab^2}$

18. $\left(\dfrac{xy^2}{x^{-1}}\right)^{-2}$

In Exercises 19–22, solve the system of linear equation.

19. $x + y = 2$
$x - y = 0$

20. $2x - 3y = -6$
$5x + 7y = 14$

21. $-x \ + 4y = 1$
$0.5x - 2y = 7$

22. $3x - 4y = 12$
$0.75x - y = 3$

In Exercises 23 and 24, factor the polynomial completely.

23. $x^4 - 1$

24. $x^2(x - 2) - 4(x - 2)$

In Exercises 25–32, solve the given equation or inequality.

25. $-2x \le 4$

26. $2x - 1 = 3$

27. $2x - 4 + 3x = 1$

28. $(x + 3)^2 = 36$

29. $x^2 = 2x + 8$

30. $2x^2 - 2x - 2 = x^2 + 2x + 10$

31. $\dfrac{x}{2} = \dfrac{1}{4}$

32. $\dfrac{3}{t} = \dfrac{9}{4}$

In Exercises 33 and 34, use a graphing utility to graph both equations and determine how many solutions the systems has.

33. $-x + y = \ 2$
$x - y = -2$

34. $8x - 6y = 24$
$4x - 2y = 10$

In Exercises 35 and 36, find an equation of the line through the given point that is (a) parallel to the given line, and (b) perpendicular to the given line. (Write your answer in slope-intercept form.)

35. Line: $y = 2x + 3$
Point: $(2, -4)$

36. Line: $2x - 3y = 4$
Point: $(6, -1)$

37. A coat is on sale for $51 which is 15% off the regular price. Find the regular price of the coat.

38. Find two integers whose sum is 30 and three times the smaller integer is twice the larger.

39. Two drivers start from the same point and travel in the same direction. One driver is traveling at 55 mph and the other at 45 mph. How long before the two are 90 miles apart?

40. A bank contains 38 coins, all dimes and quarters. The total amount of money in the bank is $7.25. How many of each coin is in the bank?

$\boxed{8.1}$ **Answers to Exercises**

1. (a) $(-1,\ 1)$ is a solution.
(b) $(2, 1)$ is not a solution.

2. (a) $(3, 4)$ is not a solution.
(b) $(1, 1)$ is a solution.

3. (a) $\left(1,\ \frac{1}{2}\right)$ is not a solution.
(b) $(0, 2)$ is a solution.

4. (a) $\left(\frac{21}{5},\ \frac{2}{5}\right)$ is a solution.
(b) $(0, 0)$ is not a solution.

5. One solution   **6.** No solutions   **7.** Infinitely many solutions   **8.** One solution

9.
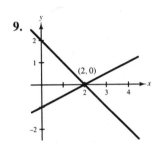

10.
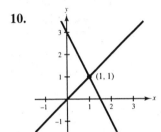

11.
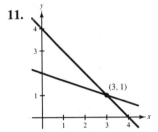

12.
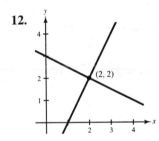

13.
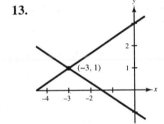

14.
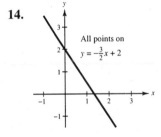

15.
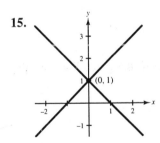

16.
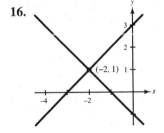

17.
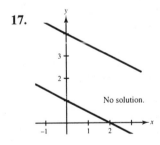

18.
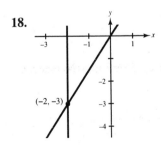

19. One solution

20. No solutions

21. Infinitely many solutions

22. One solution

23. (1, 3)

24. (2, 0)

8.2 | Answers to Exercises

1. (2, −1)

2. (−1, −2)

3. (1, 4)

4. (2, 0)

5. (−1, 4)

6. (2, −3)

7. (−3, 1)

8. (3, 1)

9. (3, 0)

10. (2, 1)

11. No solution

12. (−2, 1)

13. All (x, y) such that $x - 2y = 5$

14. No solution

15. (1, 1)

16. All (x, y) such that $\frac{1}{4}x + \frac{1}{2}y = 4$

17. (−15, −12)

18. (2, −1)

19. $a = 1$

20. $a = 6$

21. $b = 10$

22. $b = -6$

8.3 | Answers to Exercises

1. (2, −1)

2. (14, 5)

3. (1, 1)

4. $\left(2, \frac{3}{4}\right)$

5. (−5, 8)

6. (0, −1)

7. (11, 5)

8. $\left(\frac{11}{3}, 2\right)$

9. (−13, 10)

10. (−1, −1)

11. All (x, y) such that $2x + y = 3$

12. No solution

13. $\left(\frac{11}{5}, \frac{7}{5}\right)$

14. (2, −2)

15. No solution

16. All (x, y) such that $6x - 3y = 9$

17. (−1, 1)

18. $\left(\frac{32}{7}, \frac{3}{7}\right)$

19. (−1, 1)

20. $\left(1, \frac{1}{4}\right)$

21. $\left(-\frac{2}{5}, -2\right)$

22. (−3, 7)

23. $a = 2$

24. $a = 4$

25. $k = 4$

26. $k = -15$

8.4 | Answers to Exercises

1. 15 and 24

2. 13 and 39

3. 19 and 29

4. 27 nickels
23 dimes

5. 50 nickels
25 quarters

6. 45 dimes
    55 quarters

7. Length: 8 in.
    Width: 4 in.

8. Length: $12\frac{1}{2}$ ft
    Width: $10\frac{1}{2}$ ft

9. Length: 15 m
    Width: 10 m

10. $15

11. $350

12. $5000 at $7\frac{1}{2}\%$
    $5500 at $9\frac{1}{4}\%$

13. 175 adult tickets
    125 children's tickets

14. $1.14 for unleaded
    $1.28 for super unleaded

15. 16 lbs. of 80¢ seed

16. 12 liters of 20% solution

17. 1 hr

18. Rate of boat in calm water: 15 mph
    Rate of current: 3 mph

19. 45

20. $m = 1;\ b = \frac{2}{3};\ y = x + \frac{2}{3}$

Answers to Cumulative Practice Test P–8

1. $2x + 4$

2. $(2 + 3)x$

3. $(6 - 5)ab$

4. $a(x + y) + b(x + y)$

5. (e)

6. (c)

7. (a)

8. (b)

9. (d)

10. $2x + 5y - 9 = 0$

11. $x + 1 = 0$

12. $6x + 5y - 13 = 0$

13. $-4x + 4$

14. $x^2 + y^2$

15. $4x^2 - 9$

16. $-3x + 15$

17. $\dfrac{a}{b^5}$

18. $\dfrac{1}{x^4 y^4}$

19. $(1, 1)$

20. $(0, 2)$

21. No solution

22. All (x, y) such that $3x - 4y = 12$

23. $(x + 1)(x - 1)(x^2 + 1)$

24. $(x + 2)(x - 2)^2$

25. All real numbers greater than or equal to -2

26. $x = 2$

27. $x = 1$

28. $x = -9$ or $x = 3$

29. $x = -2$ or $x = 4$

30. $x = -2$ or $x = 6$

31. $x = \frac{1}{2}$

32. $t = \frac{4}{3}$

33. Infinite

34. None

35. (a) $y = 2x - 8$
    (b) $y = -\frac{1}{2}x - 3$

36. (a) $y = \frac{2}{3}x - 5$
    (b) $y = -\frac{3}{2}x + 8$

37. $60

38. 12 and 18

39. 9 hours

40. 15 dimes, 23 quarters

CHAPTER NINE
Rational Expressions and Equations

Simplifying Rational Expressions

Section Highlights

1. A fraction whose numerator and denominator are polynomials is called a **rational expression**.

2. In an rational expression, we cannot allow the variable to be any number that produces zero in the denominator. The usable values of the variable are the **domain** of the rational expression.

3. An rational expression is **simplified** or in **reduced form** if its numerator and denominator have no common factors.

4. Let a, b, and c represent real numbers, variables, or algebraic expressions such that $b \neq 0$ and $c \neq 0$. Then

$$\frac{a \cdot c}{b \cdot c} = \frac{a}{b} \cdot \frac{c}{c} = \frac{a}{b} \cdot 1 = \frac{a}{b}.$$

Therefore,

$$\frac{a \cdot c}{b \cdot c} = \frac{a}{b}.$$

EXAMPLE 1 ■   Finding the Domain of a Rational Expression

Find the domain of each of the following by identifying the excluded values.

(a) $\dfrac{3}{x-1}$

(b) $\dfrac{x}{x^2-4}$

Solution

(a) The denominator is zero when $x - 1 = 0$ or $x = 1$. Thus, the domain is all real numbers except 1.

(b) The denominator is zero when $x^2 - 4 = (x+2)(x-2) = 0$ or when $x = 2$ or $x = -2$. Thus, the domain is all real numbers except 2 and -2.

Starter Exercise 1 | *Fill in the blanks.*

Find the domain of each of the following by identifying the excluded values.

(a) $\dfrac{x+1}{x+2}$ $= x^2 + 2$

The denominator is zero when $\boxed{} = 0$, or when $x = \boxed{}$. Thus, the domain is all real numbers except $\boxed{}$.

(b) $\dfrac{x}{x^2+5x+6}$

The denominator is zero when

$$\boxed{} = \left(\boxed{}\right)\left(\boxed{}\right) = 0,$$

or when $x = \boxed{}$ or $x = \boxed{}$. Thus, the domain is all real numbers except $\boxed{}$ and $\boxed{}$.

272

EXAMPLE 2 ■ Simplifying Rational Expressions

Simplify the rational expression $\dfrac{2x^2 + 4x}{2x^2}$.

Solution

To see if there are any common factors between the numerator and denominator, we must first factor. Then we will reduce if there are any common factors between the numerator and denominator. Note that the domain is all real numbers except 0.

$$\frac{2x^2 + 4x}{2x^2} = \frac{2x(x+2)}{2x(x)}$$ Factor the numerator and denominator.

$$= \frac{2x(x+2)}{2x(x)}$$ Divide out the common factor $2x$.

$$= \frac{x+2}{x}$$ Simplified

Alternative Method:

$$\frac{2x^2 + 4x}{2x^2} = \frac{2x(x+2)}{2x(x)}$$ Factor the numerator and denominator.

$$= \frac{2x}{2x} \cdot \frac{x+2}{x}$$ Separate into product of two fractions.

$$= 1 \cdot \frac{x+2}{x}$$ $\frac{2x}{2x} = 1$

$$= \frac{x+2}{x}$$ Simplified form

Starter Exercise 2 *Fill in the blanks.*

Simplify the following rational expressions.

(a) $\dfrac{3x^2}{6x^2 + x} = \dfrac{\boxed{}(3x)}{x\left(\boxed{}+\boxed{}\right)} = \dfrac{\diagup(3x)}{\diagup x\left(\boxed{}+\boxed{}\right)} = \dfrac{3x}{\boxed{}}$, $x \neq \boxed{}$

(b) **Alternative Method:**

$$\frac{3x^2}{6x^2 + x} = \frac{x(3x)}{x\left(\boxed{}\right)} = \frac{x}{x} \cdot \frac{\boxed{}}{\boxed{}} = 1 \cdot \frac{\boxed{}}{\boxed{}} = \boxed{},$$

$$x \neq \boxed{}$$

EXAMPLE 3 ■ Simplifying Rational Expressions

Simplify the rational expression $\dfrac{x^2 - 5x - 14}{x^3 + 4x^2 + 4x}$.

Solution

$$\frac{x^2 - 5x - 14}{x^3 + 4x^2 + 4x} = \frac{(x+2)(x-7)}{x(x+2)(x+2)} \qquad \text{Factor.}$$

$$= \frac{\cancel{(x+2)}(x-7)}{x\cancel{(x+2)}(x+2)} \qquad \text{Divide out common factor } x+2.$$

$$= \frac{x-7}{x(x+2)} \qquad \text{Simplified form}$$

Alternative Method:

$$\frac{x^2 - 5x + 4}{x^3 + 4x^2 + 4x} = \frac{(x+2)(x-7)}{x(x+2)(x+2)} \qquad \text{Factor.}$$

$$= \frac{x+2}{x+2} \cdot \frac{x-7}{x(x+2)} \qquad \text{Separate fractions into a product.}$$

$$= 1 \cdot \frac{x-7}{x(x+2)} \qquad \frac{x+2}{x+2} = 1$$

$$= \frac{x-7}{x(x+2)} \qquad \text{Simplified form}$$

Starter Exercise 3   *Fill in the blanks.*

Simplify the rational expression $\dfrac{12 - x - x^2}{x^3 - 7x^2 + 12x}$.

$$\frac{12 - x - x^2}{x^3 - 7x^2 + 12x} = \frac{(3-x)\left(\boxed{}\right)}{\boxed{}(x-3)\left(x-\boxed{}\right)} = \frac{(-1)\left(x-\boxed{}\right)\left(\boxed{}\right)}{\boxed{}(x-3)\left(x-\boxed{}\right)}$$

$$= \frac{(-1)\left(x\cancel{-\boxed{}}\right)\left(\boxed{}\right)}{\boxed{}\cancel{(x-3)}\left(x-\boxed{}\right)}$$

$$= -\frac{\boxed{}}{\boxed{}\left(\boxed{}\right)}, \quad x \neq \boxed{}$$

Alternative Method:

$$\frac{12 - x - x^2}{x^3 - 7x^2 + 12x} = \frac{(3-x)\left(\boxed{}\right)}{x(x-3)\left(\boxed{}\right)} = \frac{(-1)(x-3)\left(\boxed{}\right)}{x(x-3)\left(\boxed{}\right)}$$

$$= \frac{x-3}{x-3} \cdot \frac{-1\left(\boxed{}\right)}{\boxed{}\left(\boxed{}\right)} = \boxed{} \cdot \frac{-1\left(\boxed{}\right)}{\boxed{}\left(\boxed{}\right)}$$

$$= -\frac{\boxed{}}{\boxed{}\left(\boxed{}\right)}, \quad x \neq \boxed{}, \boxed{}, \boxed{}$$

■ **Solutions to Starter Exercises** ■

1. (a) The denominator is zero when $\boxed{x+2} = 0$, or when $x = \boxed{-2}$. Thus, the domain is all real numbers except $\boxed{-2}$.

(b) The denominator is zero when $\boxed{x^2+5x+6} = \left(\boxed{x+3}\right)\left(\boxed{x+2}\right) = 0$, or when $x = \boxed{-3}$ or $x = \boxed{-2}$. Thus, the domain is all real numbers except $\boxed{-3}$ and $\boxed{-2}$.

2. $= \dfrac{\boxed{x}\,(3x)}{x\left(\boxed{6x}+\boxed{1}\right)} = \dfrac{\cancel{x}\,(3x)}{\cancel{x}\left(\boxed{6x}+\boxed{1}\right)} = \dfrac{3x}{\boxed{6x+1}},\ x \neq \boxed{0}$

Alternative Method:

$= \dfrac{x(3x)}{x\left(\boxed{x+x}\right)} = \dfrac{x}{x} \cdot \dfrac{\boxed{3x}}{6x+1} = 1 \cdot \dfrac{\boxed{3x}}{6x+1} = \boxed{\dfrac{3x}{6x+1}},\ x \neq \boxed{0}$

3. $= \dfrac{(3-x)\left(\boxed{4+x}\right)}{\boxed{x}\,(x-3)\left(x-\boxed{4}\right)} = \dfrac{(-1)\left(x-\boxed{3}\right)\left(\boxed{4+x}\right)}{\boxed{x}\,(x-3)\left(x-\boxed{4}\right)}$

$= \dfrac{(-1)\left(\cancel{x-\boxed{3}}\right)\left(\boxed{4+x}\right)}{\boxed{x}\,\cancel{(x-3)}\left(x-\boxed{4}\right)} = -\dfrac{\boxed{4+x}}{\boxed{x}\left(\boxed{x-4}\right)},\ x \neq \boxed{3}$

Alternative Method:

$= \dfrac{(3-x)\left(\boxed{4+x}\right)}{x(x-3)\left(\boxed{x-4}\right)} = \dfrac{(-1)(x-3)\left(\boxed{4+x}\right)}{x(x-3)\left(\boxed{x-4}\right)}$

$= \dfrac{x-3}{x-3} \cdot \dfrac{-1\left(\boxed{4+x}\right)}{\boxed{x}\left(\boxed{x-4}\right)} = \boxed{1} \cdot \dfrac{-1\left(\boxed{4+x}\right)}{\boxed{x}\left(\boxed{x-4}\right)}$

$= -\dfrac{\boxed{4+x}}{\boxed{x}\left(\boxed{x-4}\right)},\ x \neq \boxed{3}$

9.1 EXERCISES

In Exercises 1–4, find the domain of the given rational expression.

1. $\dfrac{5}{x+2}$ **2.** $\dfrac{x-1}{6}$ **3.** $\dfrac{x}{x^2-1}$ **4.** $\dfrac{5}{x^2+4}$

In Exercises 5–10, find the missing factor.

5. $\dfrac{x}{x+1} = \dfrac{x^2}{(x+1)\boxed{}}$ **6.** $\dfrac{2}{2x-1} = \dfrac{2\boxed{}}{2x^2-x}$

7. $\dfrac{2}{3} = \dfrac{2\boxed{}}{3(x+1)}$

8. $\dfrac{x-2}{x} = \dfrac{(x-2)\left(\boxed{}\right)}{x(x-3)}$

9. $\dfrac{x}{x+2} = \dfrac{x\boxed{}}{x^2+4x+4}$

10. $\dfrac{x+3}{x-3} = \dfrac{(x+3)\boxed{}}{x^2-9}$

In Exercises 11–30, simplify the given rational expression.

11. $\dfrac{4}{2x}$

12. $\dfrac{9x}{27}$

13. $\dfrac{x^2}{2x}$

14. $\dfrac{3z^3}{6z}$

15. $\dfrac{x(x+2)}{(x+1)(x+2)}$

16. $\dfrac{(x+6)(x-1)}{(x-1)(x+5)}$

17. $\dfrac{xy(x-7)}{x^2y(x-7)}$

18. $\dfrac{x^2-1}{x^2+2x+1}$

19. $\dfrac{4x^2-2x}{2x^3-2x}$

20. $\dfrac{x^2+7x+12}{x^2+6x+8}$

21. $\dfrac{x^2+x-6}{x^2+4x+3}$

22. $\dfrac{z^2-7z+10}{z^2-6z+5}$

23. $\dfrac{x^2+8x+12}{x^2+4x-12}$

24. $\dfrac{6+5a-a^2}{a^2-5a-6}$

25. $\dfrac{z^2+z-2}{z+z^2-2z^3}$

26. $\dfrac{6x^2+5x+1}{9x^2+9x+2}$

27. $\dfrac{x^3+6x^2+9x}{x^2-9}$

28. $\dfrac{a^3-a}{a^3+7a^2+6a}$

29. $\dfrac{x^3-3x^2+x-3}{x-3}$

30. $\dfrac{x^2-4}{x^3-x^2-4x+4}$

31. *Average Cost*   A machine shop has a setup cost of $4000 for production of a new product. The cost of producing each unit is $9.

   (a) Write a rational expression that gives the average cost per unit when x units are produced.

   (b) Determine the domain of the fraction in part (a).

   (c) Find the average cost per unit when 100 units are produced.

9.2 | Multiplying and Dividing Rational Expressions

Section Highlights

1. Let a, b, c, and d represent real numbers, variables, or algebraic expressions such that $b \neq 0$, $c \neq 0$, and $d \neq 0$. Then

   (i) $\dfrac{a}{b} \cdot \dfrac{c}{d} = \dfrac{a \cdot c}{b \cdot d}$

   (ii) $\dfrac{a}{b} \div \dfrac{c}{d} = \dfrac{a}{b} \cdot \dfrac{d}{c} = \dfrac{a \cdot d}{b \cdot c}.$

2. Sometimes division of rational expressions can be written as a compound fraction.

EXAMPLE 1 ■ Multiplying Rational Expressions

Multiply the following rational expressions.

(a) $\dfrac{x^2 y^3}{2xy} \cdot \dfrac{-4xy^2}{x^3}$

(b) $\dfrac{2x^2}{x^2 + 6x + 9} \cdot \dfrac{x^2 - 9}{6x^3}$

(c) $\dfrac{x - y}{x^2 - y^2} \cdot \dfrac{x + y}{6xy}$

Solution

(a) $\dfrac{x^2 y^3}{2xy} \cdot \dfrac{-4xy^2}{x^3} = \dfrac{-4x^3 y^5}{2x^4 y}$ Multiply numerators and denominators.

$= \dfrac{2(-2)x^3(y)(y^4)}{2(x^3)(x)(y)}$ Divide out common factors.

$= \dfrac{-2y^4}{x}, \; y \neq 0$ Simplified form

(b) $\dfrac{2x^2}{x^2 + 6x + 9} \cdot \dfrac{x^2 - 9}{6x^3} = \dfrac{2x^2(x + 3)(x - 3)}{2x^2(3x)(x + 3)^2}$ Multiply and factor.

$= \dfrac{2x^2(x + 3)(x - 3)}{2x^2(3x)(x + 3)(x + 3)}$ Divide out common factors.

$= \dfrac{x - 3}{3x(x + 3)}$ Simplified form

(c) $\dfrac{x - y}{x^2 - y^2} \cdot \dfrac{x + y}{6xy} = \dfrac{(x - y)(x + y)}{6xy(x - y)(x + y)}$ Multiply and factor.

$= \dfrac{(x - y)(x + y)}{6xy(x - y)(x + y)}$ Divide out common factors.

$= \dfrac{1}{6xy}, \; x \neq y, \; -y$ Simplified form

Starter Exercise 1   *Fill in the blanks.*

Multiply the following rational expressions.

(a) $\dfrac{2a^2 b}{15ab^2} \cdot \dfrac{5a^3}{4b^3} = \dfrac{2(5)a^{\square} b}{15(4)ab^{\square}} = \dfrac{2(5)a(a^{\square})b}{3(2)(5)ab(b^{\square})} = \boxed{}, \; a \neq 0$

(b) $\dfrac{12 - x - x^2}{x^2 - 9} \cdot \dfrac{x^2 + 2x - 3}{x + 4} = \dfrac{(3 - x)(4 + x)(x + 3)\left(\boxed{}\right)}{(x + 3)\left(\boxed{}\right)(x + 4)}$

$= \dfrac{\boxed{}(x - 3)(4 + x)(x + 3)\left(\boxed{}\right)}{(x + 3)\left(\boxed{}\right)(x + 4)}$

$= \boxed{}, \; x \neq \boxed{}, \boxed{}, \boxed{}$

(c) $\dfrac{x}{x^2 + 5x + 6} \cdot (x + 2) = \dfrac{x}{\left(\boxed{}\right)(x + 2)} \cdot \dfrac{x + 2}{1}$

$= \dfrac{x(x + 2)}{\left(\boxed{}\right)(x + 2)} = \boxed{}, \; x \neq \boxed{}$

EXAMPLE 2 ■ Multiplying Three Rational Expressions

Multiply the following rational expressions.

$$\frac{x}{x-2} \cdot \frac{x^2-x-2}{x^2-3x} \cdot \frac{x-3}{x+4}$$

Solution

$$\frac{x}{x-2} \cdot \frac{x^2-x-2}{x^2-3x} \cdot \frac{x-3}{x+4} = \frac{x(x+1)(x-2)(x-3)}{(x-2)(x)(x-3)(x+4)}$$ Multiply and factor.

$$= \frac{\cancel{x}(x+1)\cancel{(x-2)}\cancel{(x-3)}}{\cancel{(x-2)}\cancel{(x)}\cancel{(x-3)}(x+4)}$$ Divide out common factors.

$$= \frac{x+1}{x+4}, \quad x \neq 0, \ 2, \ 3$$ Simplified form

---

> **Starter Exercise 2** | *Fill in the blanks.*

Multiply the following rational expression.

$$\frac{x+2}{x+1} \cdot \frac{2x^2-3x+1}{x+2} \cdot \frac{x+1}{x-1} = \frac{(x+2)\left(\boxed{}\right)\left(\boxed{}\right)(x+1)}{(x+1)(x+2)(x-1)}$$

$$= \frac{\cancel{(x+2)}\left(\boxed{}\right)\left(\boxed{}\right)\cancel{(x+1)}}{\cancel{(x+1)}\cancel{(x+2)}\cancel{(x-1)}}$$

$$= \boxed{}, \quad x \neq \boxed{}, \boxed{}, \boxed{}$$

EXAMPLE 3 ■ Dividing Rational Expressions

Perform the following divisions.

(a) $\dfrac{2x}{x-1} \div \dfrac{4x}{x-1}$

(b) $\dfrac{2x^2+3x+1}{x+3} \div \dfrac{2x^2+x-1}{x+3}$

Solution

(a) $\dfrac{2x}{x-1} \div \dfrac{4x}{x-1} = \dfrac{2x}{x-1} \cdot \dfrac{x-1}{4x}$ Invert and multiply.

$$= \frac{2x(x-1)}{(x-1)(2)(2x)}$$ Multiply and factor.

$$= \frac{\cancel{2x}\cancel{(x-1)}}{\cancel{(x-1)}(2)\cancel{(2x)}}$$ Divide out common factors.

$$= \frac{1}{2}, \quad x \neq 1, \ 0$$ Simplified form

(b) $\dfrac{2x^2 + 3x + 1}{x + 3} \div \dfrac{2x^2 + x - 1}{x + 3} = \dfrac{2x^2 + 3x + 1}{x + 3} \cdot \dfrac{x + 3}{2x^2 + x - 1}$ Invert and multiply.

$\qquad\qquad = \dfrac{(2x + 1)(x + 1)(x + 3)}{(x + 3)(2x - 1)(x + 1)}$ Multiply and factor.

$\qquad\qquad = \dfrac{(2x + 1)\cancel{(x + 1)}\cancel{(x + 3)}}{\cancel{(x + 3)}(2x - 1)\cancel{(x + 1)}}$ Divide out common factors.

$\qquad\qquad = \dfrac{2x + 1}{2x - 1}, \quad x \neq -3, -1$ Simplified form

---

Starter Exercise 3   *Fill in the blanks.*

Perform the following divisions.

(a) $\dfrac{6x^2}{x - 3} \div \dfrac{3x}{x - 3} = \dfrac{6x^2}{x - 3} \cdot \dfrac{x - 3}{3x} = \dfrac{3x\left(\boxed{}\right)(x - 3)}{(x - 3)(3x)}$

$\qquad\qquad = \dfrac{\cancel{3x}\left(\boxed{}\right)\cancel{(x - 3)}}{\cancel{(x - 3)}\cancel{(3x)}} = \boxed{}, \quad x \neq \boxed{}, \boxed{}$

(b) $\dfrac{x - 1}{x^2 - x - 6} \div \dfrac{2x^2 + x - 3}{x + 2} = \dfrac{x - 1}{x^2 - x - 6} \cdot \dfrac{\boxed{}}{2x^2 + x - 3}$

$\qquad\qquad = \dfrac{(x - 1)\left(\boxed{}\right)}{(x - 3)\left(\boxed{}\right)(2x + 3)\left(\boxed{}\right)}$

$\qquad\qquad = \dfrac{\cancel{(x - 1)}\left(\boxed{\cancel{}}\right)}{(x - 3)\left(\boxed{\cancel{}}\right)(2x + 3)\left(\boxed{\cancel{}}\right)}$

$\qquad\qquad = \boxed{}, \quad x \neq \boxed{}, \boxed{}$

EXAMPLE 4 ■   Simplifying a Compound Fraction

Simplify the compound fraction $\dfrac{\left(\dfrac{4 - 3z - z^2}{z + 6}\right)}{\left(\dfrac{z^2 - 5z + 4}{2z + 12}\right)}$.

Solution

$\dfrac{\left(\dfrac{4 - 3z - z^2}{z + 6}\right)}{\left(\dfrac{z^2 - 5z + 4}{2z + 12}\right)} = \dfrac{4 - 3z - z^2}{z + 6} \cdot \dfrac{2z + 12}{z^2 - 5z + 4}$ Invert and multiply.

$\qquad\qquad = \dfrac{(1 - z)(4 + z)(2)(z + 6)}{(z + 6)(z - 1)(z - 4)}$ Multiply and factor.

$\qquad\qquad = \dfrac{-1\cancel{(z - 1)}(4 + z)(2)\cancel{(z + 6)}}{\cancel{(z + 6)}\cancel{(z - 1)}(z - 4)}$ Divide out common factors.

$\qquad\qquad = -\dfrac{2(4 + z)}{z - 4}, \quad z \neq -6, 1$ Simplified form

■ **Starter Exercise 4** | *Fill in the blanks.*

Simplify the following compound fraction.

$$\frac{\left(\dfrac{6x^2 + 11x - 2}{x^2 + 4x - 21}\right)}{\left(\dfrac{x + 2}{x^2 + 10x + 21}\right)} = \frac{6x^2 + 11x - 2}{x^2 + 4x - 21} \cdot \frac{x^2 + 10x + 21}{x + 2}$$

$$= \frac{(x + 2)\left(\boxed{}\right)(x + 7)\left(\boxed{}\right)}{\left(\boxed{}\right)(x + 7)(x + 2)}$$

$$= \boxed{} , \quad x \neq \boxed{} , \boxed{}$$

■ **Solutions to Starter Exercises** ■

1. (a) $= \dfrac{2(5)a^{\boxed{5}}b}{15(4)ab^{\boxed{5}}} = \dfrac{2(5)a(a^{\boxed{4}})b}{3(2)(5)ab(b^{\boxed{4}})} = \boxed{\dfrac{a^4}{3b^4}}$

(b) $= \dfrac{(3 - x)(4 + x)(x + 3)\left(\boxed{x - 1}\right)}{(x + 3)\left(\boxed{x - 3}\right)(x + 4)}$

$= \dfrac{\boxed{-1}\,(x - 3)(4 + x)(x + 3)\left(\boxed{x - 1}\right)}{(x + 3)\left(\boxed{x - 3}\right)(x + 4)}$

$= \boxed{-(x - 1)} , \quad x \neq \boxed{-3} , \boxed{-4} , \boxed{3}$

(c) $= \dfrac{x}{\left(\boxed{x + 3}\right)(x + 2)} \cdot \dfrac{x + 2}{1} = \dfrac{x(x + 2)}{\left(\boxed{x + 3}\right)(x + 2)}$

$= \boxed{\dfrac{x}{x + 3}} , \quad x \neq \boxed{-2}$

2. $= \dfrac{(x + 2)\left(\boxed{2x - 1}\right)\left(\boxed{x - 1}\right)(x + 1)}{(x + 1)(x + 2)(x - 1)}$

$= \dfrac{(x + 2)\left(\boxed{2x - 1}\right)\left(\boxed{x - 1}\right)(x + 1)}{(x + 1)(x + 2)(x - 1)}$

$= \boxed{2x - 1} , \quad x \neq \boxed{-1} , \boxed{-2} , \boxed{1}$

3. (a) $= \dfrac{3x\left(\boxed{2x}\right)(x - 3)}{(x - 3)(3x)} = \dfrac{3x\left(\boxed{2x}\right)(x - 3)}{(x - 3)(3x)} = \boxed{2x} , \quad x \neq \boxed{0} , \boxed{3}$

■ **Solutions to Starter Exercises** ■

3. —CONTINUED—

(b) $= \dfrac{x-1}{x^2-x-6} \cdot \dfrac{\boxed{x+2}}{2x^2+x-3} = \dfrac{(x-1)\left(\boxed{x+2}\right)}{(x-3)\left(\boxed{x+2}\right)(2x+3)\left(\boxed{x-1}\right)}$

$= \dfrac{(x-1)\left(\boxed{x+2}\right)}{(x-3)\left(\boxed{x+2}\right)(2x+3)\left(\boxed{x-1}\right)}$

$= \boxed{\dfrac{1}{(x-3)(2x+3)}}, \quad x \neq \boxed{-2}, \boxed{1}$

4. $= \dfrac{(x+2)\left(\boxed{6x-1}\right)(x+7)\left(\boxed{x+3}\right)}{\left(\boxed{x-3}\right)(x+7)(x+2)}$

$= \boxed{\dfrac{(6x-1)(x+3)}{x-3}}, \quad x \neq \boxed{-7}, \boxed{-2}$

9.2 EXERCISES

In Exercises 1–4, find the missing expression so that the two rational expressions are equivalent.

1. $\dfrac{2x}{3} = \dfrac{2x\left(\boxed{}\right)}{3x+6}$

2. $\dfrac{2x}{x-1} = \dfrac{2x(x-7)}{(x-1)\left(\boxed{}\right)}$

3. $\dfrac{xy}{6} = \dfrac{x^2y}{6\boxed{}}$

4. $\dfrac{3x}{x-3} = \dfrac{3x\left(\boxed{}\right)}{x^2-9}$

In Exercises 5–18, multiply the given rational expressions and simplify your answer.

5. $\dfrac{4x}{3} \cdot \dfrac{9y}{2}$

6. $\dfrac{ab^2}{4xy} \cdot \dfrac{xy^2}{ab}$

7. $\dfrac{x+2}{x} \cdot \dfrac{2x^2}{3x+6}$

8. $\dfrac{2x-1}{5x+10} \cdot (x+2)$

9. $\dfrac{5}{2x+3} \cdot \dfrac{2x+3}{5}$

10. $(4x-2) \cdot \dfrac{x+5}{2x-1}$

11. $\dfrac{1-x}{1+x} \cdot \dfrac{x+1}{x-1}$

12. $\dfrac{(5x-1)(2x-3)}{2x+7} \cdot \dfrac{x+6}{(x+7)(5x-1)}$

13. $\dfrac{z^2-4}{z^2+3z} \cdot \dfrac{z+3}{z^2-4z+4}$

14. $\dfrac{x^2+5x+6}{x+4} \cdot \dfrac{x^2+x-12}{x^2+6x+9}$

15. $\dfrac{x^2-y^2}{x^2-2xy} \cdot \dfrac{x^3-2x^2y}{2x^2+3xy+y^2}$

16. $\dfrac{a^2+2ab+b^2}{a^2+2ab-3b^2} \cdot \dfrac{a^2-3ab+2b^2}{a^2+3ab+2b^2}$

17. $\dfrac{x+1}{x+2} \cdot \dfrac{x^2+3x+2}{x^2-2x} \cdot \dfrac{x}{x+1}$

18. $\dfrac{2x^2-x-1}{x} \cdot \dfrac{x}{(x-1)(x-2)} \cdot \dfrac{x-2}{2x+1}$

In Exercises 19–26, perform the indicated division and simplify your answer.

19. $\dfrac{6x^2}{7} \div \dfrac{3x}{14}$

20. $\dfrac{2x(x+5)}{x+3} \div \dfrac{x(x+5)}{3(x+3)}$

21. $\dfrac{x^2-y^2}{2x+y} \div \dfrac{x-y}{2x+y}$

22. $\dfrac{x^2-5x-6}{x^2+4x+3} \div \dfrac{x^2-4x+4}{x^2+3x+2}$

23. $\dfrac{\left(\dfrac{x}{x+2}\right)}{\left(\dfrac{x}{x+2}\right)}$

24. $\dfrac{\left(\dfrac{2x^2}{3}\right)}{\left(\dfrac{x^2}{9}\right)}$

25. $\dfrac{\left(\dfrac{x^2-4x-12}{x^2+4x-12}\right)}{\left(\dfrac{x+2}{x-2}\right)}$

26. $\dfrac{\left(\dfrac{6x^2-xy-y^2}{4x^2+3xy-y^2}\right)}{\left(\dfrac{2x-y}{x^2-y^2}\right)}$

In Exercises 27–30, perform the indicated operations and simplify your answer.

27. $\dfrac{x^2}{5} \cdot \dfrac{5x+10}{x^2+x} \div \dfrac{x+2}{x+1}$

28. $\left(\dfrac{ab^2}{3a}\right)^2 \div \left(\dfrac{ab}{3}\right)^3$

29. $\left[\left(\dfrac{x+1}{2}\right)^2 \cdot \left(\dfrac{x-1}{3}\right)^2\right] \div \dfrac{x^2-1}{36}$

30. $\left[\left(\dfrac{x+2}{x}\right)^2 \cdot \dfrac{x^2}{x^2+4x+4}\right] \div \dfrac{2x-5}{5x+1}$

31. *Photocopy Rate*   A photocopier produces copies at a rate of 10 pages per minute.

(a) Determine the time required to copy one page.

(b) Determine the time required to copy x pages.

(c) Determine the time required to copy 100 pages.

9.3 | Adding and Subtracting Rational Expressions

Section Highlights

1. Let a, b, and c be numbers, variables, or algebraic expressions such that $c \neq 0$. Then

   (i) $\dfrac{a}{c} + \dfrac{b}{c} = \dfrac{a+b}{c}$

   (ii) $\dfrac{a}{c} - \dfrac{b}{c} = \dfrac{a-b}{c}$.

2. The **least common multiple** of two or more polynomials is the simplest polynomial that is a multiple of each of the original polynomials.

3. The **least common denominator** of two or more fractions is the least common multiple of all the denominators.

EXAMPLE 1 ■   Adding Fractions with Like Denominators

Add the fractions $\dfrac{x}{2} + \dfrac{x+1}{2}$.

Solution

$$\dfrac{x}{2} + \dfrac{x+1}{2} = \dfrac{x+(x+1)}{2} = \dfrac{2x+1}{2}$$

Starter Exercise 1   *Fill in the blanks.*

Add the following fractions.

$$\frac{2}{x-2}+\frac{x-1}{x-2}=\frac{2+\left(\boxed{}\right)}{x-2}=\boxed{}$$

EXAMPLE 2 ■   Subtracting Fractions with Like Denominators

Subtract the fractions $\dfrac{x+1}{x-6}-\dfrac{2x-3}{x-6}$.

Solution

$$\frac{x+1}{x-6}-\frac{2x-3}{x-6}=\frac{(x+1)-(2x-3)}{x-6}=\frac{-x+4}{x-6}$$

Starter Exercise 2   *Fill in the blanks.*

Subtract the following fractions.

$$\frac{x+5}{2x-1}-\frac{x+7}{2x-1}=\frac{(x+5)-(x+7)}{\boxed{}}=\boxed{}$$

EXAMPLE 3 ■   Adding Fractions and Simplifying

Add the fractions $\dfrac{x}{x^2-1}+\dfrac{1}{x^2-1}$.

Solution

$$\frac{x}{x^2-1}+\frac{1}{x^2-1}=\frac{x+1}{x^2-1}$$ Add numerators.

$$=\frac{x+1}{(x+1)(x-1)}$$ Factor completely.

$$=\frac{\cancel{x+1}}{\cancel{(x+1)}(x-1)}$$ Divide out common factors.

$$=\frac{1}{x-1},\ \ x\neq-1$$ Simplified form.

Starter Exercise 3   *Fill in the blanks.*

Add the following fractions.

$$\frac{x+1}{x^2+2x}+\frac{1}{x^2+2x}=\frac{\boxed{}}{x^2+2x}=\frac{\boxed{}}{\boxed{}(x+2)}=\frac{\boxed{\diagup}}{\boxed{}\cancel{(x+2)}}=\boxed{},x\neq\boxed{}$$

EXAMPLE 4 ■   Combining Three Fractions with Like Denominators

Perform the following operations.

$$\frac{x^2-3}{x-1}+\frac{2x+1}{x-1}-\frac{x}{x-1}$$

Solution

$$\frac{x^2-3}{x-1}+\frac{2x+1}{x-1}-\frac{x}{x-1}=\frac{(x^2-3)+(2x+1)-x}{x-1}$$

$$=\frac{x^2+x-2}{x-1}=\frac{(x-1)(x+2)}{x-1}=x+2,\ x\neq 1$$ ■

| **Starter Exercise 4** | *Fill in the blanks.* |

Perform the following operations.

$$\frac{2x^2+3}{x^2-4}-\frac{x^2+5x}{x^2-4}+\frac{10x+3}{x^2-4}=\frac{2x^2+3-(x^2+5x)+10x+3}{\boxed{}}$$

$$=\frac{x^2+\boxed{}+6}{\boxed{}}=\frac{(\boxed{})(\boxed{})}{(\boxed{})(\boxed{})}$$

$$=\frac{\boxed{}}{\boxed{}},\ x\neq\boxed{}$$

EXAMPLE 5 ■   Finding Least Common Multiples

Find the least common multiple of each of the following sets of polynomials.

(a) 6, 12, 18

(b) $2x^2,\ x^3$

(c) $2x+2,\ x+1$

(d) $x^2-x-6,\ x^2-9$

Solution

(a) $6=2\cdot 3,\ \ 12=2^2\cdot 3,\ \ 18=2\cdot 3^2$

The different factors are 2 and 3. Using the highest powers of these factors, we conclude that the least common multiple is $2^2\cdot 3^2=36$.

(b) The different factors are 2 and x. Using the highest powers of these factors, we conclude that the least common multiple is $2x^3$.

(c) $2x+2=2(x+1),\ \ x+1$

The different factors are 2 and $x+1$. Using the highest powers of these factors, we conclude that the least common multiple is $2(x+1)$.

(d) $x^2-x-6=(x-3)(x+2),\ \ x^2-9=(x+3)(x-3)$

The different factors are $x-3$, $x+3$, and $x+2$. Using the highest powers of these factors, we conclude that the least common multiple is $(x-3)(x+3)(x+2)$. ■

Starter Exercise 5 | *Fill in the blanks.*

Find the least common multiple of each of the following sets of polynomials.

(a) 5, 9, 15

$5 = 5, \quad 9 = \boxed{}^2, \quad 15 = \boxed{} \cdot \boxed{}$

Least common multiple is $3^{\boxed{}} \cdot 5 = \boxed{}$.

(b) $25x, \ 10x^2$

$25x = 5^2 x, \quad 10x^2 = \boxed{} \cdot \boxed{} \cdot x^2$

Least common multiple is $\boxed{} \cdot \boxed{} \cdot x^2 = \boxed{}$.

(c) $x^2 - x - 2, \ x^2 + x - 6$

$x^2 - x - 2 = (x - 2)(x + 1), \quad x^2 + x - 6 = \left(\boxed{}\right)(x + 3)$

Least common multiple is $\left(\boxed{}\right)\left(\boxed{}\right)\left(\boxed{}\right)$.

(d) $x^2 + 4x + 4, \ x^2 - 4$

$x^2 + 4x + 4 = (x + 2)^2, \quad x^2 - 4 = (x + 2)(x - 2)$

Least common multiple is $\left(\boxed{}\right)^2 (x - 2)$.

EXAMPLE 6 ■ Adding Fractions with Unlike Denominators

Add the following fractions.

(a) $\dfrac{2}{5x} + \dfrac{6}{x^2}$

(b) $\dfrac{-1}{x^2 + 3x + 2} + \dfrac{1}{x + 1}$

Solution

(a) The least common denominator is $5x^2$.

$$\frac{2}{5x} + \frac{6}{x^2} = \frac{2}{5x} \cdot \frac{x}{x} + \frac{6}{x^2} \cdot \frac{5}{5} \qquad \text{Find equivalent fractions that have the least common denominator.}$$

$$= \frac{2x}{5x^2} + \frac{30}{5x^2} \qquad \text{Same denominators}$$

$$= \frac{2x + 30}{5x^2} \qquad \text{Add fractions.}$$

(b) Since $x^2 + 3x + 2 = (x + 1)(x + 2)$, the least common denominator is $(x + 1)(x + 2)$.

$$\frac{-1}{x^2 + 3x + 2} + \frac{1}{x + 1} = \frac{-1}{(x + 1)(x + 2)} + \frac{1}{x + 1} \cdot \frac{x + 2}{x + 2} \qquad \text{Find equivalent fractions that have the least common denominator.}$$

$$= \frac{-1}{(x + 1)(x + 2)} + \frac{x + 2}{(x + 1)(x + 2)} \qquad \text{Same denominators}$$

$$= \frac{-1 + (x + 2)}{(x + 1)(x + 2)} \qquad \text{Add fractions.}$$

$$= \frac{\cancel{x + 1}}{\cancel{(x + 1)}(x + 2)} \qquad \text{Divide out common factors.}$$

$$= \frac{1}{x + 2}, \quad x \neq -1 \qquad \text{Simplified form}$$

■

> **Starter Exercise 6** *Fill in the blanks.*

(a) Add the fractions: $\dfrac{7}{2x^2} + \dfrac{5}{6x}$

Least common denominator is $6x^2$.

$$\frac{7}{2x^2} + \frac{5}{6x} = \frac{7}{2x^2} \cdot \frac{3}{3} + \frac{5}{6x} \cdot \frac{x}{x} = \frac{\boxed{}}{6x^2} + \frac{\boxed{}}{6x^2} = \frac{\boxed{} + \boxed{}}{6x^2}$$

(b) Add the fractions: $\dfrac{x}{x^2 - 16} + \dfrac{2}{x^2 + 2x - 8}$

Least common denominator is $(x+4)(x-4)(x-2)$.

$$\frac{x}{x^2-16} + \frac{2}{x^2+2x-8} = \frac{x}{(x+4)(x-4)} \cdot \frac{\boxed{}}{\boxed{}} + \frac{2}{(x+4)(x-2)} \cdot \frac{\boxed{}}{\boxed{}}$$

$$= \frac{\boxed{}}{(x+4)(x-4)(x-2)} + \frac{\boxed{}}{(x+4)(x-4)(x-2)}$$

$$= \frac{\boxed{} + \boxed{}}{(x+4)(x-4)(x-2)} = \frac{\boxed{}}{(x+4)(x-4)(x-2)}$$

EXAMPLE 7 ■ Subtracting Fractions with Unlike Denominators

Subtract the fractions $\dfrac{2x}{x+5} - \dfrac{3}{x-1}$.

Solution

Least common denominator is $(x+5)(x-1)$.

$$\frac{2x}{x+5} - \frac{3}{x-1} = \frac{2x}{x+5} \cdot \frac{x-1}{x-1} - \frac{3}{x-1} \cdot \frac{x+5}{x+5}$$

Find equivalent fractions that have the least common denominator.

$$= \frac{2x^2 - 2x}{(x+5)(x-1)} - \frac{3x+15}{(x-1)(x+5)}$$

Same denominators

$$= \frac{(2x^2 - 2x) - (3x+15)}{(x+5)(x-1)}$$

Subtract fractions.

$$= \frac{2x^2 - 5x - 15}{(x+5)(x-1)}$$

Simplified form

> **Starter Exercise 7** *Fill in the blanks.*

Subtract the fractions: $\dfrac{x}{x^2 + 2x + 1} - \dfrac{2}{x+1}$

Least common denominator is $(x+1)^2$.

$$\frac{x}{x^2+2x+1} - \frac{2}{x+1} = \frac{x}{(x+1)^2} - \frac{2}{x+1} \cdot \frac{x+1}{x+1}$$

$$= \frac{x}{(x+1)^2} - \frac{\boxed{}}{(x+1)^2} = \frac{x - \left(\boxed{}\right)}{(x+1)^2} = \frac{\boxed{}}{(x+1)^2}$$

EXAMPLE 8 ■ Combining Fractions with Unlike Denominators

Combine the following fractions.

$$\frac{x}{2x-2} - \frac{2x}{1-x} - \frac{3}{4x}$$

Solution

Least common denominator is $4x(x-1)$. Note that since $1-x = -1(x-1)$, we can rewrite the expression as

$$\frac{x}{2x-2} - \frac{2x}{1-x} \cdot \frac{-1}{-1} - \frac{3}{4x} = \frac{x}{2x-2} + \frac{2x}{x-1} - \frac{3}{4x}.$$

Now we simplify.

$$\frac{x}{2x-2} + \frac{2x}{x-1} - \frac{3}{4x} = \frac{x}{2(x-1)} \cdot \frac{2x}{2x} + \frac{2x}{x-1} \cdot \frac{4x}{4x} - \frac{3}{4x} \cdot \frac{x-1}{x-1} \qquad \text{Find equivalent fractions that have least common denominator.}$$

$$= \frac{2x^2}{4x(x-1)} + \frac{8x^2}{4x(x-1)} - \frac{3x-3}{4x(x-1)} \qquad \text{Same denominators}$$

$$= \frac{10x^2 - 3x + 3}{4x(x-1)} \qquad \text{Simplified form}$$

Starter Exercise 8 | *Fill in the blanks.*

Combine the fractions: $\dfrac{x^2+x-10}{x^2+5x} - \dfrac{x+1}{2x+10} + \dfrac{5}{2x}$

Least common denominator is $2x(x+5)$.

$$\frac{x^2+x-10}{x^2+5x} - \frac{x+1}{2x+10} + \frac{5}{2x} = \frac{x^2+x-10}{x(x+5)} \cdot \frac{2}{2} - \frac{x+1}{2(x+5)} \cdot \frac{x}{x} + \frac{5}{2x} \cdot \frac{x+5}{x+5}$$

$$= \frac{2x^2+2x-20}{2x(x+5)} - \frac{\boxed{}}{2x(x+5)} + \frac{5x+\boxed{}}{2x(x+5)}$$

$$= \frac{(2x^2+2x-20) - \left(\boxed{}\right) + 5x + \boxed{}}{2x(x+5)}$$

$$= \frac{\boxed{}}{2x(x+5)} = \frac{\left(\boxed{}\right)\left(\boxed{}\right)}{2x(x+5)}$$

$$= \frac{\boxed{}}{2x}, \quad x \neq -5$$

EXAMPLE 9 ■ Simplifying Compound Fractions

Simplify the following compound fractions.

(a) $\dfrac{\left(\dfrac{2}{x} - \dfrac{x}{2}\right)}{\left(1 + \dfrac{2}{x}\right)}$

(b) $\dfrac{\left(\dfrac{1}{2x+1} - 1\right)}{\left(\dfrac{x}{2x+1} - 1\right)}$

Solution

(a) $\dfrac{\left(\dfrac{2}{x} - \dfrac{x}{2}\right)}{\left(1 + \dfrac{2}{x}\right)} = \dfrac{\left(\dfrac{4}{2x} - \dfrac{x^2}{2x}\right)}{\left(\dfrac{x}{x} + \dfrac{2}{x}\right)}$ Find least common denominator.

$= \dfrac{\left(\dfrac{4 - x^2}{2x}\right)}{\left(\dfrac{x + 2}{x}\right)}$ Add fractions.

$= \dfrac{4 - x^2}{2x} \cdot \dfrac{x}{x + 2}$ Invert and multiply.

$= \dfrac{\cancel{x}(2 + \cancel{x})(x - 2)}{2\cancel{x}\cancel{(x + 2)}}$ Divide out common factors.

$= \dfrac{x - 2}{2}, \quad x \neq 0$ Simplified form

(b) This time we will multiply numerator and denominator by the least common denominator of every fraction.

$$\dfrac{\left(\dfrac{1}{2x + 1} - 1\right)}{\left(\dfrac{x}{2x + 1} - 1\right)} = \dfrac{\left(\dfrac{1}{2x + 1} - 1\right)}{\left(\dfrac{x}{2x + 1} - 1\right)} \cdot \dfrac{(2x + 1)}{(2x + 1)}$$

$$= \dfrac{\dfrac{1}{2x + 1}(2x + 1) - 1(2x + 1)}{\dfrac{x}{2x + 1}(2x + 1) - 1(2x + 1)} = \dfrac{1 - (2x + 1)}{x - (2x + 1)} = \dfrac{-2x}{-x - 1}, \quad x \neq -\dfrac{1}{2}$$

■

Starter Exercise 9   *Fill in the blanks.*

Simplify the following compound fractions.

(a) We will use the method in Example 9a.

$$\dfrac{\left(\dfrac{x}{3} - \dfrac{1}{3}\right)}{\left(1 - \dfrac{1}{2x}\right)} = \dfrac{\left(\dfrac{x}{3} - \dfrac{1}{3}\right)}{\left(\dfrac{2x}{\Box} - \dfrac{1}{2x}\right)} = \dfrac{\left(\dfrac{x - 1}{3}\right)}{\left(\dfrac{2x - 1}{2x}\right)} = \dfrac{x - 1}{3} \cdot \boxed{} = \boxed{}, \quad x \neq \Box$$

(b) We will use the method in Example 9b.

$$\dfrac{\left(\dfrac{2}{x} - 1\right)}{\left(2 - \dfrac{1}{x}\right)} = \dfrac{\left(\dfrac{2}{x} - 1\right)}{\left(2 - \dfrac{1}{x}\right)} \cdot \dfrac{x}{\Box} = \dfrac{\dfrac{2}{x}x - x}{2\Box - \dfrac{1}{x}\Box} = \dfrac{\Box - x}{\Box}, \quad x \neq \Box$$

■ **Solutions to Starter Exercises** ■

1. $= \dfrac{2 + \left(\boxed{x-1} \right)}{x-2} = \boxed{\dfrac{x+1}{x-2}}$

2. $= \dfrac{(x+5) - (x+7)}{\boxed{2x-1}} = \boxed{\dfrac{-2}{2x-1}}$

3. $= \dfrac{\boxed{x+2}}{x^2 + 2x} = \dfrac{\boxed{x+2}}{\boxed{x}\,(x+2)} = \dfrac{\cancel{x+2}}{\boxed{x}\,\cancel{(x+2)}} = \boxed{\dfrac{1}{x}}, \quad x \ne \boxed{-2}$

4. $= \dfrac{2x^2 + 3 - (x^2 + 5x) + 10x + 3}{\boxed{x^2 - 4}} = \dfrac{x^2 + \boxed{5x} + 6}{\boxed{x^2 - 4}}$

$= \dfrac{\left(\boxed{x+2} \right)\left(\boxed{x+3} \right)}{\left(\boxed{x+2} \right)\left(\boxed{x-2} \right)} = \dfrac{x+3}{x-2}, \quad x \ne \boxed{-2}$

5. (a) $5 = 5, \quad 9 = \boxed{3}^{\,2}, \quad 15 = \boxed{3} \cdot \boxed{5}$

Least common multiple is $3^{\boxed{2}} \cdot 5 = \boxed{45}$.

(b) $25x = 5^2 x, \quad 10x^2 = \boxed{2} \cdot \boxed{5} \cdot x^2$

Least common multiple is $\boxed{2} \cdot \boxed{5^2} \cdot x^2 = \boxed{50x^2}$.

(c) $x^2 - x - 2 = (x-2)(x+1), \quad x^2 + x - 6 = \left(\boxed{x-2} \right)(x+3)$

Least common multiple is $\left(\boxed{x-2} \right)\left(\boxed{x+1} \right)\left(\boxed{x+3} \right)$.

(d) Least common multiple is $\left(\boxed{x+2} \right)^2 (x-2)$.

6. (a) $= \dfrac{\boxed{21}}{6x^2} + \dfrac{\boxed{5x}}{6x^2} = \dfrac{\boxed{21} + \boxed{5x}}{6x^2}$

(b) $= \dfrac{x}{(x+4)(x-4)} \cdot \dfrac{\boxed{x-2}}{\boxed{x-2}} + \dfrac{2}{(x+4)(x-2)} \cdot \dfrac{\boxed{x-4}}{\boxed{x-4}}$

$= \dfrac{\boxed{x^2 - 2x}}{(x+4)(x-4)(x-2)} + \dfrac{\boxed{2x-8}}{(x+4)(x-4)(x-2)}$

$= \dfrac{\boxed{x^2 - 2x} + \boxed{2x-8}}{(x+4)(x-4)(x-2)} = \dfrac{\boxed{x^2 - 8}}{(x+4)(x-4)(x-2)}$

■   **Solutions to Starter Exercises**   ■

7. $= \dfrac{x}{(x+1)^2} - \dfrac{\boxed{2x+2}}{(x+1)^2} = \dfrac{x - \left(\boxed{2x+2}\right)}{(x+1)^2} = \dfrac{\boxed{-x-2}}{(x+1)^2}$

8. $= \dfrac{2x^2 + 2x - 20}{2x(x+5)} - \dfrac{\boxed{x^2+x}}{2x(x+5)} + \dfrac{5x + \boxed{25}}{2x(x+5)}$

$= \dfrac{(2x^2 + 2x - 20) - \left(\boxed{x^2+x}\right) + 5x + \boxed{25}}{2x(x+5)} = \dfrac{\boxed{x^2 + 6x + 5}}{2x(x+5)}$

$= \dfrac{\left(\boxed{x+1}\right)\left(\boxed{x+5}\right)}{2x(x+5)} = \dfrac{\boxed{x+1}}{2x}, \quad x \neq -5$

9. (a) $= \dfrac{\left(\dfrac{x}{3} - \dfrac{1}{3}\right)}{\left(\dfrac{2x}{\boxed{2x}} - \dfrac{1}{2x}\right)} = \dfrac{\left(\dfrac{x-1}{3}\right)}{\left(\dfrac{2x-1}{2x}\right)} = \dfrac{x-1}{3} \cdot \boxed{\dfrac{2x}{2x-1}}$

$= \boxed{\dfrac{2x(x-1)}{3(2x-1)}}, \quad x \neq \boxed{0}$

(b) $= \dfrac{\left(\dfrac{2}{x} - 1\right)}{\left(2 - \dfrac{1}{x}\right)} \cdot \dfrac{x}{\boxed{x}} = \dfrac{\dfrac{2}{x}x - 1 \cdot x}{2\boxed{x} - \dfrac{1}{x}\boxed{x}} = \dfrac{\boxed{2} - x}{\boxed{2x-1}}, \quad x \neq \boxed{0}$

9.3 | EXERCISES

In Exercises 1–8, perform the indicated operations and simplify your answer.

1. $\dfrac{5}{a} + \dfrac{2}{a}$ **2.** $\dfrac{4}{x^2} - \dfrac{2}{x^2}$ **3.** $\dfrac{1}{2x} - \dfrac{5}{2x}$ **4.** $\dfrac{x}{5} - \dfrac{1+x}{5}$

5. $\dfrac{2}{x-1} - \dfrac{1+x}{x-1}$ **6.** $\dfrac{3x}{x+3} + \dfrac{9}{x+3}$ **7.** $\dfrac{5z-2}{z+1} - \dfrac{4z-3}{z+1}$ **8.** $\dfrac{5x+2}{x+2} - \dfrac{3x-2}{x+2}$

In Exercises 9–12, find the least common multiple of the given polynomials.

9. $3x^2, \ x^3$ **10.** $x^2, \ x(x-1)$

11. $x^2 - 9, \ x^2 + 3x$ **12.** $x^2 - 4, \ x^2 - 4x + 4$

In Exercises 13–16, rewrite the given fractions equivalently so that they have the same denominator.

13. $\dfrac{5}{2x}, \ \dfrac{4}{x^2}$ **14.** $\dfrac{1}{2x-4}, \ \dfrac{x}{x-2}$

15. $\dfrac{1}{x^2 - 16}$, $\dfrac{2}{x^2 - 8x + 16}$

16. $\dfrac{x}{x^2 - x - 6}$, $\dfrac{2}{x^2 - 3x}$

In Exercises 17–36, perform the specified operations and simplify your answers.

17. $\dfrac{1}{2x} + \dfrac{2}{x}$

18. $\dfrac{3}{5x} - \dfrac{2}{x}$

19. $\dfrac{4}{x} - \dfrac{5}{x^2}$

20. $\dfrac{9}{2z^2} + \dfrac{6}{z}$

21. $\dfrac{2}{x - 1} + \dfrac{2}{1 - x}$

22. $\dfrac{4}{2 - x} - \dfrac{3}{x - 2}$

23. $4 - \dfrac{2}{x - 1}$

24. $\dfrac{6}{2x - 1} + 1$

25. $\dfrac{x}{x + 1} - \dfrac{2}{x + 2}$

26. $\dfrac{4}{x - 3} - \dfrac{5x}{x + 1}$

27. $\dfrac{4}{x^2 - 4} - \dfrac{x}{x - 2}$

28. $\dfrac{1}{x + 3} + \dfrac{2}{x^2 + 3x}$

29. $\dfrac{x + 4}{x} + \dfrac{x - 2}{x + 1}$

30. $\dfrac{3}{t^2 + 2t} - \dfrac{1}{t}$

31. $\dfrac{2x - 1}{x^2 + 6x + 8} - \dfrac{x - 2}{x^2 - 4x - 12}$

32. $\dfrac{x + 1}{2x} - \dfrac{1 - x}{x + 1}$

33. $\dfrac{1}{x} + \dfrac{1}{x^2 + x} - \dfrac{2}{x + 1}$

34. $\dfrac{1}{2x} - \dfrac{1}{x + 1} + \dfrac{x + 1}{x^2 + x}$

35. $u - \dfrac{2u}{u - v} + \dfrac{v}{u + v}$

36. $\dfrac{3}{2x} - \dfrac{1}{y} + \dfrac{2x}{x + y}$

In Exercises 37–40, simplify the given compound fractions.

37. $\dfrac{\left(\dfrac{2}{x^2}\right)}{\left(\dfrac{6}{x}\right)}$

38. $\dfrac{\left(1 + \dfrac{2}{x}\right)}{\left(\dfrac{2}{x}\right)}$

39. $\dfrac{\left(z - \dfrac{1}{z}\right)}{\left(1 + \dfrac{2}{z}\right)}$

40. $\dfrac{\left(\dfrac{1}{x - 1} + \dfrac{2}{x + 1}\right)}{\left(x - \dfrac{1}{x + 1}\right)}$

41. *Average of Two Numbers* Determine the average of the two real numbers given by $x/5$ and $2x/3$.

9.4 | Solving Equations

Section Highlights

1. When solving equations that contain fractions, we first multiply both sides of the equation by the least common denominator.

2. Recall from Section 9.1 that we exclude values of the variable that make the denominator zero. Sometimes, even though no mistake has been made in the solution process, a trial solution may not satisfy the original equation. Such solutions are called extraneous. This is why we always check our trial solutions in the original equation.

EXAMPLE 1 ■ An Equation Containing Constant Denominators

Solve the following equation.

$$\frac{x}{10} - \frac{2x+1}{15} = \frac{1}{5}.$$

Solution

$$\frac{x}{10} - \frac{2x+1}{15} = \frac{1}{5} \qquad \text{Given equation}$$

$$30\left[\frac{x}{10} - \frac{2x+1}{15}\right] = 30 \cdot \frac{1}{5} \qquad \begin{array}{l}\text{Multiply both sides by}\\ \text{least common denominator.}\end{array}$$

$$3x - 2(2x+1) = 6 \qquad \text{Distribute and simplify.}$$

$$3x - 4x - 2 = 6 \qquad \text{Distribute.}$$

$$-x - 2 = 6 \qquad \text{Simplify.}$$

$$-x = 8 \qquad \text{Add 2 to both sides.}$$

$$x = -8 \qquad \text{Multiply both sides by } -1.$$

Thus, the solution is $x = -8$. Check this in the original equation. ■

Starter Exercise 1 | *Fill in the blanks.*

Solve the following equation.

$$(x+1) - \frac{2x+1}{3} = \frac{4}{3}$$

$$3\left[(x+1) - \frac{2x+1}{3}\right] = 3 \cdot \frac{4}{3}$$

$$3(x+1) - \left(\boxed{}\right) = \boxed{}$$

$$3x + 3 - \boxed{} - \boxed{} = \boxed{}$$

$$\boxed{} + 2 = \boxed{}$$

$$x = \boxed{}$$

EXAMPLE 2 ■ Equations Containing Variable Denominators

Solve the following equations.

(a) $\dfrac{3}{x} + \dfrac{1}{2} = \dfrac{2}{x}$

(b) $\dfrac{2}{x+1} + \dfrac{x}{x-2} = 1$

Solution

(a) The least common denominator is $2x$. Thus, we multiply both sides of the equation by $2x$.

$$\frac{3}{x} + \frac{1}{2} = \frac{2}{x} \qquad \text{Given equation}$$

$$2x\left(\frac{3}{x} + \frac{1}{2}\right) = 2x\left(\frac{2}{x}\right) \qquad \text{Multiply both sides by } 2x.$$

$$2x\left(\frac{3}{x}\right) + 2x\left(\frac{1}{2}\right) = 2x\left(\frac{2}{x}\right) \qquad \text{Distributive Property.}$$

$$6 + x = 4 \qquad \text{Simplify.}$$

$$x = -2 \qquad \text{Solution}$$

Check -2 in the original equation.

$$\frac{3}{x} + \frac{1}{2} = \frac{2}{x} \qquad \text{Given equation}$$

$$\frac{3}{-2} + \frac{1}{2} \overset{?}{=} \frac{2}{-2} \qquad \text{Replace } x \text{ by } -2.$$

$$-\frac{3}{2} + \frac{1}{2} \overset{?}{=} -1$$

$$-\frac{2}{2} \overset{?}{=} -1$$

$$-1 = -1 \qquad \text{Solution checks.}$$

(b) The least common denominator is $(x + 1)(x - 2)$.

$$\frac{2}{x + 1} + \frac{x}{x - 2} = 1 \qquad \text{Given equation}$$

$$(x + 1)(x - 2)\left[\frac{2}{x + 1} + \frac{x}{x - 2}\right] = (x + 1)(x - 2) \cdot 1 \qquad \text{Multiply both sides by } (x + 1)(x - 2).$$

$$2(x - 2) + x(x + 1) = (x + 1)(x - 2) \qquad \text{Distribute and simplify.}$$

$$2x - 4 + x^2 + x = x^2 - x - 2$$

$$x^2 + 3x - 4 = x^2 - x - 2$$

$$4x = 2$$

$$x = \frac{1}{2}$$

Check $\frac{1}{2}$ in the original equation to verify the solution.

Starter Exercise 2 *Fill in the blanks.*

Solve the following equations.

(a)
$$\frac{1}{x-1} + 2 = \frac{3}{x-1}$$

$$(x-1)\left[\frac{1}{x-1} + 2\right] = (x-1)\frac{3}{x-1}$$

$$\boxed{} \cdot \frac{1}{x-1} + \boxed{} \cdot 2 = (x-1)\frac{3}{x-1}$$

$$\boxed{} + \boxed{} - \boxed{} = 3$$

$$\boxed{} - \boxed{} = 3$$

$$\boxed{} = 4$$

$$x = \boxed{}$$

(b)
$$\frac{1}{x} + \frac{2x}{x+1} = 2$$

$$x(x+1)\left[\frac{1}{x} + \frac{2x}{x+1}\right] = \boxed{}(x+1)\cdot 2$$

$$x+1+2\boxed{} = 2x^2 + \boxed{}$$

$$x+1 = \boxed{}$$

$$1 = \boxed{}$$

EXAMPLE 3 ■ An Equation with No Solution

Solve the equation $\dfrac{2x}{x-1} = 1 + \dfrac{2}{x-1}$.

Solution

The least common denominator is $x-1$.

$$\frac{2x}{x-1} = 1 + \frac{2}{x-1} \qquad \text{Given equation}$$

$$(x-1)\frac{2x}{x-1} = (x-1)\left(1 + \frac{2}{x-1}\right) \qquad \text{Multiply both sides by } x-1.$$

$$2x = (x-1) + 2 \qquad \text{Distribute and simplify.}$$

$$2x = x+1$$

$$x = 1$$

When you check 1 in the original equation, you find you have zero in the denominator. Therefore, 1 cannot be a solution. Therefore, the given equation has no solution. ■

Starter Exercise 3 *Fill in the blanks.*

Solve the following equation.

$$\frac{2x}{x-2} = 1 + \frac{4}{x-2}$$

$$\left(\boxed{}\right)\frac{2x}{x-2} = (x-2)\left[1 + \frac{4}{x-2}\right]$$

$$\boxed{} = (x-2) + 4$$

$$\boxed{} = \boxed{} + 2$$

$$x = \boxed{}$$

EXAMPLE 4 ■ An Equation with Two Solutions

Solve the following equation.

$$1 = \frac{1}{x - 2} - \frac{4}{x^2 - 4}$$

Solution

$1 = \dfrac{1}{x - 2} - \dfrac{4}{x^2 - 4}$	Given equation
$(x^2 - 4) \cdot 1 = (x^2 - 4)\left[\dfrac{1}{x - 2} - \dfrac{4}{x^2 - 4}\right]$	Multiply both sides by $x^2 - 4$.
$x^2 - 4 = (x + 2) - 4$	Distribute and simplify.
$x^2 - x - 2 = 0$	Standard form
$(x - 2)(x + 1) = 0$	Factor.
$x - 2 = 0 \quad \text{or} \quad x + 1 = 0$	Set each factor equal to 0.
$x = 2 \qquad\qquad x = -1$	Solve each equation.

After checking each solution in the original equation, we conclude that the solution is $x = -1$. ■

Starter Exercise 4 *Fill in the blanks.*

Solve the following equation.

$$1 = \frac{1}{x + 2} + \frac{2}{x^2 + 4x + 4}$$

$$1 = \frac{1}{x + 2} + \frac{2}{(x + 2)^2}$$

$$(x + 2)^2 \cdot 1 = \left(\boxed{}\right)^2\left[\frac{1}{x + 2} + \frac{2}{(x + 2)^2}\right]$$

$$x^2 + \boxed{} + 4 = (x + 2) + 2$$

$$x^2 + \boxed{} = 0$$

$$x\left(x + \boxed{}\right) = 0$$

$$\boxed{} = 0 \quad \text{or} \quad x + \boxed{} = 0$$

$$x = \boxed{}$$

EXAMPLE 5 ■ An Application: Average Speed

A freight train can travel 55 miles in the same amount of time a passenger train can travel 80. The average rate of the passenger train is 25 mph more than the average rate of the freight train. Find the average rate of each train.

Solution

Verbal model: | Passenger train time | = | Freight train time |

Formula: $\text{Time} = \dfrac{\text{Distance}}{\text{Rate}}$

Labels: Freight: distance $= 55$; rate $= r$
Passenger: distance $= 80$; rate $= r + 25$

Equation: $\dfrac{80}{r+25} = \dfrac{55}{r}$

$$80r = 55(r + 25)$$
$$80r = 55r + 1375$$
$$25r = 1375$$
$$r = 55$$

Thus, the average rate of the freight train is 55 mph and the average rate of the passenger train is $55 + 25 = 80$ mph. ∎

| **Starter Exercise 5** | *An Application: Work Rates—Fill in the blanks.*

Dave can sweep his driveway in 30 minutes. If Dave's brother helps, the two can sweep the driveway in 18 minutes. How long does it take Dave's brother to sweep the driveway alone?

Verbal model: | Dave's rate | + | Brother's rate | = | Together rate |

Labels: Dave: time $= 30$; rate $= 1/30$
Brother: time $= t$; rate $= 1/t$

Together: time $= 18$; rate $= 1/\boxed{}$

Equation: $\dfrac{1}{\boxed{}} + \dfrac{1}{t} = \dfrac{1}{\boxed{}}$

$$90t\left[\dfrac{1}{\boxed{}} + \dfrac{1}{t}\right] = 90t\left(\dfrac{1}{\boxed{}}\right)$$

$$\boxed{} + 90 = \boxed{}$$

$$90 = \boxed{}$$

$$\boxed{} = t$$

■ **Solutions to Starter Exercises** ■

1. $3(x+1) - \left(\boxed{2x+1}\right) = \boxed{4}$

$3x + 3 - \boxed{2x} - \boxed{1} = \boxed{4}$

$\boxed{x} + 2 = \boxed{4}$

$x = \boxed{2}$

■ **Solutions to Starter Exercises** ■

2. (a) $\boxed{(x-1)} \cdot \dfrac{1}{x-1} + \boxed{(x-1)} \cdot 2 = (x-1)\dfrac{3}{x-1}$

$$\boxed{1} + \boxed{2x} - \boxed{2} = 3$$

$$\boxed{2x} - \boxed{1} = 3$$

$$\boxed{2x} = 4$$

$$x = \boxed{2}$$

(b) $x(x+1)\left[\dfrac{1}{x} + \dfrac{2x}{x+1}\right] = \boxed{x}\,(x+1) \cdot 2$

$$x + 1 + 2\boxed{x^2} = 2x^2 + \boxed{2x}$$

$$x + 1 = \boxed{2x}$$

$$1 = \boxed{x}$$

3. $\left(\boxed{x-2}\right)\dfrac{2x}{x-2} = (x-2)\left[1 + \dfrac{4}{x-2}\right]$

$$\boxed{2x} = (x-2) + 4$$

$$\boxed{2x} = \boxed{x} + 2$$

$$x = \boxed{2}$$

4. $(x+2)^2 \cdot 1 = \left(\boxed{x+2}\right)^2\left[\dfrac{1}{x+2} + \dfrac{2}{(x+2)^2}\right]$

$$x^2 + \boxed{4x} + 4 = (x+2) + 2$$

$$x^2 + \boxed{3x} = 0$$

$$x\left(x + \boxed{3}\right) = 0$$

$$\boxed{x} = 0 \quad \text{or} \quad x + \boxed{3} = 0$$

$$x = \boxed{-3}$$

5. Rate $= \dfrac{1}{\boxed{18}}$; Equation: $\dfrac{1}{\boxed{30}} + \dfrac{1}{t} = \dfrac{1}{\boxed{18}}$

$$90t\left[\dfrac{1}{\boxed{30}} + \dfrac{1}{t}\right] = 90t\left(\dfrac{1}{\boxed{18}}\right)$$

$$\boxed{3t} + 90 = \boxed{5t}$$

$$90 = \boxed{2t}$$

$$\boxed{45} = t$$

9.4 | EXERCISES

In Exercises 1 and 2, determine whether the given value of x is a solution of the equation.

Equation   *Values*

1. $\dfrac{x}{2} - \dfrac{1}{x} = \dfrac{1}{2}$ (a) $x = -1$ (b) $x = 0$ (c) $x = 2$ (d) $x = \dfrac{1}{2}$

2. $2 + \dfrac{1}{x+1} = 3$ (a) $x = -1$ (b) $x = 0$ (c) $x = 2$ (d) $x = \dfrac{1}{2}$

In Exercises 3–30, solve the given equation.

3. $\dfrac{x}{5} = \dfrac{2}{5}$

4. $\dfrac{x}{15} = \dfrac{4}{15}$

5. $\dfrac{x}{2} = \dfrac{2}{5}$

6. $\dfrac{x}{4} = \dfrac{3}{2}$

7. $\dfrac{x+4}{12} - \dfrac{1}{6} = \dfrac{1}{4}$

8. $\dfrac{2x+1}{6} + \dfrac{1}{2} = \dfrac{1}{3}$

9. $\dfrac{1}{x} = \dfrac{2}{5}$

10. $\dfrac{3}{t} = 9$

11. $\dfrac{1}{x-4} = -\dfrac{1}{6}$

12. $-\dfrac{1}{2-t} = \dfrac{1}{2}$

13. $\dfrac{2}{x+1} = \dfrac{1}{5}$

14. $\dfrac{7}{t-1} = \dfrac{1}{6}$

15. $\dfrac{1}{12} = \dfrac{5}{x-3}$

16. $-\dfrac{2}{3} = \dfrac{2}{2x-1}$

17. $-\dfrac{1}{x+1} = \dfrac{1}{x-2}$

18. $\dfrac{2}{x-3} = \dfrac{2}{2x-1}$

19. $\dfrac{2}{2x+1} = \dfrac{-4}{x+1}$

20. $\dfrac{1}{3x-1} = \dfrac{4}{x-3}$

21. $1 - \dfrac{10}{x^2-25} = \dfrac{1}{x+5}$

22. $\dfrac{x+2}{x+1} - \dfrac{4}{x^2+3x+2} = \dfrac{12}{x^2+3x+2}$

23. $4 = \dfrac{100}{x^2}$

24. $1 + \dfrac{2}{t} = \dfrac{15}{t^2}$

25. $\dfrac{3}{z} = 1 - \dfrac{10}{z^2}$

26. $x = 2 - \dfrac{2}{x+1}$

27. $\dfrac{8}{y} = \dfrac{y}{2}$

28. $\dfrac{x}{x-3} + \dfrac{x}{x-1} = \dfrac{16}{x^2-4x+3}$

29. $\dfrac{x}{4} = \dfrac{1+2/x}{1+6/x}$

30. $\dfrac{x}{5} = \dfrac{1+2/x}{1+8/x}$

31. *Number Problem*   Find a number such that the sum of this number and its reciprocal is $\frac{26}{5}$.

32. *Number Problem*   Find a number such that the sum of the reciprocal of this number and the reciprocal of twice this number is $\frac{3}{10}$.

33. *Average Cost*   The average cost for producing x units of a product is given by

$$\text{Average cost} = \frac{1}{4} + \frac{3000}{x}.$$

Determine the number of units that must be produced to obtain an average cost of \$1.75 per unit.

34. *Work Rate*   One pump can empty a pool in 2 hours and another pump can empty the same pool in 3 hours. How long would it take both pumps working together to empty the pool?

35. A jogger can jog 5 miles in the same amount of time that a cyclist can ride 20 miles. If the average rate of the cyclist is eight more than three times the rate of the jogger, find the rate of each.

Cumulative Practice Test for Chapters P–9

In Exercises 1–4, state the property of algebra that justifies the equation.

1. $2x + 4x = (2 + 4)x$

2. $x = 1x$

3. $(x - 3x) + 4 = 4 + (x - 3x)$

4. $\frac{1}{2}(2y) = \left(\frac{1}{2} \cdot 2\right)y$

In Exercises 5–7, find an equation of the line through the given points. (Write your answer in slope-intercept form.)

5. $(2, 1), (3, 3)$

6. $(1, 2), (1, 5)$

7. $(-2, -1), (5, -1)$

In Exercises 8 and 9, solve the system of linear equations by the method of elimination.

8. $\begin{aligned} 3x - 2y &= 5 \\ 2x + 2y &= 10 \end{aligned}$

9. $\begin{aligned} 5x - 2y &= 1 \\ 3x + 5y &= 1 \end{aligned}$

In Exercises 10 and 11, solve the system of equations by the method of substitution.

10. $\begin{aligned} x + 3y &= 6 \\ x + 2y &= 1 \end{aligned}$

11. $\begin{aligned} 2x - y &= 5 \\ 3x + 2y &= 4 \end{aligned}$

In Exercises 12–14, graph the equation or inequality on the rectangular coordinate system.

12. $y = -\frac{3}{2}x + 1$

13. $x = 2$

14. $y \leq \frac{1}{2}x - 1$

In Exercises 15–22, simplify the given expression.

15. $6z - (5z + 1)$

16. $(x - 3y)(x + 3y)$

17. $\left(\dfrac{x^2 y^3}{x^{-1}}\right)^{-3}$

18. $(x - y)^2 + 2xy$

19. $5[x - (2x + 3)] - 4x$

20. $\dfrac{x^2 + 7x + 12}{x^2 - 4} \cdot \dfrac{x^2 - 2x}{x^2 + 6x + 9}$

21. $\dfrac{2}{x + 1} + \dfrac{3}{x - 1}$

22. $\dfrac{2 - x}{x^2 - x} \div \left[\dfrac{2}{x + 1} - \dfrac{x}{x^2 - 1}\right]$

In Exercises 23 and 24, find an equation of the line through the given point that is (a) parallel to the given line, and (b) perpendicular to the given line. (Write your answer in general form.)

	Line	*Point*		*Line*	*Point*
23.	$y = 3x + 1$	$(3, 1)$	**24.**	$3x - 2y = 6$	$(-6, 4)$

In Exercises 25 and 26, factor the polynomial completely.

25. $x^2(x + 2) - (x + 2)$

26. $x^4 - 16$

In Exercises 27–36, solve the given equation or inequality.

27. $1 - 2x \geq 5$

28. $1 - 2x = 5$

29. $2(x + 3) = 4$

30. $(z + 1)^2 = 4$

31. $\dfrac{z}{4} = \dfrac{1}{2}$

32. $x^2 = -2x - 1$

33. $2x^2 + x - 2 = x^2 + 2x + 4$

34. $\dfrac{9}{t} = \dfrac{3}{4}$

35. $\dfrac{x - 2}{x + 1} + \dfrac{3}{x + 1} = 0$

36. $\dfrac{1}{x - 1} = x - \dfrac{1}{x - 1}$

37. It takes Sandy 2 hours to mow the lawn with her power mower. It takes Joanne 4 hours to mow the lawn with her push mower. How long would it take Sandy and Joanne to mow the lawn together?

38. A boat can travel 24 miles in 2 hours heading downstream. It takes 3 hours for the boat to travel the same distance upstream. Find the rate of the boat in calm water and the rate of the current.

39. A total of $12,000 is invested in two funds. One fund pays $7\frac{3}{4}$% and the other $8\frac{1}{2}$% simple interest. Find the amount invested in each fund if the annual interest is $982.50.

40. How much 20% alcohol solution must be mixed with 50% alcohol solution to make 10 quarts of 38% alcohol solution?

9.1 | Answers to Exercises

1. All real numbers except -2 **2.** All real numbers **3.** All real numbers except -1 and 1

4. All real numbers **5.** x **6.** x **7.** $x + 1$

8. $x - 3$ **9.** $x + 2$ **10.** $x + 3$ **11.** $\dfrac{2}{x}$ **12.** $\dfrac{x}{3}$

13. $\dfrac{x}{2}, x \neq 0$ **14.** $\dfrac{z^2}{2}, z \neq 0$ **15.** $\dfrac{x}{x+1}, x \neq -2$ **16.** $\dfrac{x+6}{x+5}, x \neq 1$

17. $\dfrac{1}{x}, x \neq 7$ **18.** $\dfrac{x-1}{x+1}, x \neq -1$ **19.** $\dfrac{2x-1}{x^2-1}, x \neq 0$ **20.** $\dfrac{x+3}{x+2}, x \neq -4$

21. $\dfrac{x-2}{x+1}, x \neq -3$ **22.** $\dfrac{z-2}{z-1}, z \neq 5$ **23.** $\dfrac{x+2}{x-2}, x \neq -6$ **24.** $-1, a \neq -1, 6$

25. $-\dfrac{z+2}{z(1+2z)}, z \neq 1$ **26.** $\dfrac{2x+1}{3x+2}, x \neq -\dfrac{1}{3}$ **27.** $\dfrac{x^2+3x}{x-3}, x \neq -3$ **28.** $\dfrac{a-1}{a+6}, a \neq -1, 0$

29. $x^2 + 1, x \neq 3$ **30.** $\dfrac{1}{x-1}, x \neq -2, 2$ **31.** (a) $\dfrac{4000 + 9x}{x}$
(b) $0, 1, 2, \ldots$
(c) $49

9.2 | Answers to Exercises

1. $x + 2$ **2.** $x - 7$ **3.** x **4.** $x + 3$ **5.** $6xy$

6. $\dfrac{by}{4}$ **7.** $\dfrac{2x}{3}$ **8.** $\dfrac{2x-1}{5}$ **9.** 1 **10.** $2(x+5)$

11. -1 **12.** $\dfrac{(2x-3)(x+6)}{(2x+7)(x+7)}$ **13.** $\dfrac{z+2}{z(z-2)}$ **14.** $\dfrac{(x+2)(x-3)}{(x+3)}$

15. $\dfrac{x(x-y)}{2x+y}$ **16.** $\dfrac{(a+b)(a-2b)}{(a+3b)(a+2b)}$ **17.** $\dfrac{x+1}{x-2}$ **18.** 1

19. $4x$ **20.** 6   **21.** $x+y$ **22.** $\dfrac{(x-6)(x+2)(x+1)}{(x+3)(x-2)^2}$

23. 1   **24.** 6   **25.** $\dfrac{x-6}{x+6}$ **26.** $\dfrac{(3x+y)(x-y)}{4x-y}$

27. x **28.** $\dfrac{3b}{a^3}$ **29.** x^2-1 **30.** $\dfrac{5x-1}{2x-5}$

31. (a) 1/10 minute per copy
  (b) $x/10$ minutes
  (c) $100/10 = 10$ minutes

9.3	**Answers to Exercises**

1. $\dfrac{7}{a}$ **2.** $\dfrac{2}{x^2}$ **3.** $-\dfrac{2}{x}$ **4.** $-\dfrac{1}{5}$ **5.** -1

6. 3   **7.** 1   **8.** 2   **9.** $3x^3$ **10.** $x^2(x-1)$

11. $x(x+3)(x-3)$ **12.** $(x-2)^2(x+2)$ **13.** $\dfrac{5}{2x} = \dfrac{5x}{2x^2},\ \ \dfrac{4}{x^2} = \dfrac{8}{2x^2}$

14. $\dfrac{1}{2x-4},\ \ \dfrac{x}{x-2} = \dfrac{2x}{2x-4}$

15. $\dfrac{1}{x^2-16} = \dfrac{x-4}{(x-4)^2(x+4)},\ \ \dfrac{2}{x^2-8x+16} = \dfrac{2(x+4)}{(x-4)^2(x+4)}$

16. $\dfrac{x}{x^2-x-6} = \dfrac{x^2}{x(x-3)(x+2)},\ \ \dfrac{2}{x^2-3x} = \dfrac{2x+4}{x(x-3)(x+2)}$

17. $\dfrac{5}{2x}$ **18.** $-\dfrac{7}{5x}$ **19.** $\dfrac{4x-5}{x^2}$ **20.** $\dfrac{9+12z}{2z^2}$

21. 0   **22.** $-\dfrac{7}{x-2}$ **23.** $\dfrac{4x-6}{x-1}$ **24.** $\dfrac{2x+5}{2x-1}$

25. $\dfrac{x^2-2}{(x+1)(x+2)}$ **26.** $\dfrac{-5x^2+19x+4}{(x-3)(x+1)}$ **27.** $\dfrac{4-2x-x^2}{(x+2)(x-2)}$ **28.** $\dfrac{x+2}{x(x+3)}$

29. $\dfrac{2x^2+3x+4}{x(x+1)}$ **30.** $\dfrac{1-t}{t^2+2t}$ **31.** $\dfrac{x^2-15x+14}{(x+2)(x+4)(x-6)}$ **32.** $\dfrac{3x^2+1}{2x(x+1)}$

33. $\dfrac{2-x}{x(x+1)}$ **34.** $\dfrac{x+3}{2x(x+1)}$ **35.** $\dfrac{u^3-2u^2-uv^2-uv-v^2}{(u+v)(u-v)}$

36. $\dfrac{4x^2y-2x^2+xy+3y^2}{2xy(x+y)}$ **37.** $\dfrac{1}{3x}$ **38.** $\dfrac{x+2}{2}$

39. $\dfrac{z^2 - 1}{z + 2}$ **40.** $\dfrac{3x - 1}{(x - 1)(x^2 + x - 1)}$ **41.** $\dfrac{\dfrac{x}{5} + \dfrac{2x}{3}}{2} = \dfrac{13x}{30}$

9.4 | Answers to Exercises

1. (a) Yes   **2.** (a) No   **3.** $x = 2$ **4.** $x = 4$ **5.** $x = \frac{4}{5}$
  (b) No   (b) Yes
  (c) Yes   (c) No
  (d) No   (d) No

6. $x = 6$ **7.** $x = 1$ **8.** $x = -1$ **9.** $x = \frac{5}{2}$ **10.** $t = \frac{1}{3}$

11. $x = -2$ **12.** $t = 4$ **13.** $x = 9$ **14.** $t = 43$ **15.** $x = 63$

16. $x = -1$ **17.** $x = \frac{1}{2}$ **18.** $x = -2$ **19.** $x = -\frac{3}{5}$ **20.** $x = \frac{1}{11}$

21. $x = 6$ **22.** $x = -6, 2$ **23.** $x = -5, 5$ **24.** $t = -5, 3$ **25.** $z = -2, 5$

26. $x = 0$ **27.** $y = -4, 4$ **28.** $x = -2, 4$ **29.** $x = -4, 2$ **30.** $x = -5, 2$

31. 5 or $\frac{1}{5}$ **32.** 5   **33.** 2000 units   **34.** 1 hr 12 min   **35.** Jogger: 8 mph
                                                                 Cyclist: 32 mph

---

Answers to Cumulative Practice Test P–9

1. Distributive Property

2. Multiplicative Identity Property

3. Commutative Property of Addition

4. Associative Property of Multiplication

5. $y = 2x - 3$ **6.** $x = 1$ **7.** $y = -1$ **8.** $(3, 2)$

9. $\left(\frac{7}{31}, \frac{2}{31}\right)$ **10.** $(-9, 5)$ **11.** $(2, -1)$

12. 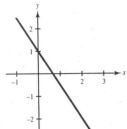   **13.** 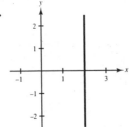   **14.**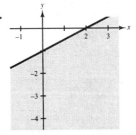

15. $z - 1$ **16.** $x^2 - 9y^2$ **17.** $\dfrac{1}{x^9 y^9}$ **18.** $x^2 - y^2$ **19.** $-9x - 15$

20. $\dfrac{x(x+4)}{(x+2)(x+3)}$

21. $\dfrac{5x+1}{x^2-1}$

22. $-\dfrac{x+1}{x}$

23. (a) $3x-y=8$
(b) $x+3y=6$

24. (a) $3x-2y=-26$
(b) $2x+3y=0$

25. $(x+1)(x-1)(x+2)$

26. $(x+2)(x-2)(x^2+4)$

27. All real numbers less than or equal to -2

28. $x=-2$

29. $x=-1$

30. $z=-3$ or $z=1$

31. $z=2$

32. $x=-1$

33. $x=-2$ or $x=3$

34. $t=12$

35. No solution

36. $x=-1$ or $x=2$

37. 1 hr, 20 min

38. Rate of boat in calm water: 10 mph
Rate of current: 2 mph

39. $5000 at $7\frac{3}{4}\%$
$7000 at $8\frac{1}{2}\%$

40. 4 qts

CHAPTER TEN
Radical Expressions and Equations

Square Roots and Radicals

Section Highlights

1. If a is a nonnegative real number and b is a real number such that $b^2 = a$, then b is called a **square root** of a.
2. Every nonnegative real number has two real square roots that are opposites of each other.
3. If a is a nonnegative real number and b is the *nonnegative* square root of a, then we can write $\sqrt{a} = b$. The number inside the radical sign is called the **radicand**. The number b is called the **principal square root** of a.
4. **Perfect squares** are numbers that are squares of integers. The square root of a nonperfect square is an *irrational* number.

EXAMPLE 1 ■ Finding the Square Roots of Numbers

Find all square roots of the numbers (a) 64 and (b) -16.

Solution

(a) The square roots of 64 are 8 and -8 since $8^2 = 64$ and $(-8)^2 = 64$.

(b) There are no square roots of -16 since there are no real numbers that multiplied times themselves equal -16.

Starter Exercise 1 | *Fill in the blanks.*

(a) Find all square roots of 36. The square roots of 36 are 6 and -6 since

$$6^2 = \boxed{} \text{ and } (-6)^2 = \boxed{}.$$

(b) Find all square roots of $\frac{9}{16}$. The square roots of $\frac{9}{16}$ are $\dfrac{\boxed{}}{4}$ and $\boxed{}$ since

$$\boxed{}^2 = \frac{9}{16} \text{ and } \boxed{}^2 = \frac{9}{16}.$$

EXAMPLE 2 ■ Finding the Principal Square Root of a Number

Find the principal square root of 25.

Solution

The number 25 has two square roots, 5 and -5. The principal square root is the positive one: $\sqrt{25} = 5$.

| Starter Exercise 2 | *Fill in the blanks.*

(a) Find the principal square root of 144. The two square roots of 144 are 12 and -12.

$$\sqrt{144} = \boxed{12}$$

(b) Find the principal square root of $\frac{1}{16}$. The two square roots of $\frac{1}{16}$ are $\frac{1}{4}$ and $\boxed{}$.

$$\sqrt{\tfrac{1}{16}} = \boxed{}$$

EXAMPLE 3 ■ Classifying Square Roots as Rational or Irrational

Classify (a) $-\sqrt{9}$ and (b) $\sqrt{15}$ as rational or irrational.

Solution

(a) $-\sqrt{9} = -3$ is rational.

(b) $\sqrt{15}$ is irrational since 15 is not a perfect square.   ■

| Starter Exercise 3 | *Fill in the blanks.*

Classify (a) $\sqrt{\tfrac{16}{25}}$ and (b) $\sqrt{0}$ as rational or irrational.

(a) $\sqrt{\tfrac{16}{25}} = \boxed{}$ is $\boxed{}$.   (b) $\sqrt{0} = \boxed{}$ is $\boxed{}$.

EXAMPLE 4 ■ Approximating a Square Root

Give a rough approximation of (a) $\sqrt{70}$ and (b) $\sqrt{160}$ without using a calculator.

Solution

(a) The number 70 lies between the two perfect squares $64 = 8^2$ and $81 = 9^2$. Thus, $\sqrt{70}$ must lie between 8 and 9. Since 70 is a little closer to 64, we approximate $\sqrt{70} \approx 8.3$.

(b) $\quad 144 < 160 < 169$ Between two perfect squares

$\Rightarrow \sqrt{144} < \sqrt{160} < \sqrt{169}$ Take the square root.

$\quad \Rightarrow 12 < \sqrt{160} < 13$ Simplify.

Since 160 is closer to 169, $\sqrt{160} \approx 12.7$.   ■

| Starter Exercise 4 | *Fill in the blanks.*

Approximate $\sqrt{105}$ without using a calculator.

$$100 < 105 < \boxed{}$$

$$\sqrt{100} < \sqrt{105} < \sqrt{\boxed{}}$$

$$\boxed{} < \sqrt{105} < \boxed{}$$

$$\sqrt{105} \approx \boxed{}$$

Solutions to Starter Exercises

1. (a) $6^2 = \boxed{36}$ and $(-6)^2 = \boxed{36}$

   (b) $\dfrac{\boxed{3}}{4}$ and $\boxed{-\frac{3}{4}}$ since

   $\boxed{\left(\frac{3}{4}\right)}^2 = \frac{9}{16}$ and

   $\boxed{\left(-\frac{3}{4}\right)}^2 = \frac{9}{16}$.

2. (a) $\sqrt{144} = \boxed{12}$

   (b) $\boxed{-\frac{1}{4}}$; $\sqrt{\frac{1}{16}} = \boxed{\frac{1}{4}}$

3. (a) $\sqrt{\frac{16}{25}} = \boxed{\frac{4}{5}}$ is $\boxed{\text{rational}}$.

   (b) $\sqrt{0} = \boxed{0}$ is $\boxed{\text{rational}}$.

4. $100 < 105 < \boxed{121}$

   $\sqrt{100} < \sqrt{105} < \sqrt{\boxed{121}}$

   $\boxed{10} < \sqrt{105} < \boxed{11}$

   $\sqrt{105} \approx \boxed{10.1}$

10.1 | EXERCISES

In Exercises 1 and 2, fill in the blank with the appropriate real number.

1. $6^2 = 36 \Rightarrow$ positive square root of 36 is _6_ .

2. $(-7)^2 = 49 \Rightarrow$ negative square root of 49 is _−7_ .

In Exercises 3–6, find both square roots of the given number, if possible. (Do not use a calculator.)

3. 4 _=2_   **4.** $\frac{49}{9}$ _$\frac{7}{3}$_   **5.** -25 **6.** -0.01

In Exercises 7–16, evaluate the given expression without a calculator, if possible.

7. $\sqrt{16}$ **8.** $\sqrt{64}$ **9.** $-\sqrt{144}$ **10.** $\sqrt{-144}$ **11.** $\sqrt{256}$

12. $\sqrt{625}$ **13.** $\sqrt{\frac{1}{64}}$ **14.** $-\sqrt{\frac{16}{49}}$ **15.** $\sqrt{-0.09}$ **16.** $\sqrt{\frac{50}{2}}$

In Exercises 17 and 18, determine whether the square root is rational or irrational.

17. $\sqrt{40}$ **18.** $\sqrt{\frac{4}{81}}$ _$\frac{2}{9}$_

In Exercises 19–26, use a calculator to approximate the expression. Round your answer to three decimal places.

19. $\sqrt{23}$ **20.** $\sqrt{512}$ **21.** $-\sqrt{172.8}$ **22.** $\sqrt{\frac{72}{5}}$

23. $\sqrt{94(83)}$ **24.** $6 - \sqrt{14}$ **25.** $\dfrac{2 - \sqrt{6}}{4}$ **26.** $\dfrac{5 + 2\sqrt{7}}{9}$

In Exercises 27 and 28, approximate the given expression without using a calculator.

27. $\sqrt{19}$ **28.** $\sqrt{96}$

29. Use a calculator to evaluate (a) $\left(\sqrt{3.7}\right)^2$ and (b) $\left(\sqrt{841}\right)^2$.

30. Use the result of Exercise 29 to determine $\left(\sqrt{a}\right)^2$, where $a \geq 0$.

10.2   Simplifying Radicals

Section Highlights

1. A (square root) radical expression is in simplest form if the following conditions are met.
    (a) All possible perfect square factors have been removed from the radicand.
    (b) All radicands are free of fractions.
    (c) No denominator contains a radical.
2. If a and b are nonnegative real numbers, then

    (a) $\sqrt{a \cdot b} = \sqrt{a} \cdot \sqrt{b}$ (b) $\sqrt{\dfrac{a}{b}} = \dfrac{\sqrt{a}}{\sqrt{b}}$.

3. If x is a real number, then $\sqrt{x^2} = |x|$.

EXAMPLE 1 ■   Simplifying Radicals

Simplify (a) $\sqrt{80}$ and (b) $\sqrt{75}$.

Solution

In each case, the first step is to find the largest perfect square factor of the radicand.

(a) The largest perfect square factor of 80 is 16.
$$\sqrt{80} = \sqrt{16 \cdot 5} = \sqrt{16} \cdot \sqrt{5} = 4\sqrt{5}$$

(b) The largest perfect square factor of 75 is 25.
$$\sqrt{75} = \sqrt{25 \cdot 3} = \sqrt{25} \cdot \sqrt{3} = 5\sqrt{3}$$

> **Starter Exercise 1**   *Fill in the blanks.*
>
> (a) Simplify $\sqrt{48}$. The largest perfect square factor of 48 is 16.
> $$\sqrt{48} = \sqrt{16 \cdot \Box} = \sqrt{16} \cdot \sqrt{\Box} = \Box \cdot \sqrt{\Box}$$
>
> (b) Simplify $\sqrt{294}$. The largest perfect square factor of 294 is $\Box$.
> $$\sqrt{294} = \sqrt{\Box \cdot \Box} = \sqrt{\Box} \cdot \sqrt{\Box} = \Box \cdot \sqrt{\Box}$$

EXAMPLE 2 ■   Simplifying Radicals Involving Variable Factors

Simplify (a) $\sqrt{9x^2}$ and (b) $\sqrt{72x^5}$.

Solution

(a) $\sqrt{9x^2} = \sqrt{9} \cdot \sqrt{x^2} = 3|x|$

(b) $\sqrt{72x^5} = \sqrt{36 \cdot (x^2)^2 \cdot 2x}$

$\qquad\qquad = \sqrt{36}\sqrt{(x^2)^2} \cdot \sqrt{2x} = 6x^2\sqrt{2x}$ ■

Starter Exercise 2   *Fill in the blanks.*

(a) Simplify $\sqrt{16x^2}$.

$$\sqrt{16x^2} = \sqrt{16} \cdot \sqrt{\Box} = \Box|\Box|$$

(b) Simplify $\sqrt{49x^5}$.

$$\sqrt{49x^5} = \sqrt{\Box} \cdot \sqrt{(x^2)^{\Box}} \cdot \sqrt{\Box} = \Box\sqrt{\Box}$$

(c) Simplify $\sqrt{27x^3}$.

$$\sqrt{27x^3} = \sqrt{\Box x^2 \cdot 3x} = \sqrt{\Box} \cdot \sqrt{\Box} \cdot \sqrt{3x} = \Box$$

EXAMPLE 3 ■   Simplifying Radicals Involving Fractions

Simplify the following radicals.

(a) $\sqrt{\dfrac{48x}{3}}$

(b) $\sqrt{\dfrac{3x^4}{27y^2}}$

Solution

(a) $\sqrt{\dfrac{48x}{3}} = \sqrt{16x} = \sqrt{16}\sqrt{x} = 4\sqrt{x}$

(b) $\sqrt{\dfrac{3x^4}{27y^2}} = \sqrt{\dfrac{x^4}{9y^2}} = \dfrac{\sqrt{(x^2)^2}}{\sqrt{9}\sqrt{y^2}} = \dfrac{x^2}{3|y|}$ ■

Starter Exercise 3   *Fill in the blanks.*

(a) Simplify $\sqrt{\dfrac{24x}{25}}$.

$$\sqrt{\dfrac{24x}{25}} = \dfrac{\sqrt{\Box \cdot 6x}}{\sqrt{25}} = \dfrac{\Box\sqrt{6x}}{\Box}$$

(b) Simplify $\sqrt{\dfrac{125xy^5}{45x^3y}}$.

$$\sqrt{\dfrac{125xy^5}{45x^3y}} = \sqrt{\dfrac{25y^4}{\Box}} = \dfrac{\sqrt{\Box}\sqrt{(y^2)^2}}{\sqrt{\Box}\sqrt{\Box}} = \Box$$

EXAMPLE 4 ■ Rationalizing Denominators

Simplify $\dfrac{3}{\sqrt{45}}$ by rationalizing denominators.

Solution

$$\frac{3}{\sqrt{45}} = \frac{3}{\sqrt{45}} \cdot \frac{\sqrt{5}}{\sqrt{5}} = \frac{3\sqrt{5}}{15} = \frac{\sqrt{5}}{5}$$

Starter Exercise 4 *Fill in the blanks.*

(a) Simplify $\sqrt{\dfrac{11}{12}}$ by rationalizing the denominator.

$$\sqrt{\frac{11}{12}} = \frac{\sqrt{11}}{\sqrt{12}} \cdot \frac{\sqrt{\square}}{\sqrt{\square}} = \frac{\sqrt{33}}{\square}$$

(b) Simplify $\dfrac{10}{\sqrt{20}}$ by rationalizing the denominator.

$$\frac{10}{\sqrt{20}} = \frac{10}{\sqrt{20}} \cdot \boxed{} = \frac{10\sqrt{\square}}{\square} = \boxed{}$$

EXAMPLE 5 ■ Rationalizing Denominators Containing Variable Factors

Write the radical $\sqrt{\dfrac{2}{x}}$ in simplest form.

Solution

$$\sqrt{\frac{2}{x}} = \frac{\sqrt{2}}{\sqrt{x}} \cdot \frac{\sqrt{x}}{\sqrt{x}} = \frac{\sqrt{2x}}{x}$$

Starter Exercise 5 *Fill in the blanks.*

(a) Write in simplest form: $\sqrt{\dfrac{1}{9x}}$

$$\sqrt{\frac{1}{9x}} = \frac{1}{\sqrt{9x}} \cdot \frac{\sqrt{x}}{\sqrt{x}} = \boxed{}$$

(b) Write in simplest form: $\sqrt{\dfrac{20x^5y}{15x^2y^2}}$

$$\sqrt{\frac{20x^5y}{15x^2y^2}} = \sqrt{\frac{4x^3}{\square}} = \frac{\sqrt{4x^3}}{\sqrt{\square}} \cdot \boxed{} = \frac{\sqrt{\square}}{\square}$$

$$= \boxed{} = \boxed{}$$

■ **Solutions to Starter Exercises** ■

1. (a) $\sqrt{48} = \sqrt{16 \cdot \boxed{3}} = \sqrt{16} \cdot \sqrt{\boxed{3}} = \boxed{4} \cdot \sqrt{\boxed{3}}$

(b) The largest perfect square factor of 294 is $\boxed{49}$.

$\sqrt{294} = \sqrt{\boxed{49} \cdot \boxed{6}} = \sqrt{\boxed{49}} \cdot \sqrt{\boxed{6}} = \boxed{7} \cdot \sqrt{\boxed{6}}$

2. (a) $\sqrt{16x^2} = \sqrt{16} \cdot \sqrt{\boxed{x^2}} = \boxed{4} \,|\,\boxed{x}\,|$

(b) $\sqrt{49x^5} = \sqrt{\boxed{49}} \cdot \sqrt{(x^2)^{\boxed{2}}} \cdot \sqrt{\boxed{x}} = \boxed{7x^2}\sqrt{\boxed{x}}$

(c) $\sqrt{27x^3} = \sqrt{\boxed{9} \, x^2 \cdot 3x} = \sqrt{\boxed{9}} \cdot \sqrt{\boxed{x^2}} \cdot \sqrt{3x} = \boxed{3x\sqrt{3x}}$

3. (a) $\sqrt{\dfrac{24x}{25}} = \dfrac{\sqrt{\boxed{4} \cdot 6x}}{\sqrt{25}} = \dfrac{\boxed{2}\sqrt{6x}}{\boxed{5}}$

(b) $\sqrt{\dfrac{125xy^5}{45x^3y}} = \sqrt{\dfrac{25y^4}{9x^2}} = \dfrac{\sqrt{\boxed{25}}\sqrt{(y^2)^2}}{\sqrt{\boxed{9}}\sqrt{\boxed{x^2}}} = \boxed{\dfrac{5y^2}{3|x|}}$

4. (a) $\sqrt{\dfrac{11}{12}} = \dfrac{\sqrt{11}}{\sqrt{12}} \cdot \dfrac{\sqrt{\boxed{3}}}{\sqrt{\boxed{3}}} = \dfrac{\sqrt{33}}{\boxed{6}}$

(b) $\dfrac{10}{\sqrt{20}} = \dfrac{10}{\sqrt{20}} \cdot \boxed{\dfrac{\sqrt{5}}{\sqrt{5}}} = \dfrac{10\sqrt{\boxed{5}}}{\boxed{10}} = \boxed{\sqrt{5}}$

5. (a) $= \boxed{\dfrac{\sqrt{x}}{3x}}$

(b) $= \sqrt{\dfrac{4x^3}{3y}} = \dfrac{\sqrt{4x^3}}{\sqrt{3y}} \cdot \boxed{\dfrac{\sqrt{3y}}{\sqrt{3y}}} = \dfrac{\sqrt{\boxed{12x^3y}}}{3y}$

$= \boxed{\dfrac{\sqrt{4x^2 \cdot 3xy}}{3y}} = \boxed{\dfrac{2x\sqrt{3xy}}{3y}}$

10.2 EXERCISES

In Exercises 1–32, simplify the expression. Remember to rationalize denominators and use absolute values when necessary.

1. $\sqrt{12}$ **2.** $\sqrt{18}$ **3.** $\sqrt{50}$ **4.** $\sqrt{72}$ **5.** $\sqrt{200}$

6. $\sqrt{256}$ **7.** $\sqrt{20,000}$ **8.** $\sqrt{980}$ **9.** $\sqrt{9x^2}$ **10.** $\sqrt{16x^4}$

11. $\sqrt{36a^2b^2}$ **12.** $\sqrt{49x^2y^4}$ **13.** $\sqrt{8a^6}$ **14.** $\sqrt{12x^4}$ **15.** $\sqrt{4x^5}$

16. $\sqrt{18x^3}$ **17.** $\sqrt{24x^7}$ **18.** $\sqrt{72x^3}$ **19.** $\sqrt{50x^2y^3}$ **20.** $\sqrt{288x^3y^5}$

21. $\sqrt{\dfrac{50}{2}}$ **22.** $\sqrt{\dfrac{48}{3}}$ **23.** $\dfrac{1}{\sqrt{3}}$ **24.** $\dfrac{2}{\sqrt{10}}$ **25.** $\sqrt{\dfrac{3x}{5}}$

26. $\sqrt{\dfrac{2x}{7y}}$ **27.** $\sqrt{\dfrac{4x}{3}}$ **28.** $\sqrt{\dfrac{2x^2y}{4xy^3}}$ **29.** $\dfrac{10x}{\sqrt{20x^3}}$ **30.** $\dfrac{5z}{\sqrt{50xz^2}}$

31. $\sqrt{\dfrac{12uv^3}{3u^2v}}$ **32.** $\sqrt{\dfrac{45x^3z}{18x^4z^2}}$

In Exercises 33 and 34, use a calculator to approximate the expressions to three decimal places.

33. $\dfrac{1+\sqrt{6}}{\sqrt{5}}$ **34.** $\dfrac{\sqrt{2}-3\sqrt{5}}{\sqrt{7}}$

10.3 Operations with Radical Expressions

Section Highlights

1. $\sqrt{a}+\sqrt{b}$ **does NOT equal** $\sqrt{a+b}$.
2. $a+b$ and $a-b$ are called conjugates of each other. Recall that $(a+b)(a-b)=a^2-b^2$.

EXAMPLE 1 ■ Simplifying Radical Expressions

Simplify the radical expressions (a) $5\sqrt{2}+3\sqrt{2}$ and (b) $9\sqrt{7}-5\sqrt{3}+2\sqrt{7}$.

Solution

(a) $5\sqrt{2} + 3\sqrt{2} = (5+3)\sqrt{2}$ Distributive Property

$\qquad\qquad = 8\sqrt{2}$ Simplify

(b) $9\sqrt{7} - 5\sqrt{3} + 2\sqrt{7} = 9\sqrt{7} + 2\sqrt{7} - 5\sqrt{3}$ Commutative Property

$\qquad\qquad\qquad = (9+2)\sqrt{7} - 5\sqrt{3}$ Distributive Property

$\qquad\qquad\qquad = 11\sqrt{7} - 5\sqrt{3}$ Simplify. ∎

Starter Exercise 1   *Fill in the blanks.*

(a) Simplify the radical expression $17\sqrt{3} - 6\sqrt{3}$.

$$17\sqrt{3} - 6\sqrt{3} = \left(\boxed{} - \boxed{}\right)\sqrt{3} = \boxed{}$$

(b) Simplify the radical expression $\sqrt{6} + 3\sqrt{5} + 7\sqrt{6} - 2\sqrt{5}$.

$$\sqrt{6} + 3\sqrt{5} + 7\sqrt{6} - 2\sqrt{5} = \sqrt{6} + \boxed{} + 3\sqrt{5} - 2\sqrt{5}$$

$$= \left(\boxed{} + \boxed{}\right)\sqrt{6} + \left(\boxed{}\right)\sqrt{5} = \boxed{}$$

(c) Simplify the radical expression $\sqrt{48} + 5\sqrt{27}$.

$$\sqrt{48} + 5\sqrt{27} = \boxed{}\sqrt{3} + 5\left(\boxed{}\sqrt{3}\right)$$

$$= \boxed{}\sqrt{3} + \boxed{}\sqrt{3} = \left(\boxed{} + \boxed{}\right)\boxed{} = \boxed{}$$

EXAMPLE 2 ■   Simplifying Radical Expressions with Variable Radicands

Simplify the radical expressions (a) $1 + 2\sqrt{x} - \sqrt{9} + 3\sqrt{x}$ and (b) $7x\sqrt{8x} - 2\sqrt{2x^3}$.

Solution

(a) $1 + 2\sqrt{x} - \sqrt{9} + 3\sqrt{x} = 1 + 2\sqrt{x} - 3 + 3\sqrt{x}$ Simplify radicals.

$\qquad\qquad\qquad = 1 - 3 + 2\sqrt{x} + 3\sqrt{x}$ Commutative Property

$\qquad\qquad\qquad = 1 - 3 + (2+3)\sqrt{x}$ Distributive Property

$\qquad\qquad\qquad = -2 + 5\sqrt{x}$ Simplify.

(b) $7x\sqrt{8x} - 2\sqrt{2x^3} = 7x\left(2\sqrt{2x}\right) - 2x\sqrt{2x}$ Simplify radicals.

$\qquad\qquad\qquad = 14x\sqrt{2x} - 2x\sqrt{2x}$ Simplify terms.

$\qquad\qquad\qquad = (14-2)x\sqrt{2x}$ Distributive Property

$\qquad\qquad\qquad = 12x\sqrt{2x}$ Simplify. ∎

Starter Exercise 2 | *Fill in the blanks.*

(a) Simplify $10\sqrt{z} + z - 8\sqrt{z}$.

$$10\sqrt{z} + z - 8\sqrt{z} = 10\sqrt{z} - \boxed{} + z = \left(\boxed{} - \boxed{}\right)\sqrt{z} + z = \boxed{}$$

(b) Simplify $\sqrt{9z} + \sqrt{4z}$.

$$\sqrt{9z} + \sqrt{4z} = \boxed{}\sqrt{z} + \boxed{}\sqrt{z} = \left(\boxed{} + \boxed{}\right)\sqrt{z} = \boxed{}$$

(c) Simplify $\sqrt{63x^5} - \sqrt{28x^5}$.

$$\sqrt{63x^5} - \sqrt{28x^5} = \boxed{}\sqrt{7x} - \boxed{}\sqrt{7x} = \left(\boxed{} - \boxed{}\right)\boxed{} = \boxed{}$$

EXAMPLE 3 ■ Multiplying Radicals

Multiply the following and simplify your answer.

(a) $\sqrt{x}\left(\sqrt{x} + \sqrt{3x^3}\right)$

(b) $\left(2 - \sqrt{3}\right)\left(2 + \sqrt{3}\right)$

Solution

(a) $\sqrt{x}\left(\sqrt{x} + \sqrt{3x^3}\right) = \sqrt{x}\sqrt{x} + \sqrt{x}\sqrt{3x^3}$ Distributive Property

$$= \sqrt{x^2} + \sqrt{3x^4}$$ Multiply radicands.

$$= x + x^2\sqrt{3}$$ Simplify.

(b) $\left(2 - \sqrt{3}\right)\left(2 + \sqrt{3}\right) = 2^2 - \left(\sqrt{3}\right)^2$ Special Product Formula

$$= 4 - 3$$

$$= 1$$

Starter Exercise 3 | *Fill in the blanks.*

(a) Multiply and simplify: $\sqrt{2x}\left(\sqrt{2x^3} + \sqrt{8x}\right)$

$$\sqrt{2x}\left(\sqrt{2x^3} + \sqrt{8x}\right) = \sqrt{2x}\sqrt{2x^3} + \boxed{}\sqrt{8x} = \sqrt{\boxed{}} + \sqrt{\boxed{}} = \boxed{} + \boxed{}$$

(b) Multiply and simplify: $\left(\sqrt{x} + \sqrt{3}\right)\left(\sqrt{x} - \sqrt{3}\right)$

$$\left(\sqrt{x} + \sqrt{3}\right)\left(\sqrt{x} - \sqrt{3}\right) = \left(\sqrt{x}\right)^2 - \boxed{} = \boxed{} - \boxed{}$$

EXAMPLE 4 ■   Simplifying Quotients Involving Radicals

Rationalize the denominator and simplify $\dfrac{\sqrt{2}}{\sqrt{x}+\sqrt{6}}$.

Solution

$$\frac{\sqrt{2}}{\sqrt{x}+\sqrt{6}} = \frac{\sqrt{2}}{\sqrt{x}+\sqrt{6}} \cdot \frac{\sqrt{x}-\sqrt{6}}{\sqrt{x}-\sqrt{6}} \qquad \text{Multiply by } \frac{\text{conjugate of denominator}}{\text{conjugate of denominator}}.$$

$$= \frac{\sqrt{2x}-\sqrt{12}}{\left(\sqrt{x}\right)^2 - \left(\sqrt{6}\right)^2} = \frac{\sqrt{2x}-2\sqrt{3}}{x-6}$$

■

Starter Exercise 4   *Fill in the blanks.*

(a) Rationalize the denominator and simplify: $\dfrac{2}{\sqrt{6}+\sqrt{2}}$

$$\frac{2}{\sqrt{6}+\sqrt{2}} = \frac{2}{\sqrt{6}+\sqrt{2}} \cdot \frac{\boxed{}-\boxed{}}{\boxed{}-\boxed{}} = \frac{\boxed{}-\boxed{}}{\boxed{}^2 - \boxed{}^2}$$

$$= \frac{\boxed{}-\boxed{}}{\boxed{}-\boxed{}} = \frac{\boxed{}-\boxed{}}{\boxed{}} = \frac{2\left(\boxed{}-\boxed{}\right)}{2\cdot\boxed{}} = \boxed{}$$

(b) Rationalize the denominator and simplify: $\dfrac{x}{\sqrt{x}-\sqrt{y}}$

$$\frac{x}{\sqrt{x}-\sqrt{y}} = \frac{x}{\sqrt{x}-\sqrt{y}} \cdot \frac{\boxed{}}{\boxed{}} = \frac{\boxed{}}{\boxed{}^2 - \boxed{}^2} = \boxed{}$$

■ **Solutions to Starter Exercises** ■

1. (a) $= \left(\boxed{17} - \boxed{6}\right)\sqrt{3} = \boxed{11\sqrt{3}}$

(b) $= \sqrt{6} + \boxed{7\sqrt{6}} + 3\sqrt{5} - 2\sqrt{5} = \left(\boxed{1} + \boxed{7}\right)\sqrt{6} + \left(\boxed{3-2}\right)\sqrt{5}$

$= \boxed{8\sqrt{6}+\sqrt{5}}$

(c) $= \boxed{4}\sqrt{3} + 5\left(\boxed{3}\sqrt{3}\right) = \boxed{4}\sqrt{3} + \boxed{15}\sqrt{3} = \left(\boxed{4} + \boxed{15}\right)\boxed{\sqrt{3}}$

$= \boxed{19\sqrt{3}}$

2. (a) $= 10\sqrt{z} - \boxed{8\sqrt{z}} + z = \left(\boxed{10} - \boxed{8}\right)\sqrt{z} + z = \boxed{2\sqrt{z}+z}$

(b) $= \boxed{3}\sqrt{z} + \boxed{2}\sqrt{z} = \left(\boxed{3} + \boxed{2}\right)\sqrt{z} = \boxed{5\sqrt{z}}$

(c) $= \boxed{3x^2}\sqrt{7x} - \boxed{2x^2}\sqrt{7x} = \left(\boxed{3} - \boxed{2}\right)\boxed{x^2\sqrt{7x}} = \boxed{x^2\sqrt{7x}}$

■ **Solutions to Starter Exercises** ■

3. (a) $= \sqrt{2x}\sqrt{2x^3} + \boxed{\sqrt{2x}}\;\boxed{\sqrt{8x}} = \sqrt{\boxed{4x^4}} + \sqrt{\boxed{16x^2}} = \boxed{2x^2} + \boxed{4x}$

(b) $= \left(\sqrt{x}\right)^2 - \boxed{(\sqrt{3})^2} = \boxed{x} - \boxed{3}$

4. (a) $= \dfrac{2}{\sqrt{6}+\sqrt{2}} \cdot \dfrac{\boxed{\sqrt{6}} - \boxed{\sqrt{2}}}{\boxed{\sqrt{6}} - \boxed{\sqrt{2}}} = \dfrac{\boxed{2\sqrt{6}} - \boxed{2\sqrt{2}}}{\boxed{(\sqrt{6})}^2 - \boxed{(\sqrt{2})}^2}$

$= \dfrac{\boxed{2\sqrt{6}} - \boxed{2\sqrt{2}}}{\boxed{6} - \boxed{2}} = \dfrac{\boxed{2\sqrt{6}} - \boxed{2\sqrt{2}}}{\boxed{4}}$

$= \dfrac{2\left(\boxed{\sqrt{6}} - \boxed{\sqrt{2}}\right)}{2 \cdot \boxed{2}} = \dfrac{\sqrt{6}-\sqrt{2}}{2}$

(b) $= \dfrac{x}{\sqrt{x}-\sqrt{y}} \cdot \dfrac{\boxed{\sqrt{x}+\sqrt{y}}}{\boxed{\sqrt{x}+\sqrt{y}}} = \dfrac{\boxed{x\sqrt{x}+x\sqrt{y}}}{\boxed{(\sqrt{x})}^2 - \boxed{(\sqrt{y})}^2} = \dfrac{\boxed{x\sqrt{x}+x\sqrt{y}}}{x-y}$

10.3 EXERCISES

In Exercises 1–10, simplify the radical expression.

1. $2 - \sqrt{4}$
2. $12\sqrt{3} - 10\sqrt{3}$
3. $\frac{1}{2}\sqrt{7} + \frac{1}{4}\sqrt{7}$
4. $\sqrt{27} - \sqrt{48}$

5. $\sqrt{5} + \sqrt{45} - \sqrt{125}$
6. $7\sqrt{x} - 5\sqrt{x}$
7. $5\sqrt{2x} - \sqrt{8x}$

8. $\sqrt{112x^3} + x\sqrt{175x}$
9. $5\sqrt{x^3y^5} - xy^2\sqrt{xy}$
10. $\sqrt{x^2y^3} - 5\sqrt{xy^2}$

In Exercises 11–22, perform the specified multiplication and simplify the product.

11. $\sqrt{2} \cdot \sqrt{8}$
12. $\sqrt{6}(2 - \sqrt{5})$
13. $\sqrt{5}(\sqrt{15} - 2)$

14. $(\sqrt{3} + \sqrt{7})(\sqrt{3} - \sqrt{7})$
15. $(\sqrt{x} + \sqrt{2})(\sqrt{x} - \sqrt{2})$
16. $(\sqrt{6} + \sqrt{5})^2$

17. $(\sqrt{3} + 1)(\sqrt{5} - 2)$
18. $\sqrt{x}(\sqrt{x} + \sqrt{2})$
19. $\sqrt{x}(\sqrt{2x^3} + \sqrt{2x})$

20. $(\sqrt{2x} + \sqrt{x})(\sqrt{2x} - \sqrt{x})$
21. $(\sqrt{3} + \sqrt{x})^2$
22. $(\sqrt{x} + 5)(2\sqrt{x} - 1)$

In Exercises 23–26, find the conjugate of the expression. Then find the product of the expression and its conjugate.

23. $1 + \sqrt{2}$ **24.** $2 - \sqrt{3}$ **25.** $\sqrt{x} + 5$ **26.** $\sqrt{x} + \sqrt{14}$

In Exercises 27–32, rationalize the denominator and simplify.

27. $\dfrac{1}{\sqrt{3} - 1}$ **28.** $\dfrac{2}{\sqrt{5} + 4}$ **29.** $\dfrac{2x}{\sqrt{x} + \sqrt{2}}$

30. $\left(\sqrt{2} + 1\right) \div \left(\sqrt{2} - 6\right)$ **31.** $\left(\sqrt{x} - 2\right) \div \left(\sqrt{x} + 3\right)$ **32.** $\left(\sqrt{x} + \sqrt{6}\right) \div \left(\sqrt{x} - \sqrt{6}\right)$

In Exercises 33 and 34, rewrite the expression as a single fraction and simplify.

33. $2 - \dfrac{1}{\sqrt{2}}$ **34.** $\dfrac{1}{\sqrt{3}} - \sqrt{12}$

In Exercises 35 and 36, insert the correct symbol ($<$, $>$, or $=$) between the two numbers. You may want to use a calculator.

35. $\sqrt{2} + \sqrt{3}$ ☐ $\sqrt{2 + 3}$ **36.** 7 ☐ $\sqrt{3^2 + 4^2}$

10.4   Solving Equations and Applications

Section Highlights

1. If $a = b$, then $a^2 = b^2$. Note that $a = b$ and $a^2 = b^2$ are not equivalent. Thus, all solutions of $a^2 = b^2$ must be checked in $a = b$ to determine whether they are extraneous.

2. **Pythagorean Theorem** If a and b are the lengths of the legs of a right triangle and c is the length of the hypotenuse, then

$$c^2 = a^2 + b^2$$
$$c = \sqrt{a^2 + b^2}.$$

EXAMPLE 1 ■ Solving a Radical Equation Having One Radical

Solve (a) $\sqrt{2x + 1} = -3$ and (b) $x + \sqrt{x} = 6$.

Solution

(a) $\sqrt{2x + 1} = -3$ Given equation   **Check:** $\sqrt{2x + 1} = -3$

$\left(\sqrt{2x + 1}\right)^2 = (-3)^2$ Square both sides. $\sqrt{2(4) + 1} \overset{?}{=} -3$

$2x + 1 = 9$ Simplify. $\sqrt{8 + 1} \overset{?}{=} -3$

$2x = 8$ $\sqrt{9} \overset{?}{=} -3$

$x = 4$ Solution $3 \neq -3$

Therefore, there is no solution.

(b)
$$x + \sqrt{x} = 6 \qquad \text{Given equation}$$
$$\sqrt{x} = 6 - x \qquad \text{Isolate radical.}$$
$$\left(\sqrt{x}\right)^2 = (6 - x)^2 \qquad \text{Square both sides.}$$
$$x = 36 - 12x + x^2 \qquad \text{Simplify.}$$
$$0 = x^2 - 13x + 36 \qquad \text{Standard form}$$
$$0 = (x - 9)(x - 4) \qquad \text{Factor.}$$
$$x - 9 = 0 \text{ or } x - 4 = 0 \qquad \text{Set both factors equal to 0.}$$
$$x = 9 \text{ or } \qquad x = 4 \qquad \text{Solutions}$$

Try checking these solutions. When you do, you will find that 4 is a valid solution, but 9 is not. ∎

Starter Exercise 1   *Fill in the blanks.*

(a) Solve $\sqrt{3x - 2} = 2$.
$$\sqrt{3x - 2} = 2$$
$$\left(\sqrt{3x - 2}\right)^2 = 2^2$$
$$\boxed{} - \boxed{} = \boxed{}$$
$$\boxed{} = \boxed{}$$
$$x = \boxed{}$$

(b) Solve $36 + 5\sqrt{x} = x$.
$$36 + 5\sqrt{x} = x$$
$$5\sqrt{x} = x - 36$$
$$\left(5\sqrt{x}\right)^2 = (x - 36)^2$$
$$25x = \boxed{}$$
$$0 = \boxed{}$$
$$0 = \left(\boxed{}\right)\left(\boxed{}\right)$$
$$\vdots$$

EXAMPLE 2 ■   Solving Radical Equations Having Two Radicals

Solve $\sqrt{x + 6} = \sqrt{2x - 3}$.

Solution
$$\sqrt{x + 6} = \sqrt{2x - 3} \qquad \text{Given equation}$$
$$\left(\sqrt{x + 6}\right)^2 = \left(\sqrt{2x - 3}\right)^2 \qquad \text{Square both sides.}$$
$$x + 6 = 2x - 3 \qquad \text{Simplify.}$$
$$x = 2x - 9$$
$$-x = -9$$
$$x = 9 \qquad \text{Solution}$$

After checking 9 in the original equation, you can see that the solution is 9. ∎

Starter Exercise 2   *Fill in the blanks.*

(a) Solve $\sqrt{5x+6}-\sqrt{3x+4}=0$.

$$\sqrt{5x+6}-\sqrt{3x+4}=0$$
$$\sqrt{5x+6}=\boxed{}$$
$$\left(\sqrt{5x+6}\right)^2=\left(\boxed{}\right)^2$$
$$5x+6=\boxed{}$$
$$5x=\boxed{}$$
$$\boxed{}=\boxed{}$$
$$x=\boxed{}$$

(b) Solve $\sqrt{x+1}+2\sqrt{4x-1}=0$.

$$\sqrt{x+1}+2\sqrt{4x-1}=0$$
$$\sqrt{x+1}=-2\sqrt{4x-1}$$
$$\left(\sqrt{x+1}\right)^2=\boxed{}$$
$$x+1=\boxed{}$$
$$x=\boxed{}$$
$$-15x=\boxed{}$$
$$x=\boxed{}$$

Conclusion: $\boxed{}$

EXAMPLE 3 ■   Pythagorean Theorem

The length of one leg of a right triangle is 3 inches, the other is 4 inches. Find the length of the hypotenuse.

Solution

In $c=\sqrt{a^2+b^2}$, replace a by 3 and b by 4.

$$c=\sqrt{3^2+4^2}$$
$$c=\sqrt{25}$$
$$c=5$$

The length of the hypotenuse is 5 inches.

Starter Exercise 3   *Fill in the blanks.*

A ladder is 12 feet long and leaning against a wall with its base 3 feet from the wall. How high up on the wall does the ladder reach? Note that the ladder, wall, and ground form a right triangle. Use the Pythagorean Theorem with $c=12$, $b=3$, and find a.

$$12^2=a^2+3^2$$
$$\boxed{}=a^2+\boxed{}$$
$$\boxed{}=a^2$$
$$\boxed{}\approx a$$

■ **Solutions to Starter Exercises** ■

1. (a) $\boxed{3x} - \boxed{2} = \boxed{4}$

$\qquad\qquad \boxed{3x} = \boxed{6}$

$\qquad\qquad\quad x = \boxed{2}$

(b) $\qquad 25x = \boxed{x^2 - 72x + 1296}$

$\qquad\qquad 0 = \boxed{x^2 - 97x + 1296}$

$\qquad\qquad 0 = \left(\boxed{x - 16}\right)\left(\boxed{x - 81}\right)$

$\qquad x - 16 = 0 \ \text{ or } x - 81 = 0$

$\qquad\quad x = 16 \text{ or } \qquad x = 81$

After checking, we see only 81 is a solution.

2. (a) $\qquad \sqrt{5x+6} = \boxed{\sqrt{3x+4}}$

$\qquad \left(\sqrt{5x+6}\right)^2 = \left(\boxed{\sqrt{3x+4}}\right)^2$

$\qquad\qquad 5x + 6 = \boxed{3x + 4}$

$\qquad\qquad\quad 5x = \boxed{3x - 2}$

$\qquad\qquad \boxed{2x} = \boxed{-2}$

$\qquad\qquad\quad x = \boxed{-1}$

(b) $\left(\sqrt{x+1}\right)^2 = \boxed{\left(-2\sqrt{4x-1}\right)}$

$\qquad\quad x + 1 = \boxed{4(4x-1)}$

$\qquad\qquad x = \boxed{16x - 5}$

$\qquad\quad -15x = \boxed{-5}$

$\qquad\qquad x = \boxed{\tfrac{1}{3}}$

Conclusion: $\boxed{\text{No solution}}$

3. $\boxed{144} = a^2 + \boxed{9}$

$\quad \boxed{135} = a^2$

$\boxed{11.62} \approx a$

11.62 feet high

10.4 | EXERCISES

In Exercises 1–16, solve the given equation.

1. $\sqrt{x} = 5$

2. $\sqrt{t} - 3 = 0$

3. $\sqrt{t} + 7 = 0$

4. $\sqrt{y+7} = 10$

5. $\sqrt{5x} = 15$

6. $\sqrt{2x-3} = 9$

7. $\sqrt{4t+6} = 2$

8. $10 - \sqrt{x} = 7$

9. $2\sqrt{x+1} = 6$

10. $\sqrt{x^2+1} = x + 1$

11. $\sqrt{x+3} = \sqrt{2x+1}$

12. $\sqrt{6x-1} = 2\sqrt{x}$

13. $14 + 5\sqrt{x} = x$

14. $x - 2\sqrt{x} = 3$

15. $x - 2 = \sqrt{8-x}$

16. $\sqrt{3x+7} = x - 1$

In Exercises 17–20, find the length x of the unknown side of the right triangle. (Round your answer to two decimal places.)

	Hypotenuse	Legs		Hypotenuse	Legs
17.	16	4 and x	**18.**	x	2 and 5
19.	2	x and 1	**20.**	5	4 and x

21. *Height of a Ladder*   A ladder is 20 feet long, and it is leaning against a house at a point 16 feet above the ground. How far from the house is the base of the ladder?

22. *Distance Between Two Points*   Find the length of the line segment connecting the points $(2, 4)$ and $(7, 8)$ in the rectangular coordinate system.

In Exercises 23 and 24, use the equation for the velocity of a freely-falling object $(v = \sqrt{2gh})$.

23. *Velocity of an Object*   An object is dropped from a height of 25 feet. Find the velocity of the object when it strikes the ground.

24. *Height of an Object*   An object strikes the ground with a velocity of 48 feet per second. Find the height from which the object was dropped.

25. *Height*   The time t in seconds for a freely-falling object to fall d feet is given by $t = \sqrt{d/16}$. A boy drops a rock from a bridge and notes that it hits the water after approximately 4 seconds. Estimate the height of the bridge.

26. *Pendulum Length*   The time t in seconds for a pendulum of length L feet to move through one complete cycle (its period) is given by

$$t = 2\pi\sqrt{\frac{L}{32}}.$$

How long is the pendulum of a grandfather clock that has a period of 2.4 seconds?

27. *Demand*   The demand equation for a certain product is given by $p = 30 - \sqrt{x - 1}$ where x is the number of units demanded per day and p is the price per unit. Find the demand when the price is $25.50.

Cumulative Practice Test for Chapters P–10

In Exercises 1–6, match each equation with the property listed below that best justifies the equation.

(a) Multiplicative Identity Property
(c) Commutative Property of Addition

(b) Distributive Property
(d) Associative Property of Multiplication

1. $2(3x + 5) = 2(3x) + 2(5)$

2. $2(3x) + 2(5) = (2 \cdot 3)x + 2(5)$

3. $(2x + 3) + 6x = 6x + (2x + 3)$

4. $5x - 3x = (5 - 3)x$

5. $x = 1x$

6. $x + 7x = 7x + x$

In Exercises 7–16, simplify the given expression.

7. $\dfrac{2x}{x+1} - \dfrac{4}{x-1}$

8. $(2x-3)^2$

9. $\sqrt{\dfrac{45x^3}{5x}}$

10. $\dfrac{1}{\sqrt{5}+2}$

11. $\dfrac{x+2}{x^2-9} \div \dfrac{x^2-4}{x^2+6x+9}$

12. $-2[3x-5(x+2)]-7x$

13. $\sqrt{147x^3} - x\sqrt{27x} + \sqrt{3x^3}$

14. $\sqrt{x^2+4x+4}$

15. $\dfrac{\sqrt{2x}}{\sqrt{8x^3}}$

16. $(x^2+3x-1)(x-5)$

In Exercises 17 and 18, find an equation of the line that passes through the given points.

17. $(3, -1), (5, 3)$

18. $(9, 12), (9, 15)$

In Exercises 19 and 20, factor the given polynomial completely.

19. $x^4 - 1$

20. $2x^3 - 7x^2 - 15x$

In Exercises 21–32, solve the given equation.

21. $(x-2)^2 = 9$

22. $\dfrac{1}{x-1} = x - \dfrac{1}{x-1}$

23. $\dfrac{2}{x} = \dfrac{4}{8}$

24. $2\sqrt{x} - 1 = 9$

25. $5[x - (2x-3)] = 12$

26. $x^2 = 3x + 28$

27. $x + 3\sqrt{x} = 10$

28. $\dfrac{2}{x+1} = \dfrac{1}{x-1}$

29. $\dfrac{1}{2}x - \dfrac{1}{3} = \dfrac{3}{4}x + \dfrac{1}{12}$

30. $\sqrt{2x+1} - \sqrt{x-3} = 0$

31. $\sqrt{5x-1} + \sqrt{6x+7} = 0$

32. $4x^2 + 3x + 12 = 3x^2 - 4x$

In Exercises 33 and 34, solve the given system of equations.

33. $3x + 2y = 6$
$\ \dfrac{1}{2}x + \dfrac{1}{3}y = 1$

34. $2x - 4y = 7$
$\ 3x - y = 1$

35. The sum of three consecutive odd integers is 75. Find the integers.

36. A plane can fly 280 miles, with the wind, in two hours. Against the wind, it would take the plane three hours to fly 300 miles. What is the rate of the wind and the rate of the plane in calm air?

37. Find the length of the diagonal of a square with sides of length 5.

38. A book sells for $20.80. The bookstore uses a 30% markup rate. What is the wholesale cost of the book?

39. A total of $15,000 was invested in two funds paying $7\frac{1}{2}\%$ and 8% simple interest. How much was invested in each fund if the yearly interest earned on both funds totaled $1160?

40. Bob is 50 feet due north of Jennifer. Sam is 30 feet due west of Jennifer. How far apart are Bob and Sam?

10.1 | Answers to Exercises

1. 6
2. −7
3. 2 and −2
4. $\frac{7}{3}$ and −$\frac{7}{3}$
5. No square roots

6. No square roots
7. 4
8. 8
9. −12
10. Not a real number

11. 16
12. 25
13. $\frac{1}{8}$
14. −$\frac{4}{7}$
15. Not a real number

16. 5
17. Irrational
18. Rational
19. 4.796
20. 22.627

21. −13.145
22. 3.795
23. 88.329
24. 2.258
25. −0.112

26. 1.144
27. 4.4
28. 9.8
29. (a) 3.7
(b) 841
30. a

10.2 | Answers to Exercises

1. $2\sqrt{3}$
2. $3\sqrt{2}$
3. $5\sqrt{2}$
4. $6\sqrt{2}$
5. $10\sqrt{2}$

6. 16
7. $100\sqrt{2}$
8. $14\sqrt{5}$
9. $3|x|$
10. $4x^2$

11. $6|a|\,|b|$
12. $7|x|y^2$
13. $2|a^3|\sqrt{2}$
14. $2x^2\sqrt{3}$
15. $2x^2\sqrt{x}$

16. $3x\sqrt{2x}$
17. $2x^3\sqrt{6x}$
18. $6x\sqrt{2x}$
19. $5|x|y\sqrt{2y}$
20. $12xy^2\sqrt{2xy}$

21. 5
22. 4
23. $\frac{\sqrt{3}}{3}$
24. $\frac{2\sqrt{10}}{5}$
25. $\frac{\sqrt{15x}}{5}$

26. $\frac{\sqrt{14xy}}{7y}$
27. $\frac{2\sqrt{3x}}{3}$
28. $\frac{\sqrt{2x}}{2|y|}$
29. $\frac{\sqrt{5x}}{x}$
30. $\frac{z\sqrt{2x}}{2x|z|}$

31. $\frac{2|v|\sqrt{u}}{u}$
32. $\frac{\sqrt{10xz}}{2xz}$
33. 1.543
34. −2.001

10.3 | Answers to Exercises

1. 0
2. $2\sqrt{3}$
3. $\frac{3}{4}\sqrt{7}$
4. $-\sqrt{3}$
5. $-\sqrt{5}$

6. $2\sqrt{x}$
7. $3\sqrt{2x}$
8. $9x\sqrt{7x}$
9. $4xy^2\sqrt{xy}$
10. $xy\sqrt{y}-5y\sqrt{x}$

11. 4
12. $2\sqrt{6}-\sqrt{30}$
13. $5\sqrt{3}-2\sqrt{5}$
14. −4

15. $x - 2$ **16.** $11 + 2\sqrt{30}$ **17.** $\sqrt{15} - 2\sqrt{3} + \sqrt{5} - 2$ **18.** $x + \sqrt{2x}$

19. $x^2\sqrt{2} + x\sqrt{2}$ **20.** x **21.** $9 + 2\sqrt{3x} + x$ **22.** $2x + 9\sqrt{x} - 5$

23. $1 - \sqrt{2}; \; -1$ **24.** $2 + \sqrt{3}; \; 1$ **25.** $\sqrt{x} - 5; \; x - 25$ **26.** $\sqrt{x} - \sqrt{14}; \; x - 14$

27. $\dfrac{\sqrt{3} + 1}{2}$ **28.** $-\dfrac{2\sqrt{5} - 8}{11}$ **29.** $\dfrac{2x\sqrt{x} - 2x\sqrt{2}}{x - 2}$ **30.** $-\dfrac{8 + 7\sqrt{2}}{34}$

31. $\dfrac{x - 5\sqrt{x} + 6}{x - 9}$ **32.** $\dfrac{x + 2\sqrt{6x} + 6}{x - 6}$ **33.** $\dfrac{4 - \sqrt{2}}{2}$ **34.** $-\dfrac{5\sqrt{3}}{3}$

35. $>$ **36.** $>$

10.4	**Answers to Exercises**

1. $x = 25$ **2.** $x = 9$ **3.** No solution   **4.** $y = 93$ **5.** $x = 45$

6. $x = 42$ **7.** $x = -\frac{1}{2}$ **8.** $x = 9$ **9.** $x = 8$ **10.** $x = 0$

11. $x = 2$ **12.** $x = \frac{1}{2}$ **13.** $x = 49$ **14.** $x = 9$ **15.** $x = 4$

16. $x = 6$ **17.** $x \approx 15.49$ **18.** $x \approx 5.39$ **19.** $x \approx 1.73$ **20.** $x = 3$

21. 12 ft   **22.** ≈ 6.40 **23.** 40 ft per sec   **24.** 36 ft   **25.** 256 ft

26. 4.67 ft   **27.** 21.25 units

	Answers to Cumulative Practice Test P–10

1. (b)   **2.** (d)   **3.** (c)   **4.** (b)   **5.** (a)

6. (c)   **7.** $\dfrac{2x^2 - 6x - 4}{x^2 - 1}$ **8.** $4x^2 - 12x + 9$ **9.** $3|x|$ **10.** $\sqrt{5} - 2$

11. $\dfrac{x + 3}{(x - 3)(x - 2)}$ **12.** $-3x + 20$ **13.** $5x\sqrt{3x}$ **14.** $|x + 2|$

15. $\dfrac{1}{2|x|}$ **16.** $x^3 - 2x^2 - 16x + 5$ **17.** $y = 2x - 7$ **18.** $x = 9$

19. $(x + 1)(x - 1)(x^2 + 1)$ **20.** $x(2x + 3)(x - 5)$ **21.** $x = -1$ or $x = 5$ **22.** $x = -1$ or $x = 2$

23. $x = 4$ **24.** $x = 25$ **25.** $x = \dfrac{3}{5}$ **26.** $x = -4$ or $x = 7$

27. $x = 4$

28. $x = 3$

29. $x = \dfrac{-5}{3}$

30. $x = -4$

31. No solution

32. $x = -4$ or $x = -3$

33. all (x, y) such that $3x + 2y = 6$

34. $\left(-\dfrac{3}{10}, -\dfrac{19}{10}\right)$

35. $23, 25, 27$

36. Plane: 120 mph
Wind: 20 mph

37. $5\sqrt{2} \approx 7.07$

38. $16

39. $8000 at $7\frac{1}{2}\%$
$7000 at 8%

40. $10\sqrt{34} \approx 58.31$ ft

CHAPTER ELEVEN
Quadratic Functions and Equations

11.1 Solution by Extracting Square Roots

Section Highlights

1. A **quadratic equation** in x is an equation that can be written equivalently in the standard form $ax^2 + bx + c = 0$, where a, b, and c are real numbers with $a \neq 0$.

2. Let u be a real number, a variable, or an algebraic expression. The equation $u^2 = d$, where $d > 0$, has exactly two solutions: $u = \sqrt{d}$ and $u = -\sqrt{d}$ (sometimes written $u = \pm\sqrt{d}$).

EXAMPLE 1 ■ Solving Quadratic Equations by Factoring

Solve the quadratic equation $4x^2 + 4x + 1 = 0$.

Solution

$$4x^2 + 4x + 1 = 0 \qquad \text{Given equation}$$
$$(2x + 1)^2 = 0 \qquad \text{Factor.}$$
$$2x + 1 = 0 \qquad \text{Set factor equal to 0.}$$
$$x = -\tfrac{1}{2} \qquad \text{Solve.}$$

Check this in the original equation. ■

Starter Exercise 1 | *Fill in the blanks.*

Solve the equation $x^2 + 8x = x - 12$ by factoring.

$$x^2 + 8x = x - 12$$
$$x^2 + \boxed{} + 12 = 0$$
$$\left(x + \boxed{}\right)\left(x + \boxed{}\right) = 0$$
$$x + \boxed{} = 0 \ \text{ or } x + \boxed{} = 0$$
$$x = \boxed{} \text{ or } \quad x = \boxed{}$$

EXAMPLE 2 ■ Solving Quadratic Equations by Extracting Square Roots

Solve the following equations by extracting the square roots.

(a) $x^2 = 16$

(b) $3(2x - 1)^2 - 9 = 0$

Solution

(a) $x^2 = 16$ Given equation

     $x = \pm\sqrt{16}$ Extract square roots.

     $x = \pm 4$ Simplify.

(b) $3(2x - 1)^2 - 9 = 0$ Given equation

     $3(2x - 1)^2 = 9$

     $(2x - 1)^2 = 3$

     $2x - 1 = \pm\sqrt{3}$ Extract square roots.

     $2x = 1 \pm \sqrt{3}$ Add 1 to both sides.

     $x = \dfrac{1 \pm \sqrt{3}}{2}$ Multiply both sides by $\frac{1}{2}$.

■

Starter Exercise 2   *Fill in the blanks.*

(a) Solve $2x^2 = 6$ by extracting square roots.

     $2x^2 = 6$

     $x^2 = \boxed{}$

     $x = \pm \boxed{}$

(b) Solve $5(2x + 7)^2 - 10 = 0$ by extracting square roots.

     $5(2x + 7)^2 - 10 = 0$

     $5(2x + 7)^2 = \boxed{}$

     $(2x + 7)^2 = \boxed{}$

     $2x + 7 = \boxed{}$

     $2x = \boxed{}$

     $x = \dfrac{\boxed{}}{2}$

■ **Solutions to Starter Exercises** ■

1. $x^2 + \boxed{7x} + 12 = 0$

     $\left(x + \boxed{3}\right)\left(x + \boxed{4}\right) = 0$

     $x + \boxed{3} = 0$ or $x + \boxed{4} = 0$

     $x = \boxed{-3}$ or $x = \boxed{-4}$

2. (a) $x^2 = \boxed{3}$

     $x = \pm \boxed{\sqrt{3}}$

     (b) $5(2x + 7)^2 = \boxed{10}$

     $(2x + 7)^2 = \boxed{2}$

     $2x + 7 = \boxed{\pm\sqrt{2}}$

     $2x = \boxed{-7 \pm \sqrt{2}}$

     $x = \dfrac{\boxed{-7 \pm \sqrt{2}}}{2}$

| 11.1 | **EXERCISES** |

In Exercises 1–12, solve the given equation by factoring.

1. $x^2 + 5x = 0$

2. $3x^2 - 9x = 0$

3. $x^2 - 16 = 0$

4. $16x^2 - 49 = 0$

5. $x^2 - 5x + 6 = 0$

6. $x^2 + 6x + 9 = 0$

7. $9x^2 - 12x + 4 = 0$

8. $(x + 7)(x - 1) = -15$

9. $(5 - x)(x + 3) = 7$

10. $3x^2 = 8x + 3$

11. $x^2 + 5x = 6 - 5x^2$

12. $2t(t - 1) + 3(t - 1) = 0$

In Exercises 13–26, solve the quadratic equation by extracting square roots.

13. $x^2 = 25$

14. $x^2 = 169$

15. $4x^2 = 49$

16. $z^2 - 81 = 0$

17. $2x^2 - 72 = 0$

18. $y^2 + 7 = 0$

19. $(x + 1)^2 = 64$

20. $(x - 3)^2 = 121$

21. $(x - 2)^2 = 3$

22. $(x + 6)^2 = 12$

23. $(2x - 1)^2 = 8$

24. $3(x - 2)^2 = 9$

25. $4(2x - 1)^2 - 12 = 0$

26. $71(23x + 6)^2 + 15 = 0$

27. *Falling Time* The height h (in feet) of an object dropped from a tower 100 feet high is given by $h = 100 - 16t^2$, where t is the time in seconds. How long does it take the object to reach the ground?

28. *Revenue* The revenue R (in dollars) when x units of a certain product are sold is given by $R = x\left(50 - \frac{1}{4}x\right)$, $0 < x < 100$. Determine the number of units that must be sold to produce a revenue of $2500.

11.2 | Solution by Completing the Square

Section Highlights

1. To **complete the square** for the expression $x^2 + bx$, we add $(b/2)^2$, which is the square of half the coefficient of x. Consequently,

$$x^2 + bx + \left(\frac{b}{2}\right)^2 = \left(x + \frac{b}{2}\right)^2.$$

EXAMPLE 1 ■   Constructing a Perfect Square Trinomial

What term should be added to the expression $x^2 - 6x$ so that it becomes a perfect square trinomial?

Solution

The coefficient of x is -6. The square of half of -6 is $(-3)^2 = 9$. Hence, 9 is the number to be added to the expression to make it a perfect square trinomial.

$$x^2 - 6x + 9 = (x - 3)^2$$

■

| **Starter Exercise 1** | *Fill in the blanks.* |

Add the same number to both sides of $x^2 + 8x = 1$ to make the left side a perfect square trinomial.

$$b = 8 \Rightarrow \frac{b}{2} = \boxed{} \Rightarrow \left(\frac{b}{2}\right)^2 = \boxed{}$$

$$x^2 + 8x + \boxed{} = 1 + \boxed{}$$

$$\left(x + \boxed{}\right)^2 = \boxed{}$$

EXAMPLE 2 ■   Completing the Square: Leading Coefficient is 1

Solve the equation $x^2 - 12x + 1 = 0$ by completing the square.

Solution

$$
\begin{array}{ll}
x^2 - 12x + 1 = 0 & \text{Given equation} \\
x^2 - 12x = -1 & \text{Add } -1 \text{ to both sides.} \\
x^2 - 12x + (-6)^2 = -1 + (-6)^2 & \text{Add } (-6)^2 \text{ to both sides.} \\
(x - 6)^2 = 35 & \text{Binomial squared} \\
x - 6 = \pm\sqrt{35} & \text{Extract square roots.} \\
x = 6 \pm \sqrt{35} & \text{Solutions}
\end{array}
$$

■

| **Starter Exercise 2** | *Fill in the blanks.* |

(a) Solve $x^2 - 2x = 8$ by completing the square.

$$x^2 - 2x = 8$$

$$x^2 - 2x + (-1)^2 = 8 + \boxed{}$$

$$\left(x - \boxed{}\right)^2 = \boxed{}$$

$$x - \boxed{} = \pm\sqrt{\boxed{}}$$

$$x = \boxed{}$$

(b) Solve $x^2 - 3x - 6 = 0$ by completing the square.

$$x^2 - 3x - 6 = 0$$

$$x^2 - 3x = 6$$

$$x^2 - 3x + \boxed{} = 6 + \boxed{}$$

$$\left(x - \boxed{}\right)^2 = \boxed{}$$

$$x - \boxed{} = \pm\sqrt{\boxed{}}$$

$$x = \boxed{} \pm \boxed{} = \boxed{}$$

EXAMPLE 3 ■ Completing the Square: Leading Coefficient is Not 1

Solve the equation $3x^2 - x - 1 = 0$ by completing the square.

Solution

$$3x^2 - x - 1 = 0 \qquad \text{Given equation}$$

$$3x^2 - x = 1 \qquad \text{Add 1 to both sides.}$$

$$x^2 - \frac{1}{3}x = \frac{1}{3} \qquad \text{Multiply both sides by } \frac{1}{3}.$$

$$x^2 - \frac{1}{3}x + \left(-\frac{1}{6}\right)^2 = \frac{1}{3} + \left(-\frac{1}{6}\right)^2 \qquad \text{Add } \left(-\frac{1}{6}\right)^2 \text{ to both sides.}$$

$$\left(x - \frac{1}{6}\right)^2 = \frac{13}{36} \qquad \text{Binomial squared}$$

$$x - \frac{1}{6} = \pm\frac{\sqrt{13}}{6} \qquad \text{Extract square roots.}$$

$$x = \frac{1}{6} \pm \frac{\sqrt{13}}{6} = \frac{1 \pm \sqrt{13}}{6}$$

■ Starter Exercise 3   *Fill in the blanks.*

Solve $2x^2 + 15x = -\frac{9}{2}$ by completing the square.

$$2x^2 + 15x = -\frac{9}{2}$$

$$x^2 + \frac{15}{2}x = \boxed{}$$

$$x^2 + \frac{15}{2}x + \boxed{} = \boxed{} + \boxed{}$$

$$\left(x + \boxed{}\right)^2 = \boxed{}$$

$$x + \boxed{} = \pm \boxed{}$$

$$x = \boxed{} \pm \boxed{}$$

$$x = \boxed{} \quad \text{or} \quad x = \boxed{}$$

EXAMPLE 4 ■ A Quadratic Equation with No Real Solution

Solve the equation $x^2 + 2x + 8 = 0$ by completing the square.

Solution

$$x^2 + 2x + 8 = 0$$

$$x^2 + 2x = -8$$

$$x^2 + 2x + 1^2 = -8 + 1^2$$

$$(x + 1)^2 = -7$$

Since the square of a real number cannot be negative, we conclude that there are no real solutions.

■ Starter Exercise 4   *Fill in the blanks.*

Solve $x^2 + 6x + 10 = 0$ by completing the square.

$$x^2 + 6x + 10 = 0$$

$$x^2 + 6x = -10$$

$$x^2 + 6x + \boxed{} = -10 + \boxed{}$$

$$\left(x + \boxed{}\right)^2 = \boxed{}$$

Conclusion: $\boxed{}$

■   Solutions to Starter Exercises   ■

1.
$$b = 8 \Rightarrow \frac{b}{2} = \boxed{4} \Rightarrow \left(\frac{b}{2}\right)^2 = \boxed{16}$$

$$x^2 + 8x + \boxed{16} = 1 + \boxed{16}$$

$$\left(x + \boxed{4}\right)^2 = \boxed{17}$$

2. (a) $x^2 - 2x + (-1)^2 = 8 + \boxed{(-1)^2}$

$$\left(x - \boxed{1}\right)^2 = \boxed{9}$$

$$x - \boxed{1} = \pm\sqrt{\boxed{9}}$$

$$x = \boxed{\pm 3}$$

(b) $x^2 - 3x + \boxed{\left(\dfrac{3}{2}\right)^2} = 6 + \boxed{\left(\dfrac{3}{2}\right)^2}$

$$\left(x - \boxed{\dfrac{3}{2}}\right)^2 = \boxed{\dfrac{33}{4}}$$

$$x - \boxed{\dfrac{3}{2}} = \pm \boxed{\dfrac{\sqrt{33}}{2}}$$

$$x = \boxed{\dfrac{3}{2}} \pm \boxed{\dfrac{\sqrt{33}}{2}}$$

$$= \boxed{\dfrac{3 \pm \sqrt{33}}{2}}$$

■ **Solutions to Starter Exercises** ■

3.
$$x^2 + \frac{15}{2}x = \boxed{-\frac{9}{4}}$$

$$x^2 + \frac{15}{2}x + \boxed{\left(\frac{15}{4}\right)^2} = \boxed{-\frac{9}{4}} + \boxed{\left(\frac{15}{4}\right)^2}$$

$$\left(x + \boxed{\left(\frac{15}{4}\right)}\right)^2 = \boxed{\frac{189}{16}}$$

$$x + \boxed{\frac{15}{4}} = \pm \boxed{\frac{\sqrt{189}}{4}}$$

$$x = \boxed{-\frac{15}{4}} \pm \boxed{\frac{\sqrt{189}}{4}}$$

$$x = \boxed{-\frac{15 - \sqrt{189}}{4}} \text{ or } x = \boxed{-\frac{15 + \sqrt{189}}{4}}$$

4. $x^2 + 6x + \boxed{(3)^2} = -10 + \boxed{3^2}$

$$\left(x + \boxed{3}\right)^2 = \boxed{-1}$$

Conclusion: $\boxed{\text{No solution}}$

11.2 EXERCISES

In Exercises 1–4, determine the constant that must be added to the expression to make it a perfect square trinomial.

1. $x^2 + 2x + \boxed{}$

2. $z^2 - 6z + \boxed{}$

3. $y^2 + \frac{1}{2}y + \boxed{}$

4. $x^2 - \frac{1}{3}x + \boxed{}$

In Exercises 5–22, solve the equation by completing the square.

5. $x^2 - 8x = 0$

6. $z^2 + 10z = 0$

7. $x^2 + 6x + 8 = 0$

8. $x^2 - 2x - 8 = 0$

9. $x^2 - 4x + 4 = 0$

10. $x^2 - 2x + 1 = 0$

11. $y^2 - 2y = 7$

12. $z^2 + 5z = 10$

13. $x^2 + 5x + 1 = 0$

14. $u^2 - 3u - 6 = 0$

15. $z^2 + 7z - 3 = 0$

16. $x^2 - x - 9 = 0$

17. $2x^2 + 4x - 6 = 0$

18. $3x^2 - 12x - 9 = 0$

19. $3z^2 - 5z + 1 = 0$

20. $2y^2 + 6y - 3 = 0$

21. $5z^2 - 2z - 1 = 0$

22. $4z^2 - 6z - 3 = 0$

In Exercises 23 and 24, solve the equation.

23. $\frac{x}{3} + \frac{1}{x} = 2$

24. $\sqrt{x + 1} = x - 3$

25. Find two consecutive positive integers such that the sum of their squares is 145.

In Exercises 26–30, use a graphing utility to graph the following functions and estimate their x-intercepts. Compare your answers to Exercises 5–9, respectively.

26. $f(x) = x^2 - 8x$

27. $f(x) = x^2 + 10x$

28. $f(x) = x^2 + 6x + 8$

29. $f(x) = x^2 - 2x - 8$

30. $f(x) = x^2 - 4x + 4$

11.3   Solution by the Quadratic Formula

Section Highlights

1. **The Quadratic Formula**

   The solutions of a quadratic equation in the standard form $ax^2 + bx + c = 0$, $a \neq 0$ are given by the **Quadratic Formula**

   $$x = \frac{-b \pm \sqrt{b^2 - 4ac}}{2a}.$$

   Memory Aid: "Minus b, plus or minus the square root of b squared minus $4ac$, all divided by $2a$."

2. When using the Quadratic Formula, remember that before the formula can be applied, you must first write the quadratic equation in standard form.

EXAMPLE 1 ■   **Using the Quadratic Formula: Two Distinct Solutions**

Use the Quadratic Formula to solve the equation $2x^2 + 3x - 1 = 0$.

Solution

$2x^2 + 3x - 1 = 0$ Given equation in standard form: $a = 2$, $b = 3$, $c = -1$

$x = \dfrac{-b \pm \sqrt{b^2 - 4ac}}{2a}$ Quadratic Formula

$x = \dfrac{-3 \pm \sqrt{3^2 - 4(2)(-1)}}{2(2)}$ Substitute.

$x = \dfrac{-3 \pm \sqrt{17}}{4}$ Solutions

Starter Exercise 1 *Fill in the blanks.*

Use the Quadratic Formula to solve $3x^2 - 2x - 4 = 0$.

$3x^2 - 2x - 4 = 0$

$$x = \frac{\boxed{} \pm \sqrt{\boxed{}^2 - 4(3)\left(\boxed{}\right)}}{2(3)}$$

$$x = \frac{\boxed{} \pm \sqrt{\boxed{}}}{\boxed{}} = \frac{\boxed{} \pm 2\sqrt{\boxed{}}}{\boxed{}}$$

$$x = \frac{\cancel{2}\left(\boxed{} \pm \sqrt{\boxed{}}\right)}{\cancel{2} \cdot \boxed{}}$$

$$x = \boxed{}$$

EXAMPLE 2 ■ Using the Quadratic Formula: One Repeated Solution

Use the Quadratic Formula to solve the equation $9x^2 - 12x + 4 = 0$.

Solution

$9x^2 - 12x + 4 = 0$ Given equation in standard form: $a = 9$, $b = -12$, $c = 4$

$x = \dfrac{-b \pm \sqrt{b^2 - 4ac}}{2a}$ Quadratic Formula

$x = \dfrac{-(-12) \pm \sqrt{(-12)^2 - 4(9)(4)}}{2(9)}$ Substitute.

$x = \dfrac{12 \pm \sqrt{0}}{18} = \dfrac{2}{3}$ Solution

Starter Exercise 2 *Fill in the blanks.*

Use the Quadratic Formula to solve the equation $16x^2 - 24x + 9 = 0$.

$$x = \frac{-b \pm \sqrt{b^2 - 4ac}}{2a}$$

$$x = \frac{-(-24) \pm \sqrt{(-24)^2 - 4(16)\left(\boxed{}\right)}}{2\left(\boxed{}\right)}$$

$$x = \frac{24 \pm \sqrt{\boxed{}}}{\boxed{}} = \boxed{}$$

EXAMPLE 3 ■ Using the Quadratic Formula: No Real Solution

Use the Quadratic Formula to solve the equation $4x^2 + x + 6 = 0$.

Solution

$$4x^2 + x + 6 = 0 \qquad \text{Given equation in standard form: } a = 4, b = 1, c = 6$$

$$x = \frac{-b \pm \sqrt{b^2 - 4ac}}{2a} \qquad \text{Quadratic Formula}$$

$$x = \frac{-1 \pm \sqrt{1^2 - 4(4)(6)}}{2(4)}$$

$$x = \frac{-1 \pm \sqrt{-95}}{8}$$

Since $\sqrt{-95}$ is not a real number, we conclude that the original equation has no real solution. ■

Starter Exercise 3 | *Fill in the blanks.*

Use the Quadratic Formula to solve the equation $3x^2 + 2x + 3 = 0$.

$$3x^2 + 2x + 3 = 0$$

$$x = \frac{-2 \pm \sqrt{2^2 - 4(\boxed{})(\boxed{})}}{2(\boxed{})}$$

$$x = \frac{-2 \pm \sqrt{\boxed{}}}{\boxed{}}$$

Conclusion: $\boxed{}$

■ Solutions to Starter Exercises ■

1. $$x = \frac{\boxed{-(-2)} \pm \sqrt{\boxed{(-2)}^2 - 4(3)\left(\boxed{-4}\right)}}{2(3)}$$

$$x = \frac{\boxed{2} \pm \sqrt{\boxed{52}}}{\boxed{6}} = \frac{\boxed{2} \pm 2\sqrt{\boxed{13}}}{\boxed{6}}$$

$$x = \frac{\cancel{2}\left(\boxed{1} \pm \sqrt{\boxed{13}}\right)}{\cancel{2} \cdot \boxed{3}}$$

$$x = \boxed{\frac{1 \pm \sqrt{13}}{3}}$$

■ **Solutions to Starter Exercises** ■

2. $x = \dfrac{-(-24) \pm \sqrt{(-24)^2 - 4(16)\left(\boxed{9}\right)}}{2\left(\boxed{16}\right)}$

$x = \dfrac{24 \pm \sqrt{\boxed{0}}}{\boxed{32}} = \boxed{\dfrac{3}{4}}$

3. $x = \dfrac{-2 \pm \sqrt{2^2 - 4\left(\boxed{3}\right)\left(\boxed{3}\right)}}{2\left(\boxed{3}\right)}$

$x = \dfrac{-2 \pm \sqrt{\boxed{-32}}}{\boxed{6}}$

Conclusion: $\boxed{\text{No solution}}$

11.3 | EXERCISES

In Exercises 1–4, use the discriminant to determine the number of real solutions of the quadratic equation.

1. $x^2 - 3x - 7 = 0$

2. $2x^2 - 3x - 4 = 0$

3. $x^2 + 2x + 1 = 0$

4. $3x^2 - x + 10 = 0$

In Exercises 5–20, use the Quadratic Formula to solve the equation.

5. $x^2 - 3x - 10 = 0$

6. $z^2 + 6z - 27 = 0$

7. $x^2 - x + 1 = 0$

8. $2x^2 - 3x + 1 = 0$

9. $3x^2 - x - 6 = 0$

10. $5x^2 + 7x + 1 = 0$

11. $10x^2 - 2x - 1 = 0$

12. $x^2 - 7x + 12 = 0$

13. $2z^2 - 3z - 5 = 0$

14. $4z^2 + 3z + 1 = 0$

15. $10x^2 + 9x + 2 = 0$

16. $2y^2 - 5y - 4 = 0$

17. $3x^2 - 6x + 2 = 0$

18. $\frac{1}{2}x^2 - \frac{1}{4}x - \frac{1}{2} = 0$

19. $0.2x^2 - 0.6x - 0.2 = 0$

20. $0.1x^2 + 0.3x - 0.2 = 0$

In Exercises 21–26, solve the quadratic equation by the most convenient method.

21. $x^2 = 16$

22. $z^2 - 3z = 0$

23. $2x(x - 1) - 7(x - 1) = 0$

24. $(x - 2)^2 - 25 = 0$

25. $x^2 + 4x - 9 = 0$

26. $6x^2 - 5x - 8 = 0$

▦ In Exercises 27 and 28, use a calculator to solve the equation. (Round your answer to three decimal places.)

27. $123x^2 - 37x - 54 = 0$

28. $0.01x^2 + 3.1x + 0.7 = 0$

In Exercises 29 and 30, solve the equation.

29. $\sqrt{4x + 3} = x + 1$

30. $\dfrac{x}{2} - \dfrac{1}{x} = 5$

31. *Number Problem* Find two consecutive odd positive integers whose product is 143.

11.4 Graphing Quadratic Functions

Section Highlights

1. The graph of the quadratic function $y = ax^2 + bx + c$ is called a **parabola**.
   (i) If $a > 0$, then the parabola opens upward.
   (ii) If $a < 0$, then the parabola opens downward.

2. The **vertex** of the parabola given by $y = ax^2 + bx + c$ occurs at the point whose x-coordinate is

   $$x = -\frac{b}{2a}.$$

   To find the y-coordinate of the vertex, substitute the x-coordinate in the equation $y = ax^2 + bx + c$.

3. To sketch the parabola given by $y = ax^2 + bx + c$, we suggest the following steps.
   (i) Use the leading coefficient test to determine whether the parabola opens upward or downward.
   (ii) Find and plot the y-intercept of the parabola and the vertex.
   (iii) Find and plot the x-intercepts of the parabola (if any).
   (iv) Make up a table of values that includes a few additional points on the parabola.
   (v) Complete the graph of the parabola with a smooth, cup-shaped curve.

EXAMPLE 1 ■ Using the Leading Coefficient Test

Determine whether the graph of the quadratic function $y = x^2 + 3x - 4$ opens upward or downward.

Solution

The coefficient of x^2 (the leading coefficient) is $a = 1$. Hence, the parabola opens upward.

Starter Exercise 1   *Fill in the blanks.*

(a) Determine whether the graph of $y = 2x^2 - 6x - 1$ opens upward or downward.

$a = 2$

Conclusion: ☐

(b) Determine whether the graph of $y = 7 + 2x - x^2$ opens upward or downward.

$a = $ ☐

Conclusion: ☐

EXAMPLE 2 ■ Finding the Vertex of a Parabola

Find the vertex of the parabola $y = 2x^2 + 4x - 1$.

Solution

$$a = 2 \text{ and } b = 4 \Rightarrow -\frac{b}{2a} = -\frac{4}{2(2)} = -1$$

Thus, the x-coordinate of the vertex is -1. To find the y-coordinate, we replace x by -1.

$$y = 2x^2 + 4x - 1 = 2(-1)^2 + 4(-1) - 1 = -3$$

Therefore, the vertex is at $(-1, \ -3)$. ■

Starter Exercise 2 | *Fill in the blanks.*

Find the vertex of the parabola $y = -3x^2 + 2x - 5$.

$$a = -3 \text{ and } b = \boxed{} \Rightarrow -\frac{b}{2a} = -\frac{\boxed{}}{2(-3)} = \boxed{}$$

$$y = -3x^2 + 2x - 5 = -3\left(\boxed{}\right)^2 + 2\left(\boxed{}\right) - 5 = \boxed{}$$

Coordinates of the vertex: $\left(\boxed{}, \ \boxed{}\right)$

EXAMPLE 3 ■ Sketching a Parabola: Two x-intercepts

Sketch the graph of the quadratic function $y = x^2 + 2x - 3$.

Solution

Leading coefficient test: Since $a = 1$, the parabola opens upward.

y-intercept: $(0, \ -3)$

x-intercepts: $0 = x^2 + 2x - 3 = (x + 3)(x - 1), \quad (-3, \ 0) \text{ and } (1, 0)$

Vertex: $x = -\dfrac{b}{2a} = -\dfrac{2}{2(1)} = -1$

$y = (-1)^2 + 2(-1) - 3 = -4, \quad (-1, \ -4)$

Table of values:

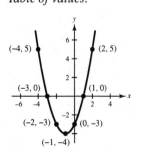

x	-4	-2	2
$y = x^2 + 2x - 3$	5	-3	5
$(x, \ y)$	$(-4, \ 5)$	$(-2, \ -3)$	$(2, 5)$

Starter Exercise 3   *Fill in the blanks.*

Sketch the graph of the quadratic function $y = -\frac{1}{2}x^2 + x + 4$.

Leading coefficient test   Since $a = -\frac{1}{2}$, the parabola opens downward.

y-intercept: $\left(0, \boxed{}\right)$

x-intercepts: $0 = -\frac{1}{2}x^2 + x + 4 \Rightarrow 0 = x^2 - 2x - 8$

$$\Rightarrow 0 = (x-4)(x+2), \quad \left(\boxed{}, 0\right) \text{ and } \left(\boxed{}, 0\right)$$

Vertex: $x = -\dfrac{b}{2a} = -\dfrac{1}{2\left(\boxed{}\right)} = \boxed{}$

$$y = -\frac{1}{2}x^2 + x + 4 = -\frac{1}{2}(1)^2 + 1 + 4 = 4\frac{1}{2}, \quad \left(\boxed{}, \boxed{}\right)$$

Table of values:

x		-3	-1	2	3
$y = -\frac{1}{2}x^2 + x + 4$					
$(x,\ y)$					

EXAMPLE 4 ■   Sketching a Parabola: One x-intercept

Sketch the graph of the quadratic function $y = x^2 + 2x + 1$.

Solution

Leading coefficient test: Since $a = 1$, the parabola opens upward.

y-intercept: $(0,\ 1)$

x-intercept: $0 = x^2 + 2x + 1 = (x+1)^2, \quad (-1,\ 0)$

Vertex: $x = -\dfrac{b}{2a} = -\dfrac{2}{2(1)} = -1$

$$y = (-1)^2 + 2(-1) + 1 = 0, \quad (-1,\ 0)$$

Table of values:

x	-4	-3	-2	1	2
$y = x^2 + 2x + 1$	9	4	1	4	9
$(x,\ y)$	$(-4,\ 9)$	$(-3,\ 4)$	$(-2,\ 1)$	$(1,\ 4)$	$(2,\ 9)$

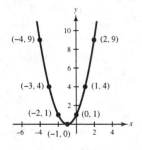

■

Starter Exercise 4 *Fill in the blanks.*

Sketch the graph of the quadratic function $y = -2x^2 + 4x - 3$.

Leading coefficient test: Since $a = -2$, the parabola opens downward.

y-intercept: $\left(0, \boxed{}\right)$

x-intercept: $0 = -2x^2 + 4x - 3 \Rightarrow$ No real solutions

$\boxed{}$

Vertex: $x = \boxed{} = \boxed{} = \boxed{}$

$y = \boxed{}, \quad \left(\boxed{}, \boxed{}\right)$

Table of values:

x	-1	2	3
$y = -2x^2 + 4x - 3$			
$(x, \; y)$			

📆 EXAMPLE 5 ■ Graphing Calculator Problem

Use a graphing calculator to sketch the graphs of the following parabolas on the same screen.

(a) $y = x^2$ (b) $y = x^2 + 2$ (c) $y = x^2 - 2$ (d) $y = (x - 2)^2$

Solution

To sketch all four graphs on the same screen, press the $\boxed{\text{Y=}}$ key, enter the equations on the lines labeled $y_1 =$, $y_2 =$, $y_3 =$, and $y_4 =$, and press $\boxed{\text{GRAPH}}$. You should obtain the graph shown on the right. Now compare the graphs in parts (b), (c), and (d) to the graph in part (a).

(a)
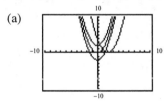

(b) The graph of $y = x^2 + 2$ is similar to the graph of $y = x^2$, except the graph has been shifted up two units.

(c) The graph of $y = x^2 - 2$ is similar to the graph of $y = x^2$, except the graph is shifted down two units.

(d) The graph of $y = (x - 2)^2$ is similar to the graph of $y = x^2$, except the graph is shifted to the right two units. ■

■   **Solutions to Starter Exercises**   ■

1. (a) Conclusion: $\boxed{\text{opens upward}}$ (b) $a = \boxed{-1}$

Conclusion: $\boxed{\text{opens downward}}$

2. $b = \boxed{2} \Rightarrow -\dfrac{b}{2a} = -\dfrac{\boxed{2}}{2(-3)} = \boxed{\dfrac{1}{3}}$

$y = -3x^2 + 2x - 5 = -3\left(\boxed{\dfrac{1}{3}}\right)^2 + 2\left(\boxed{\dfrac{1}{3}}\right) - 5 = \boxed{-4\tfrac{2}{3}}$

Coordinates of the vertex: $\left(\boxed{\dfrac{1}{3}}, \boxed{-4\tfrac{2}{3}}\right)$

3. *y-intercept:* $\left(0, \boxed{4}\right)$

x-intercepts: $\left(\boxed{-2}, 0\right)$ and $\left(\boxed{4}, 0\right)$

Vertex: $x = -\dfrac{b}{2a} = -\dfrac{1}{2\left(\boxed{-1/2}\right)} = \boxed{1}$

$\left(\boxed{1}, \boxed{4\tfrac{1}{2}}\right)$

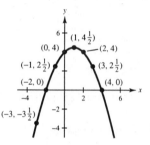

Table of values:

x	-3	-1	2	3
$y = -\tfrac{1}{2}x^2 + x + 4$	$-3\tfrac{1}{2}$	$2\tfrac{1}{2}$	4	$2\tfrac{1}{2}$
$(x,\ y)$	$\left(-3,\ -3\tfrac{1}{2}\right)$	$\left(-1,\ 2\tfrac{1}{2}\right)$	$(2,\ 4)$	$\left(3,\ 2\tfrac{1}{2}\right)$

4. *y-intercept:* $\left(0, \boxed{-3}\right)$

x-intercept: $\boxed{\text{No } x\text{-intercept}}$

Vertex: $x = \boxed{-\dfrac{b}{2a}} = \boxed{-\dfrac{4}{2(-2)}} = \boxed{1}$

$y = \boxed{-2(1)^2 + 4(1) - 3 = -1}, \left(\boxed{1}, \boxed{-1}\right)$

Table of values:

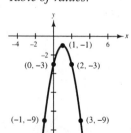

x	-1	2	3
$y = -2x^2 + 4x - 3$	-9	-3	-9
$(x,\ y)$	$(-1,\ -9)$	$(2,\ -3)$	$(3,\ -9)$

11.4 | EXERCISES

In Exercises 1–4, match the equation with the correct graph.

(a)
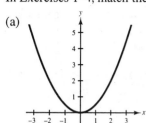

(b)
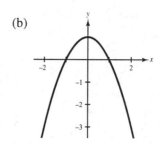

(c)
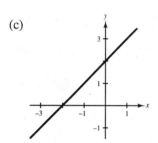

(d)
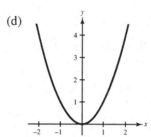

1. $y = x + 2$ **2.** $y = x^2$ **3.** $y = 1 - x^2$ **4.** $y = \frac{1}{2}x^2$

In Exercises 5–8, determine whether the graph of the quadratic function opens upward or downward.

5. $y = x^2 + x + 1$ **6.** $y = -2x^2 - 3x + 2$

7. $y = 2 - 3x - 7x^2$ **8.** $y = 14x^2 - 1$

In Exercises 9–12, find the intercepts of the graph of the given function.

9. $y = 1 - x^2$ **10.** $y = x^2 - 5x$

11. $y = x^2 + 7x + 6$ **12.** $y = x^2 - 2x + 1$

In Exercises 13–16, find the vertex of the parabola.

13. $y = x^2 + 4$ **14.** $y = x^2 - 2x + 8$

15. $y = -2x^2 + 4x$ **16.** $y = 3x^2 - 5x + 1$

In Exercises 17–28, sketch the graph of the quadratic function. Identify the vertex of the parabola.

17. $y = x^2 + 2$ **18.** $y = -x^2 - 1$ **19.** $y = x^2 + 2x$ **20.** $y = -x^2 - 2x$

21. $y = (x - 1)^2$ **22.** $y = x^2 + 2x + 1$ **23.** $y = -x^2 - 4x + 6$ **24.** $y = x^2 - 2x + 2$

25. $y = -\frac{1}{2}(x - 1)^2 + 2$ **26.** $y = \frac{1}{2}x^2 + 2x + 3$ **27.** $y = -(x + 1)^2$ **28.** $y = -2(x^2 - 2x - 2)$

In Exercises 29 and 30, expand the right side of the equation, find the vertex, and then describe how the vertex can be determined from the given completed square form of the equation.

29. $y = (x + 1)^2 + 1$ **30.** $y = -(x - 2)^2 + 3$

In Exercises 31 and 32, complete the square for the right side of the equation and use this form to find the vertex of the parabola.

31. $y = x^2 - 8x + 1$ **32.** $y = x^2 + 4x + 1$

33. Use a graphing calculator to sketch the graphs of the following parabolas on the same screen. Describe the effect of changing the constant term.

(a) $y = x^2$ (b) $y = x^2 + 1$ (c) $y = x^2 - 2$ (d) $y = x^2 - 8$

11.5 Applications of Quadratic Equations

Section Highlights

Guidelines for Solving Word Problems

1. Search for the *hidden equality*–the two essential expressions said to be equal or known to be equal. (A sketch may be helpful.)
2. Write a *verbal model* that equates these two essential expressions.
3. Assign *labels* to the fixed quantities and the variable quantities.
4. Rewrite the verbal model as an *algebraic equation* using the assigned labels.
5. *Solve* the algebraic equation.
6. *Check* to see that your solution satisfies the word problem as stated.

EXAMPLE 1 ■ Area

The length of a rectangular room is 5 feet longer than its width. The area of the room is 374 square feet. Find the dimensions of the room.

Solution

Verbal model: | Width | · | Length | = | Area |

Labels: Width = w
Length = $w + 5$
Area = 374

Equation:
$$w(w + 5) = 374$$
$$w^2 + 5w - 374 = 0$$
$$(w + 22)(w - 17) = 0$$
$$w + 22 = 0 \quad \text{or} \quad w - 17 = 0$$
$$w = -22 \text{ or} \quad w = 17$$

The width cannot be negative, so we disregard −22. If the width is 17, the length is 17 + 5 = 22. Thus, the dimensions of the room are 17 feet by 22 feet. ■

Starter Exercise 1 | *Fill in the blanks.*

A group rents a cabin at a ski resort for $300 for a weekend. If one more person goes along, the price per person is reduced by $10. How many people were in the original group?

Verbal model: | Cost per skier | · | Number of skiers | = | 300 |

Labels: Original number of skiers $= x$

New number of skiers $=$ [   ]

Original cost per skier $= \dfrac{300}{x}$

New cost per skier $= \dfrac{300}{x} - 10$

Equation: $\left(\dfrac{300}{x} - 10\right)\left(\boxed{}\right) = \boxed{}$

$\left(\dfrac{300 - 10x}{x}\right)\left(\boxed{}\right) = \boxed{}$

$(300 - 10x)\left(\boxed{}\right) = \boxed{}$

$\boxed{} = \boxed{}$

$\vdots$

There were [   ] people in the original group.

EXAMPLE 2 ■   Work

Bob takes 10 minutes longer than Nancy to sweep the driveway. Together they can sweep the driveway in 15 minutes. How long does it take Nancy to sweep the driveway alone?

Solution

Verbal model: | Rate for Bob | + | Rate for Nancy | = | Rate for both |

Labels: Both: time $= 15$, rate $= \dfrac{1}{15}$

Nancy: time $= t$, rate $= \dfrac{1}{t}$

Bob: time $= t + 10$, rate $= \dfrac{1}{t + 10}$

Equation: $\dfrac{1}{t + 10} + \dfrac{1}{t} = \dfrac{1}{15}$

$15t + 15(t + 10) = t(t + 10)$

$15t + 15t + 150 = t^2 + 10t$

$0 = t^2 - 20t - 150$

$0 = (t - 50)(t + 30)$

$t - 50 = 0 \quad \text{or} \quad t + 30 = 0$

$t = 50 \text{ or} \quad t = -30$

Choosing the positive value, we see it takes Nancy 50 minutes to sweep the driveway.

■

Starter Exercise 2 | *Fill in the blanks—Pythagorean Theorem*

Ranger Station 1 is due north of you. Ranger Station 2 is due east of you and 10 miles farther away from you than Ranger Station 1. The two stations are 50 miles apart. How far are you from Ranger Station 2?

Verbal model: $\boxed{\text{Distance to Station 1}}^2 + \boxed{\text{Distance to Station 2}}^2 = \boxed{50}^2$

Labels: Distance to Station 1 $= x$

Distance to Station 2 $= \boxed{}$

Equation: $x^2 + \left(\boxed{}\right)^2 = 50^2$

$x^2 + \boxed{} + 20x + \boxed{} = 2500$

$\boxed{} = 0$

$\vdots$

Choosing the positive value for x, the distance to Ranger Station 2 is $\boxed{}$ miles.

■   **Solutions to Starter Exercises**   ■

1. New number of skiers $= \boxed{x+1}$

$\left(\dfrac{300}{x} - 10\right)\left(\boxed{x+1}\right) = \boxed{300}$

$\left(\dfrac{300-10x}{x}\right)\left(\boxed{x+1}\right) = \boxed{300}$

$(300 - 10x)\left(\boxed{x+1}\right) = \boxed{300x}$

$\boxed{-10x^2 - 10x + 300} = \boxed{0}$

$\boxed{x^2 + x - 30} = \boxed{0}$

$\boxed{(x-5)(x+6)} = \boxed{0}$

$\boxed{x - 5 = 0}$ or $\boxed{x + 6 = 0}$

$\boxed{x = 5}$ or $\boxed{x = -6}$

There were $\boxed{5}$ people in the original group.

■ **Solutions to Starter Exercises** ■

2. Distance to Station 2 = $\boxed{x + 10}$

$$x^2 + \left(\boxed{x + 10} \right)^2 = 50^2$$

$$x^2 + \boxed{x^2} + 20x + \boxed{100} = 2500$$

$$\boxed{2x^2 + 20x - 2400} = 0$$

$$\boxed{x^2 + 10x - 1200} = \boxed{0}$$

$$\left(\boxed{x + 40} \right)\left(\boxed{x - 30} \right) = \boxed{0}$$

$$\boxed{x + 40 = 0} \text{ or } \boxed{x - 30 = 0}$$

$$\boxed{x = -40} \text{ or } \boxed{x = 30}$$

Choosing the positive value for x, the distance to Ranger Station 2 is $\boxed{x + 10 = 30 + 10 = 40}$ miles.

11.5 | EXERCISES

In Exercises 1–4, find two positive integers satisfying the given requirement.

1. The product of two consecutive integers is 240.

2. The product of two consecutive even integers is 168.

3. The sum of the squares of two consecutive integers is 85.

4. The sum of the squares of two consecutive odd integers is 202.

In Exercises 5 and 6, find the time necessary for an object to fall to ground level from an initial height of h_0 feet if its height h at any time t is given by $h = h_0 - 16t^2$.

5. $h_0 = 64$

6. $h_0 = 576$

In Exercises 7–12, find the dimensions of the rectangle and complete the table where l and w represent the length and width of a rectangle, respectively.

	Width	Length	Perimeter	Area
7.	w	$w + 6$	68 ft	
8.	w	$2w$	60 in	
9.	w	$3w$		432 ft^2
10.	$l - 2$	l		143 m^2
11.	$\frac{1}{2}l$	l	42 yds	
12.	$0.6l$	l		15 in^2

13. *Dimensions of a Triangle*   The height of a triangle is twice its base and the area is 25 square inches. Find the dimensions of the triangle.

14. *Dimensions of a Triangle*   The height of a triangle is 60% of its base and the area is 480 m^2. Find the dimensions of the triangle.

15. *Dimensions of a Rectangle*   The perimeter of a rectangle is 56 inches and the length of the diagonal is 20 inches. Find the dimensions of the rectangle.

16. *Ticket Price*   A community organization charters a fishing boat for $350. If the organization can get four more people to sign up for the trip, the price per person will drop by $10. How many people originally signed up for the trip?

17. *Distance*   A driver traveled 715 miles. If he would have increased his speed by 10 mph, he could have decreased his time by two hours. What was his original driving speed?

18. *Distance*   A bus traveled the first 100 miles of a trip at one speed and the last 165 miles at a speed 5 mph faster. The entire trip took five hours. What were the two average speeds?

19. *Work Rate*   Working together, two machines can complete a task in 8 minutes. It takes one machine twelve more minutes than the other, working alone. How long does it take each machine to complete the task alone?

20. *Work Rate*   An old printer takes five more minutes than the new printer to print a report. Working together, the two printers can print the report in six minutes. How long does it take the new printer to print the report alone?

Cumulative Practice Test for Chapters P–11

In Exercises 1–6, answer true or false.

1. $x(x + 3) = x^2 + 3$

2. $2x^2 + 3x = 5x^3$

3. $x^{-1} = \dfrac{1}{x}$, if $x \neq 0$

4. $\dfrac{x^m}{x^n} = x^{m-n}$, if $x \neq 0$

5. $(2x)(3) = (2 \cdot 3)(x \cdot 3)$

6. $\dfrac{x + y}{x} = y$

In Exercises 7–12, simplify the given expression.

7. $\dfrac{x}{x + 1} + \dfrac{2}{x - 1}$

8. $\dfrac{1}{\sqrt{3} + 1}$

9. $\sqrt{\dfrac{45x^3}{5x}}$

10. $(x + 2)^2$

11. $2[3x - (x + 6)] - 7$

12. $\sqrt{147x^3} - x\sqrt{27x} + \sqrt{3x^3}$

In Exercises 13–16, match the equation with the graphs below.

13. $2x - 3y = 9$ **14.** $x = 4$ **15.** $y - x^2 = 0$ **16.** $y = x^2 - 2x + 3$

(a)
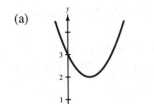

(b)
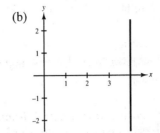

(c)
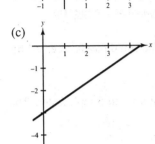

(d)
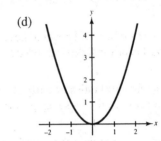

In Exercises 17–26, solve the given equation.

17. $5(x - 2) = x - 4$ **18.** $\dfrac{2}{x} = \dfrac{8}{4}$ **19.** $2x^2 = x + 6$

20. $x^2 + 4x + 1 = 0$ **21.** $\sqrt{2x - 1} - \sqrt{x + 3} = 0$ **22.** $\sqrt{2x + 3} + 9 = 0$

23. $\dfrac{1}{x} = 2 + \dfrac{1}{x + 1}$ **24.** $x - 3\sqrt{x} = 10$ **25.** $\dfrac{1}{x - 2} = \dfrac{3}{x + 2}$ **26.** $(x + 6)^2 = 4$

In Exercises 27 and 28, find an equation of the line, in general form, that passes through (1, 2) and is (a) parallel and (b) perpendicular to the line given by the equation.

27. $x + y = 1$ **28.** $x - 2y = 5$

In Exercises 29 and 30, factor the given polynomial completely.

29. $x^2(x + 2) - (x + 2)$ **30.** $x^4 - 16$

In Exercises 31–33, solve the system of linear equations.

31. $\begin{aligned} 2x - y &= 4 \\ 4x + 2y &= 1 \end{aligned}$ **32.** $\begin{aligned} 2x - 3y &= 1 \\ 4x - 6y &= 0 \end{aligned}$ **33.** $\begin{aligned} x + 5y &= 1 \\ 3x + 15y &= 3 \end{aligned}$

34. The product of two consecutive integers is 182. Find the two integers.

35. A watch is on sale for $64, which is 20% off the regular price. What is the regular price of the watch?

36. A total of $20,000 is invested in two funds which pay $7\frac{1}{2}\%$ and $8\frac{3}{4}\%$ simple interest. If the total interest earned each year was $1600, how much was invested in each fund?

37. The diagonal of a square is 18 inches. Find the length of the sides of the square.

38. How many liters of 8% acid solution must be mixed with 14% acid solution to make 8 liters of $9\frac{1}{2}$% acid solution?

39. Sailing downstream, a boat can travel 34 miles in two hours. Heading upstream, it can travel only 26 miles in the same time. Find the rate of the current.

40. Machine A takes twice as long as Machine B to complete a task. Working together, the machines can complete the task in 8 minutes. How long does it take each machine, working alone, to complete the task?

11.1	**Answers to Exercises**

1. $x = -5$ or $x = 0$ **2.** $x = 0$ or $x = 3$ **3.** $x = \pm 4$ **4.** $x = \pm \frac{7}{4}$

5. $x = 2$ or $x = 3$ **6.** $x = -3$ **7.** $x = \frac{2}{3}$ **8.** $x = -4$ or $x = -2$

9. $x = 4$ or $x = -2$ **10.** $x = -\frac{1}{3}$ or $x = 3$ **11.** $x = -\frac{3}{2}$ or $x = \frac{2}{3}$ **12.** $t = -\frac{3}{2}$ or $t = 1$

13. $x = \pm 5$ **14.** $x = \pm 13$ **15.** $x = \pm \frac{7}{2}$ **16.** $x = \pm 9$

17. $x = \pm 6$ **18.** No solution   **19.** $x = -9$ or $x = 7$ **20.** $x = -8$ or $x = 14$

21. $2 \pm \sqrt{3}$ **22.** $-6 \pm 2\sqrt{3}$ **23.** $\dfrac{1 \pm 2\sqrt{2}}{2}$ **24.** $2 \pm \sqrt{3}$

25. $x = \dfrac{1 \pm \sqrt{3}}{2}$ **26.** No solution   **27.** $2\frac{1}{2}$ sec   **28.** 100 units

---

11.2	**Answers to Exercises**

1. 1   **2.** 9   **3.** $\frac{1}{16}$ **4.** $\frac{1}{36}$

5. $x = 0$ or $x = 8$ **6.** $z = -10$ or $z = 0$ **7.** $x = -4$ or $x = -2$ **8.** $x = -2$ or $x = 4$

9. $x = 2$ **10.** $x = 1$ **11.** $y = 1 \pm 2\sqrt{2}$ **12.** $z = \dfrac{-5 \pm \sqrt{65}}{2}$

13. $x = \dfrac{-5 \pm \sqrt{21}}{2}$ **14.** $u = \dfrac{3 \pm \sqrt{33}}{2}$ **15.** $z = \dfrac{-7 \pm \sqrt{61}}{2}$ **16.** $x = \dfrac{1 \pm \sqrt{37}}{2}$

17. $x = -3, 1$ **18.** $x = 2 \pm \sqrt{7}$ **19.** $z = \dfrac{5 \pm \sqrt{13}}{6}$ **20.** $y = \dfrac{-3 \pm \sqrt{15}}{2}$

21. $z = \dfrac{1 \pm \sqrt{6}}{5}$ **22.** $z = \dfrac{3 \pm \sqrt{21}}{4}$ **23.** $x = 3 \pm \sqrt{6}$ **24.** $x = \dfrac{7 + \sqrt{17}}{2}$

25. 8 and 9

11.3 | Answers to Exercises

1. Two   **2.** Two   **3.** One   **4.** None   **5.** $x = -2$ or $x = 5$

6. $z = -9$ or $z = 3$ **7.** No real solution   **8.** $x = \dfrac{1}{2}$ or $x = 1$ **9.** $x = \dfrac{1 \pm \sqrt{73}}{6}$

10. $x = \dfrac{-7 \pm \sqrt{29}}{10}$ **11.** $x = \dfrac{1 \pm \sqrt{11}}{10}$ **12.** $x = 3$ or $x = 4$

13. $z = -1$ or $z = \dfrac{5}{2}$ **14.** No real solution   **15.** $x = -\dfrac{1}{2}$ or $x = -\dfrac{2}{5}$

16. $y = \dfrac{5 \pm \sqrt{57}}{4}$ **17.** $x = \dfrac{3 \pm \sqrt{3}}{3}$ **18.** $x = \dfrac{1 \pm \sqrt{17}}{4}$

19. $x = \dfrac{3 \pm \sqrt{13}}{2}$ **20.** $x = \dfrac{-3 \pm \sqrt{17}}{2}$ **21.** $x = \pm 4$

22. $x = 0$ or $x = 3$ **23.** $x = 1$ or $x = \dfrac{7}{2}$ **24.** $x = -3$ or $x = 7$

25. $x = -2 \pm \sqrt{13}$ **26.** $x = \dfrac{5 \pm \sqrt{217}}{12}$ **27.** $x \approx -0.529$ or $x \approx 0.830$

28. $x \approx -309.774$ or $x \approx -0.226$ **29.** $x = 1 \pm \sqrt{3}$ **30.** $x = 5 \pm 3\sqrt{3}$

31. 11 and 13

11.4 | Answers to Exercises

1. (c)   **2.** (d)   **3.** (b)   **4.** (a)   **5.** Upward

6. Downward   **7.** Downward   **8.** Upward   **9.** (0, 1), (1, 0), (−1, 0)

10. (0, 0), (5, 0)   **11.** (0, 6), (−1, 0), (−6, 0)   **12.** (0, 1), (1, 0)

13. (0, 4)   **14.** (1, 7)   **15.** (1, 2)   **16.** $\left(\dfrac{5}{6}, -\dfrac{13}{12}\right)$

17.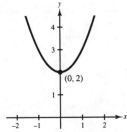

18.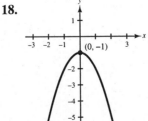

19.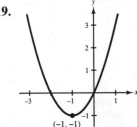

20.

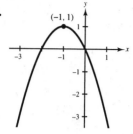

21.

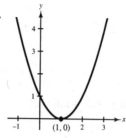

22.

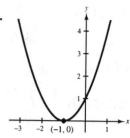

23.

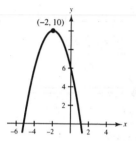

24.

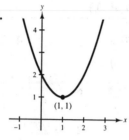

25.

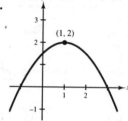

26.

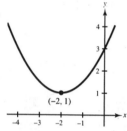

27.

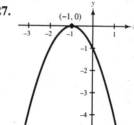

28.

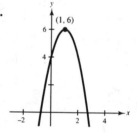

29. $(-1, 1)$

30. $(2, 3)$; If the equation is in the form $y = a(x - h)^2 + k$, then the vertex is (h, k).

31. $(4, -15)$

32. $(-2, -3)$

33.

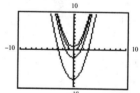

| 11.5 | **Answers to Exercises** |

1. 15 and 16   **2.** 12 and 14   **3.** 6 and 7   **4.** 9 and 11   **5.** 2 seconds

6. 6 seconds   **7.** 14 ft by 20 ft, 280 sq ft   **8.** 10 in. by 20 in., 200 sq in.

9. 12 ft by 36 ft, 96 ft   **10.** 11 in. by 13 in., 48 in.   **11.** 7 yd by 14 yd, 98 sq yd

12. 3 in. by 5 in., 16 in.

13. Base: 5 in.
Height: 10 in.

14. Base: 40 m
Height: 24 m

15. 12 in. by 16 in.

16. 10

17. 55 mph

18. 50 mph and 55 mph

19. 12 minutes and 24 minutes

20. 10 minutes

Answers to Cumulative Practice Test P–11

1. False

2. False

3. True

4. True

5. False

6. False

7. $\dfrac{x^2 + x + 2}{x^2 - 1}$

8. $\dfrac{\sqrt{3} - 1}{2}$

9. $3x$

10. $x^2 + 4x + 4$

11. $4x - 19$

12. $5x\sqrt{3x}$

13. (c)

14. (b)

15. (d)

16. (a)

17. $x = \frac{3}{2}$

18. $x = 1$

19. $x = -\frac{3}{2}$ or $x = 2$

20. $x = -2 \pm \sqrt{3}$

21. $x = 4$

22. No solution

23. $x = \dfrac{-1 \pm \sqrt{3}}{2}$

24. $x = 25$

25. $x = 4$

26. $x = -8$ or $x = -4$

27. (a) $x + y = 3$
(b) $x - y = -1$

28. (a) $x - 2y = -3$
(b) $2x + y = 0$

29. $(x + 1)(x - 1)(x + 2)$

30. $(x + 2)(x - 2)(x^2 + 4)$

31. $\left(\frac{9}{8}, -\frac{7}{4}\right)$

32. No solution

33. All (x, y) such that $x + 5y = 1$

34. 13, 14

35. $80

36. $8000 at $8\frac{3}{4}\%$
$12,000 at $7\frac{1}{2}\%$

37. $9\sqrt{2}$ in.

38. 6 liters

39. 2 mph

40. 24 min; 12 min

Warm-Up Exercises

On the following pages you will find a set of Warm-Up Exercises for each section of the textbook. These will help you review the skills and problem-solving techniques you will need to complete the exercise sets in the textbook. You will find answers for each set of Warm-Ups in the back of this *Study Guide*.

Chapter 1

Section 1.1

Warm-Up

The following warm-up exercises involve skills that were covered in earlier sections. You will use these skills in the exercise set for this section.

In Exercises 1-10, perform the indicated operations.

1. $10+(-10)$

2. $3-(-1)$

3. $5+|-3|$

4. $-4-(-10)$

5. $(6-3)+2$

6. $5+(4 \cdot 3)$

7. $\dfrac{10+8}{9}$

8. $\dfrac{|15-25|}{5}$

9. $\dfrac{27-13}{2}$

10. $\dfrac{3(7-2)}{5}$

Section 1.2

Warm-Up

The following warm-up exercises involve skills that were covered in earlier sections. You will use these skills in the exercise set for this section.

In Exercises 1-10, perform the indicated operations.

1. $10 \cdot (-10)$

2. $3 \cdot 1$

3. $4 \cdot |-3|$

4. $8 \div (-2)$

5. $-6 \cdot 2$

6. $(3-2) \cdot 9$

7. $\dfrac{2 \cdot 2}{3 \cdot 3}$

8. $\dfrac{|(-2) \cdot 8|}{4}$

9. $\dfrac{(-3) \cdot (-4)}{2}$

10. $-\dfrac{(-2)(5)}{10}$

Section 1.3

Warm-Up

The following warm-up exercises involve skills that were covered in earlier sections. You will use these skills in the exercise set for this section.

In Exercises 1–10, perform the indicated operations and write the answer in reduced form.

1. $16(-4)$ **2.** $-25(-3)$ **3.** $\dfrac{-49}{7}$ **4.** $-\dfrac{0}{32}$

5. $\dfrac{5}{16} - \dfrac{3}{16}$ **6.** $-2 + \dfrac{3}{2}$ **7.** $\left(-\dfrac{4}{3}\right)\left(-\dfrac{9}{16}\right)$ **8.** $\dfrac{7}{8} \div \dfrac{3}{16}$

9. $\dfrac{3/4}{-5/8}$ **10.** $\dfrac{\dfrac{1}{3} + \dfrac{5}{6}}{\dfrac{5}{12}}$

Section 1.4

Warm-Up

The following warm-up exercises involve skills that were covered in earlier sections. You will use these skills in the exercise set for this section.

In Exercises 1–10, evaluate the given expression.

1. $3^2 - (-4)$ **2.** $(-5)^2 + 3$

3. 9.3×10^6 **4.** $6.6 \div 10^3$

5. $\dfrac{|12 - 4^2|}{2}$ **6.** $\dfrac{-|7 + 3^2|}{4}$

7. $3 + 2(6 + 10)$ **8.** $-50 - 4(3 - 8)$

9. $(-4)^2 - (30 \div 5)$ **10.** $(8 \cdot 9) + (-4)^3$

Chapter 2

Section 2.1

Warm-Up

The following warm-up exercises involve skills that were covered in earlier sections. You will use these skills in the exercise set for this section.

In Exercises 1–6, evaluate the given expression.

1. $2(4 - 3^2)$

2. $(3 \div 9) - (12 \div 6)$

3. $(36 \div 3^2) + 4^2$

4. $(4 \cdot 2^3) + 10$

5. $\dfrac{120}{2^3 + 4^2}$

6. $-2[3 - (12 - 3)]$

In Exercises 7–10, state the property of real numbers that justifies the given statement.

7. $-5(3 + 6) = (-5)3 + (-5)6$

8. $2(-3) = -3(2)$

9. $4 + (13 - 5) = (4 + 13) - 5$

10. $15\left(\dfrac{1}{15}\right) = 1$

Section 2.2

Warm-Up

The following warm-up exercises involve skills that were covered in earlier sections. You will use these skills in the exercise set for this section.

In Exercises 1 and 2, rewrite the given expression in exponential form.

1. $3z \cdot 3z \cdot 3z \cdot 3z$

2. $x \cdot 8 \cdot y \cdot y \cdot 8 \cdot 8 \cdot x \cdot x$

In Exercises 3 and 4, rewrite the given expression as a product of factors.

3. $3^2 x^4 y^3$

4. $5(uv)^4$

In Exercises 5–10, simplify the given expression.

5. $v^2 \cdot v^3$

6. $(u^3)^2$

7. $(-2x)^2 x^4$

8. $-y^2(-2y)^3$

9. $5z^3(z^2)^2$

10. $(a + 3)^2(a + 3)^5$

Section 2.3

Warm-Up

The following warm-up exercises involve skills that were covered in earlier sections. You will use these skills in the exercise set for this section.

In Exercises 1–6, use the Distributive Property to expand the given expression.

1. $-4(2x - 5)$ **2.** $10(3t - 4)$

3. $x(-2xy + y^3)$ **4.** $-z(xz - 2y^2)$

5. $-\dfrac{3}{4}(12 - 8x)$ **6.** $\dfrac{2}{3}(-15x - 12z)$

In Exercises 7–10, simplify the given expression by combining like terms.

7. $4s - 6t + 7s + t$ **8.** $2x^2 - 4 + 5 - 3x^2$

9. $\dfrac{5x}{3} - \dfrac{2x}{3} - 4$ **10.** $3x^2y + xy - xy^2 - 6xy$

Section 2.4

Warm-Up

The following warm-up exercises involve skills that were covered in earlier sections. You will use these skills in the exercise set for this section.

In Exercises 1–4, simplify the given expression.

1. $-3(3x - 2y) + 5y$ **2.** $3v - (4 - 5v)$

3. $-y^2(y^2 + 4) + 6y^2$ **4.** $5t(2 - t) + t^2$

In Exercises 5–10, evaluate the algebraic expression for the specified values of the variables. (If not possible, state the reason.)

5. $x^2 - y^2$, $x = 4, y = 3$ **6.** $4s + st$, $s = 3, t = -4$

7. $\dfrac{x}{x^2 + y^2}$, $x = 0, y = 3$ **8.** $\dfrac{z^2 + 2}{x^2 - 1}$, $x = 1, z = 1$

9. $\dfrac{a}{1 - r}$, $a = 2, r = \dfrac{1}{2}$ **10.** $2l + 2w$, $l = 3, w = 1.5$

Section 2.5

| Warm-Up | *The following warm-up exercises involve skills that were covered in earlier sections. You will use these skills in the exercise set for this section.* |

In Exercises 1–4, simplify the algebraic expression.

1. $(-2y^2)^3$

2. $(3a^2)(4ab)$

3. $3x - 2(x - 5)$

4. $x(x - 3) - (x^2 + 6x)$

In Exercises 5 and 6, evaluate the expression for the specified value of the variable.

Expression   *Values*

5. $x^2 - 2x + 1$ (a) $x = 1$ (b) $x = 3$

6. $\dfrac{1}{x^2 + 1}$ (a) $x = 0$ (b) $x = 2$

In Exercises 7–10, translate the phrase into an algebraic expression. (Let x represent the arbitrary number.)

7. Four more than twice a number

8. A number is decreased by 10, and the result is doubled.

9. A number is decreased by five, and the result is squared.

10. A number is increased by 25, and the sum is halved.

Chapter 3

Section 3.1

| Warm-Up | *The following warm-up exercises involve skills that were covered in earlier sections. You will use these skills in the exercise set for this section.* |

In Exercises 1–6, perform the indicated operation.

1. $2(-3) + 9$

2. $-10(6 - 2)$

3. $4 - \dfrac{5}{2}$

4. $\dfrac{|18 - 25|}{6}$

5. $\left(-\dfrac{7}{12}\right)\left(\dfrac{3}{28}\right)$

6. $\dfrac{4}{3} \div \dfrac{5}{6}$

In Exercises 7–10, determine whether the given value of the variable is a solution of the equation.

7. $6x - 5 = 0, \quad x = \dfrac{5}{6}$

8. $4 - 3x = 0, \quad x = \dfrac{4}{3}$

9. $3(x - 4) = x, \quad x = 6$

10. $x + 6 = 2(3x + 1), \quad x = -3$

Section 3.2

Warm-Up

The following warm-up exercises involve skills that were covered in earlier sections. You will use these skills in the exercise set for this section.

In Exercises 1–10, solve the given equation.

1. $-10x = 1{,}000$

2. $15x = 60$

3. $-\dfrac{x}{10} = 1{,}000$

4. $\dfrac{x}{15} = 60$

5. $-\dfrac{10}{x} = 1{,}000$

6. $\dfrac{15}{x} = 60$

7. $0.35x = 70$

8. $0.60x = 24$

9. $125(1 - r) = 100$

10. $3{,}050(1 - r) = 1{,}830$

Section 3.3

Warm-Up

The following warm-up exercises involve skills that were covered in earlier sections. You will use these skills in the exercise set for this section.

In Exercises 1–10, solve the linear equation and check your solution.

1. $-t = 6$

2. $8 - z = 3$

3. $50 - z = 15$

4. $x - 6 = 3x + 10$

5. $2x - 5 = x + 9$

6. $6x + 8 = 8 - 2x$

7. $2x + \dfrac{3}{2} = \dfrac{3}{2}$

8. $-10x + \dfrac{2}{3} = \dfrac{7}{3} - 5x$

9. $\dfrac{x}{6} + \dfrac{x}{3} = 1$

10. $\dfrac{x}{5} + \dfrac{1}{5} = \dfrac{7}{10}$

Section 3.4

Warm-Up

The following warm-up exercises involve skills that were covered in earlier sections. You will use these skills in the exercise set for this section.

In Exercises 1–4, write the fraction in reduced form.

1. $\dfrac{15}{25}$ **2.** $\dfrac{36}{42}$ **3.** $\dfrac{33}{123}$ **4.** $\dfrac{28}{42}$

In Exercises 5–10, solve the given equation.

5. $\dfrac{x}{3} = \dfrac{4}{9}$ **6.** $\dfrac{t}{16} = \dfrac{1}{4}$ **7.** $\dfrac{n}{5} = \dfrac{8}{25}$ **8.** $\dfrac{x}{2} = \dfrac{15}{4}$

9. $\dfrac{6}{x} = \dfrac{3}{5}$ **10.** $\dfrac{5}{8} = \dfrac{4}{x}$

Section 3.5

Warm-Up

The following warm-up exercises involve skills that were covered in earlier sections. You will use these skills in the exercise set for this section.

In Exercises 1–4, place the correct inequality symbol ($<$ or $>$) between the two real numbers.

1. $-\dfrac{1}{2}$ -7 **2.** $-\dfrac{1}{3}$ $-\dfrac{1}{6}$

3. -2 -3 **4.** -6 $-\dfrac{13}{2}$

In Exercises 5–10, solve the equation.

5. $-2n = 5$ **6.** $16 + 2l = 64$

7. $\dfrac{9 + x}{3} = 15$ **8.** $20 - \dfrac{9}{x} = 2$

9. $4 - 3(1 - x) = 7$ **10.** $6(t - 6) = 0$

Chapter 4

Section 4.1

Warm-Up

The following warm-up exercises involve skills that were covered in earlier sections. You will use these skills in the exercise set for this section.

In Exercises 1–6, place the correct inequality symbol ($<$ or $>$) between the two real numbers, and plot each number on the real number line.

1. -4 [____] 3 2. $\dfrac{8}{3}$ [____] 2 3. -2 [____] -6

4. -8 [____] 0 5. $\dfrac{15}{4}$ [____] $-\dfrac{1}{2}$ 6. $\dfrac{2}{5}$ [____] $\dfrac{15}{16}$

In Exercises 7–10, evaluate the given expression.

7. $4 - |-3|$ 8. $-10 - (4 - 18)$

9. $\dfrac{3 - (5 - 20)}{4}$ 10. $\dfrac{|3 - 18|}{3}$

Section 4.2

Warm-Up

The following warm-up exercises involve skills that were covered in earlier sections. You will use these skills in the exercise set for this section.

In Exercises 1–4, plot the given points in a rectangular coordinate system.

1. $(-6, 4)$, $(-3, -4)$ 2. $(4, 6)$, $(8, -2)$

3. $\left(\dfrac{7}{2}, \dfrac{9}{2}\right)$, $\left(\dfrac{4}{3}, -3\right)$ 4. $\left(-\dfrac{3}{4}, -\dfrac{7}{4}\right)$, $\left(-1, \dfrac{5}{2}\right)$

In Exercises 5 and 6, solve the given equation.

5. $\dfrac{5}{6}x - 7 = 0$ 6. $16 - \dfrac{2}{3}x = 0$

In Exercises 7–10, find the unknown coordinate of the solution point of the given equation.

7. $y = \dfrac{4}{5}x + 2$, $x = 15$ 8. $y = 3 - \dfrac{5}{6}x$, $x = 12$

9. $y = 8 - 0.75x$, $y = -1$ 10. $y = 2 + 0.6x$, $y = 4.4$

Section 4.3

Warm-Up

Warm-Up

The following warm-up exercises involve skills that were covered in earlier sections. You will use these skills in the exercise set for this section.

In Exercises 1–10, solve for y in terms of x.

1. $3x + y = 4$ **2.** $x - y = 0$

3. $5 - y + x = 0$ **4.** $4 - y + x = 0$

5. $2x + 2y = 10$ **6.** $3x - 3y = 15$

7. $2x + 3y = 2$ **8.** $4x - 5y = -2$

9. $3x + 4y - 5 = 0$ **10.** $-2x - 3y + 6 = 0$

Section 4.4

Warm-Up

The following warm-up exercises involve skills that were covered in earlier sections. You will use these skills in the exercise set for this section.

In Exercises 1–4, solve the given equation.

1. $3x - 42 = 0$ **2.** $64 - 16x = 0$

3. $2 - 3x = 14 + x$ **4.** $7 + 5x = 7x - 1$

In Exercises 5–10, solve the given percent problem.

5. What is 62% of 25?   **6.** What is $\frac{1}{2}$% of 6,000?

7. 300 is what percent of 150?   **8.** 600 is what percent of 900?

9. 145.6 is 32% of what number?   **10.** 2 is 0.8% of what number?

Section 4.5

The following warm-up exercises involve skills that were covered in earlier sections. You will use these skills in the exercise set for this section.

In Exercises 1–10, solve the given equation.

1. $14 - 2x = x + 2$

2. $2(x + 1) = 0$

3. $5[1 + 2(x + 3)] = 6 - 3(x - 1)$

4. $2 - 5(x - 1) = 2[x + 10(x - 1)]$

5. $\dfrac{x}{3} + 5 = 8$

6. $\dfrac{3x}{4} + \dfrac{1}{2} = 8$

7. $\dfrac{x}{3} + \dfrac{x}{2} = \dfrac{1}{3}$

8. $\dfrac{2}{x} + \dfrac{2}{5} = 1$

9. $\dfrac{3}{x} + \dfrac{4}{3} = 1$

10. $\dfrac{x}{x + 1} - \dfrac{1}{2} = \dfrac{4}{3}$

Chapter 5

Section 5.1

The following warm-up exercises involve skills that were covered in earlier sections. You will use these skills in the exercise set for this section.

In Exercises 1–4, use the Distributive Property to expand the given expression.

1. $10(x - 1)$

2. $4(3 - 2z)$

3. $-\dfrac{1}{2}(4 - 6x)$

4. $-25(2x - 3)$

In Exercises 5 and 6, list the terms of the algebraic expression.

5. $10x - 3y + 4$

6. $-2r + 8s - 6$

In Exercises 7 and 8, give the coefficient of the term.

7. $\dfrac{3}{4}x$

8. $-8y$

In Exercises 9 and 10, simplify the expression by combining like terms.

9. $8y - 2x + 7x - 10y$

10. $\dfrac{5}{6}x - \dfrac{2}{3}x + 8$

Section 5.2

Warm-Up

The following warm-up exercises involve skills that were covered in earlier sections. You will use these skills in the exercise set for this section.

In Exercises 1 and 2, rewrite the product in exponential form.

1. $(-3)(-3)(-3)(-3)$

2. $x \cdot x \cdot x$

In Exercises 3–6, rewrite the expression as a repeated multiplication.

3. $\left(\dfrac{4}{5}\right)^4$

4. $(4.5)^5$

5. $(x^2)^3$

6. $(y^4)^2$

In Exercises 7–10, simplify the algebraic expression.

7. $2(x - 4) + 5x$

8. $4(3 - y) + 2(y + 1)$

9. $-3(z - 2) - (z - 6)$

10. $(u - 2) - 3(2u + 1)$

Section 5.3

Warm-Up

The following warm-up exercises involve skills that were covered in earlier sections. You will use these skills in the exercise set for this section.

In Exercises 1–4, write the fraction in reduced form.

1. $\dfrac{8}{12}$

2. $\dfrac{18}{144}$

3. $\dfrac{60}{150}$

4. $\dfrac{175}{42}$

In Exercises 5–10, find the product and simplify.

5. $-2x^2(5x^3)$

6. $4y(3y^2 - 1)$

7. $(2z + 1)(2z - 1)$

8. $(x + 4)(2x - 5)$

9. $(x + 7)^2$

10. $(x + 1)(x^2 - x + 1)$

Section 5.4

Warm-Up	*The following warm-up exercises involve skills that were covered in earlier sections. You will use these skills in the exercise set for this section.*

In Exercises 1–6, evaluate the given quantity.

1. 4^3 **2.** $(-2)^5$ **3.** $\left(-\dfrac{2}{3}\right)^2$ **4.** -8^2

5. $25 - 3^2 \cdot 2$ **6.** $(12 - 9)^3 \cdot 4 + 36$

In Exercises 7–10, simplify the given expression. (Assume that each denominator is not zero.)

7. $x^2 \cdot x^3$ **8.** $(ab)^4 \div (ab)^3$ **9.** $\dfrac{u^4 v^2}{uv}$ **10.** $(y^2 z^3)(z^2)$

Chapter 6

Section 6.1

Warm-Up	*The following warm-up exercises involve skills that were covered in earlier sections. You will use these skills in the exercise set for this section.*

In Exercises 1–10, find the product.

1. $2(5 - 15)$ **2.** $-3(8 + 6)$

3. $12(2x - 3)$ **4.** $7(4 - 3x)$

5. $-6(10 - 7x)$ **6.** $-2y(y + 1)$

7. $-3t(t + 2)$ **8.** $8xy(xy - 3)$

9. $(2 - x)(2 + x)$ **10.** $(x + 4)^2$

Section 6.2

Warm-Up

The following warm-up exercises involve skills that were covered in earlier sections. You will use these skills in the exercise set for this section.

In Exercises 1–10, find the product.

1. $-4(x - 6)$

2. $6(3x + 8)$

3. $y(y + 2)$

4. $-a^2(a - 1)$

5. $(x - 2)(x - 5)$

6. $(t + 3)(t + 6)$

7. $(u - 8)(u + 3)$

8. $(v - 1)(v - 6)$

9. $(z - 3)(z + 1)$

10. $(x - 4)(x + 7)$

Section 6.3

Warm-Up

The following warm-up exercises involve skills that were covered in earlier sections. You will use these skills in the exercise set for this section.

In Exercises 1 and 2, rewrite the given polynomial by factoring out the indicated fraction.

1. $\frac{1}{3}x + \frac{5}{9} = \frac{1}{9}(\qquad)$

2. $\frac{5}{8}x - \frac{3}{2} = \frac{1}{8}(\qquad)$

In Exercises 3–10, factor the polynomial completely.

3. $6x + 12$

4. $4x - 12$

5. $x^2y - xy^2$

6. $6x^3 - 3x^2 + 9x$

7. $x^2 + x - 42$

8. $x^2 + 13x + 42$

9. $x^3 - x^2 - 42x$

10. $2x^2y + 8xy - 64y$

Section 6.4

Warm-Up

The following warm-up exercises involve skills that were covered in earlier sections. You will use these skills in the exercise set for this section.

In Exercises 1–6, find the product.

1. $-12x(x - 3)$
2. $3z(-2z + 5)$
3. $(x + 10)(x - 10)$
4. $(y + a)(y - a)$
5. $(x + 5)^2$
6. $(3x - 2)^2$

In Exercises 7–10, factor the polynomial completely.

7. $10x^2 + 70$
8. $16x - 4x^2$
9. $3x^2 - 5x + 2$
10. $2x^2 - x - 1$

Section 6.5

Warm-Up

The following warm-up exercises involve skills that were covered in earlier sections. You will use these skills in the exercise set for this section.

In Exercises 1–4, solve the given equation.

1. $4x - 48 = 0$
2. $3x + 9 = 0$
3. $\frac{x}{4} + \frac{x}{3} = \frac{1}{3}$
4. $\frac{3}{x} + \frac{3}{2} = 2$

In Exercises 5–10, factor the given polynomial.

5. $3x^2 + 7x$
6. $4x^2 - 25$
7. $16 - (x - 11)^2$
8. $x^2 + 7x - 18$
9. $10x^2 + 13x - 3$
10. $6x^2 - 73x + 12$

Chapter 7

Section 7.1

Warm-Up

The following warm-up exercises involve skills that were covered in earlier sections. You will use these skills in the exercise set for this section.

In Exercises 1–6, simplify the given expression.

1. $2(x + h) - 3 - (2x - 3)$ **2.** $3(y + t) + 5 - (3y + 5)$

3. $2(x - 4)^2 - 5$ **4.** $-3(x + 2)^2 + 10$

5. $4 - 2[3 + 4(x + 1)]$ **6.** $5x + x[3 - 2(x - 3)]$

In Exercises 7–10, evaluate the expression at the given values of the variable. (If not possible, state the reason.)

7. $3x^2 - 4$ (a) $x = 0$ (b) $x = -2$ (c) $x = 2$ (d) $x = 5$

8. $9 - (x - 3)^2$ (a) $x = 0$ (b) $x = 3$ (c) $x = 6$ (d) $x = 2.5$

9. $\dfrac{x + 4}{x - 3}$ (a) $x = -4$ (b) $x = 3$ (c) $x = 4$ (d) $x = 2$

10. $x + \dfrac{1}{x}$ (a) $x = 1$ (b) $x = -1$ (c) $x = \dfrac{1}{2}$ (d) $x = 10$

Section 7.2

Warm-Up

The following warm-up exercises involve skills that were covered in earlier sections. You will use these skills in the exercise set for this section.

In Exercises 1–4, evaluate the given expression.

1. $\dfrac{4 - 2}{7 - 3}$ **2.** $\dfrac{-2 - (-5)}{10 - 1}$ **3.** $\dfrac{4 - (-5)}{-3 - (-1)}$ **4.** $\dfrac{-5 - 8}{0 - (-3)}$

In Exercises 5 and 6, find $-1/m$ for the given value of m.

5. $m = -\dfrac{5}{4}$ **6.** $m = \dfrac{7}{8}$

In Exercises 7–10, solve for y in terms of x.

7. $2x - 3y = 5$ **8.** $4x + 2y = 0$

9. $y - (-4) = 3[x - (-1)]$ **10.** $y - 7 = \dfrac{2}{3}(x - 3)$

Section 7.3

Warm-Up

The following warm-up exercises involve skills that were covered in earlier sections. You will use these skills in the exercise set for this section.

In Exercises 1–4, determine the slope of the line passing through the given points.

1. $(-3, 2)$, $(5, 0)$ **2.** $(-10, -5)$, $(4, 10)$

3. $(-4, -4)$, $(-4, 6)$ **4.** $\left(\dfrac{2}{3}, \dfrac{1}{2}\right)$, $\left(\dfrac{5}{6}, \dfrac{3}{4}\right)$

In Exercises 5 and 6, sketch the graph of the lines through the given point with the indicated slope. Make all four sketches in the same rectangular coordinate system.

	Point	Slope			

5. $(2, 3)$ (a) 0 (b) 1 (c) 2 (d) $-\dfrac{1}{3}$

6. $(-4, 1)$ (a) 3 (b) -3 (c) $\dfrac{1}{2}$ (d) undefined

In Exercises 7–10, write the equation in the form $y = mx + b$.

7. $y - 8 = -\dfrac{2}{5}(x - 4)$ **8.** $y + 3 = \dfrac{4}{3}(x - 5)$

9. $y - (-1) = \dfrac{3 - (-1)}{2 - 4}(x - 4)$ **10.** $y - 5 = \dfrac{3 - 5}{0 - 2}(x - 2)$

Section 7.4

Warm-Up

The following warm-up exercises involve skills that were covered in earlier sections. You will use these skills in the exercise set for this section.

In Exercises 1–6, sketch the line represented by the given equation.

1. $x + 5 = 0$ **2.** $x + 2y = 0$

3. $y = -\dfrac{1}{3}x + 3$ **4.** $y = \dfrac{1}{5}x - 2$

5. $2x + y - 1 = 0$ **6.** $2y - 5 = 0$

In Exercises 7–10, determine whether the given points are solution points of the equation.

	Equation	Points	

7. $8x + 3y - 15 = 0$ (a) $(1, 1)$ (b) $(0, 5)$

8. $4x - 5y + 20 = 0$ (a) $(5, 8)$ (b) $(0, 0)$

9. $\dfrac{1}{2}x - 2y + 6 = 0$ (a) $(8, 11)$ (b) $\left(2, \dfrac{7}{2}\right)$

10. $x + \dfrac{4}{5}y - 8 = 0$ (a) $\left(2, \dfrac{15}{2}\right)$ (b) $\left(1, \dfrac{5}{2}\right)$

Chapter 8

Section 8.1

The following warm-up exercises involve skills that were covered in earlier sections. You will use these skills in the exercise set for this section.

In Exercises 1–6, sketch the graph of the given equation.

1. $y = -\frac{1}{3}x + 6$ **2.** $y = 2(x - 3)$ **3.** $y - 1 = 3(x + 2)$

4. $y + 2 = \frac{1}{4}(x - 1)$ **5.** $2x + y = 4$ **6.** $5x - 2y = 3$

In Exercises 7 and 8, find an equation of the line passing through the two points.

7. $(-1, 3),\ (4, 8)$ **8.** $(2, 6),\ (5, 1)$

In Exercises 9 and 10, determine the slope of the given line.

9. $3x + 6y = 4$ **10.** $7x - 4y = 10$

Section 8.2

Warm-Up

The following warm-up exercises involve skills that were covered in earlier sections. You will use these skills in the exercise set for this section.

In Exercises 1–6, solve the given equation.

1. $x - (x + 2) = 8$ **2.** $y + (2y + 3) = 4$

3. $y - 3(4y - 2) = 1$ **4.** $x + 6(3 - 2x) = 4$

5. $3x + \frac{1}{2}(6x + 5) = \frac{3}{2}$ **6.** $4y - \frac{4}{5}(3y - 10) = 8$

In Exercises 7–10, solve the system of linear equations by graphing.

7. $-4x + 3y = 8$
    $-x + 5y = 2$ **8.** $2x + y = 3$
                        $3x - 2y = -6$

9. $x + y = 5$
    $2x - 3y = 0$ **10.** $-2x + 3y = 10$
                       $5x - 2y = 8$

Section 8.3

Warm-Up

The following warm-up exercises involve skills that were covered in earlier sections. You will use these skills in the exercise set for this section.

In Exercises 1–4, perform the indicated operations and simplify.

1. $(3x + 2y) - 2(x + y)$ **2.** $(-10u + 3v) + 5(2u - 8v)$

3. $x^2 + (x - 3)^2 + 6x$ **4.** $y^2 - (y + 1)^2 + 2y$

In Exercises 5 and 6, solve the given equation.

5. $3x + (x - 5) = 19$ **6.** $3t - 2(t + 1) = 4$

In Exercises 7–10, determine whether the lines represented by the pair of equations are parallel, perpendicular, or neither.

7. $2x - 3y = -10$
$3x + 2y = 11$

8. $x - 3y = 2$
$6x + 2y = 4$

9. $4x - 12y = 5$
$-2x + 6y = 3$

10. $5x + y = 2$
$3x + 2y = 1$

Section 8.4

Warm-Up

The following warm-up exercises involve skills that were covered in earlier sections. You will use these skills in the exercise set for this section.

In Exercises 1–6, write an algebraic expression that represents the given statement.

1. The amount of money (in dollars) represented by m nickels and n quarters

2. The amount of income tax due on a taxable income of I dollars (The income tax rate is 13%.)

3. The amount of time required to travel 250 miles at an average speed of r miles per hour

4. The discount on the price of a product with a list price of L dollars and a discount rate of 15%

5. The perimeter of a rectangle of length l and width $l/2$

6. The sum of a number n and two and one-half times that number

In Exercises 7–10, solve the given system of linear equations.

7. $x + y = 25$
$y = 10$

8. $2x - 3y = 4$
$6x = -12$

9. $x + y = 32$
$x - y = 24$

10. $2r - s = 5$
$r + 2s = 10$

Chapter 9

Section 9.1

| Warm-Up | *The following warm-up exercises involve skills that were covered in earlier sections. You will use these skills in the exercise set for this section.* |

In Exercises 1–4, fill in the missing numerator or denominator.

1. $-\dfrac{3}{4} = \dfrac{}{12}$ 2. $\dfrac{5}{8} = \dfrac{30}{}$

3. $\dfrac{5}{3} = \dfrac{-35}{}$ 4. $-\dfrac{3}{7} = \dfrac{}{42}$

In Exercises 5–8, write the fraction in reduced form.

5. $\dfrac{30}{45}$ 6. $\dfrac{36}{84}$

7. $\dfrac{84}{98}$ 8. $\dfrac{63}{99}$

In Exercises 9 and 10, factor the given polynomial.

9. $x^3 - 3x^2 + 4x$ 10. $10x^4 + 3x^3 - 4x^2$

Section 9.2

| Warm-Up | *The following warm-up exercises involve skills that were covered in earlier sections. You will use these skills in the exercise set for this section.* |

In Exercises 1–6, multiply or divide, as indicated, and simplify your answer.

1. $\dfrac{5}{16} \cdot \dfrac{4}{5}$ 2. $\dfrac{7}{12} \cdot \dfrac{9}{14}$ 3. $-\dfrac{12}{35} \cdot \dfrac{-25}{54}$

4. $\dfrac{-225}{-448} \cdot \dfrac{28}{-105}$ 5. $\dfrac{16}{3} \div \dfrac{32}{45}$ 6. $\dfrac{-7}{45} \div \dfrac{2}{3}$

In Exercises 7–10, factor the given polynomial.

7. $4x^3 - x$ 8. $9x^2 - 4$

9. $15x^2 - 11x - 14$ 10. $4x^2 - 28x + 49$

Section 9.3

Warm-Up

The following warm-up exercises involve skills that were covered in earlier sections. You will use these skills in the exercise set for this section.

In Exercises 1–10, perform the indicated operations and simplify your answer.

1. $\dfrac{7}{9} + \dfrac{2}{9}$ **2.** $\dfrac{3}{32} + \dfrac{5}{32}$ **3.** $\dfrac{11}{15} - \dfrac{2}{15}$

4. $\dfrac{3}{10} - \dfrac{8}{10}$ **5.** $\dfrac{3}{7} + \dfrac{3}{14}$ **6.** $\dfrac{5}{24} + \dfrac{3}{16}$

7. $-\dfrac{2}{3} + \dfrac{46}{75}$ **8.** $\dfrac{22}{5} - \dfrac{8}{35}$ **9.** $\dfrac{3}{4} - \dfrac{7}{8} + \dfrac{7}{12}$

10. $\dfrac{5}{9} - \dfrac{2}{3} - \dfrac{5}{18}$

Section 9.4

Warm-Up

The following warm-up exercises involve skills that were covered in earlier sections. You will use these skills in the exercise set for this section.

In Exercises 1–10, solve the given equation.

1. $5x = 8$ **2.** $-2x = 15$

3. $16x - 3 = 29$ **4.** $4 - 32x = 0$

5. $(x - 4)(x + 10) = 0$ **6.** $(2x + 3)(x - 9) = 0$

7. $x(x - 8) = 0$ **8.** $x(8 - x) = 16$

9. $x^2 + x - 56 = 0$ **10.** $t^2 - 5t = 0$

Chapter 10

Section 10.1

Warm-Up

The following warm-up exercises involve skills that were covered in earlier sections. You will use these skills in the exercise set for this section.

In Exercises 1–10, evaluate the given expression.

1. 15^2 **2.** 12^2 **3.** $(-5)^2$ **4.** -5^2

5. $\left(\dfrac{2}{3}\right)^2$ **6.** $\left(\dfrac{4}{5}\right)^2$ **7.** $\dfrac{3}{4}\left(\dfrac{4}{3}\right)^2$ **8.** $\dfrac{5}{6}\left(\dfrac{3}{5}\right)^2$

9. $-\left(\dfrac{1}{3}\right)^2\left(-\dfrac{1}{3}\right)^2$ **10.** $\left(-\dfrac{1}{5}\right)^2\left(\dfrac{1}{5}\right)^2$

Section 10.2

Warm-Up

The following warm-up exercises involve skills that were covered in earlier sections. You will use these skills in the exercise set for this section.

In Exercises 1–4, evaluate the square root, if possible. (Do not use a calculator.)

1. $\sqrt{10{,}000}$ **2.** $-\sqrt{196}$ **3.** $-\sqrt{\dfrac{169}{25}}$ **4.** $\sqrt{-\dfrac{9}{4}}$

In Exercises 5–10, simplify the given expression.

5. $(2x)^2(2x)^3$ **6.** $(-4x^2y)^3$ **7.** $\dfrac{32x^3y^2}{2xy^3}$ **8.** $\dfrac{8y^2z^{-2}}{z^2}$

9. $\dfrac{9xy^{-2}}{3x^{-2}y}$ **10.** $(u^2v)^3(u^2v)^{-3}$

Section 10.3

Warm-Up

The following warm-up exercises involve skills that were covered in earlier sections. You will use these skills in the exercise set for this section.

In Exercises 1–10, perform the indicated multiplication and simplify the result by combining any like terms.

1. $4(2 - x) + 6x$

2. $3(2x + 1) + 2(x + 1)$

3. $17 - 5(x - 2)$

4. $32 - 3(2x + 10)$

5. $(x + 6)^2$

6. $(2x - 3)^2$

7. $(5x - 3)(5x + 3)$

8. $(1 - 3x)(1 + 3x)$

9. $x(x - 7) - 3x(x + 2)$

10. $4x(x + 6) + 3x(10 - 4x)$

Section 10.4

Warm-Up

The following warm-up exercises involve skills that were covered in earlier sections. You will use these skills in the exercise set for this section.

In Exercises 1–6, solve the given equation.

1. $x + 6 = 32$

2. $7x = 18$

3. $6x - 5 = 13$

4. $45 - 8x = 21$

5. $3(x - 5)(x + 11) = 0$

6. $-2(4 - x)(16 + x) = 0$

In Exercises 7–10, simplify the given expression.

7. $\sqrt{500}$

8. $\sqrt{20} - \sqrt{5}$

9. $2\sqrt{18}\sqrt{32}$

10. $\dfrac{2}{\sqrt{2}}$

Chapter 11

Section 11.1

Warm-Up

The following warm-up exercises involve skills that were covered in earlier sections. You will use these skills in the exercise set for this section.

In Exercises 1–6, factor the given expression.

1. $9x^2 - 25$

2. $4t^2 - 12t + 9$

3. $2x^2 - 8x - 10$

4. $4s^2 - 9$

5. $3x^2 - 11x + 10$

6. $4x^3 - 12x^2 - 16x$

In Exercises 7–10, solve the given equation.

7. $5y + 4 = 0$

8. $2s - 3 = 0$

9. $(x + 7)(2x - 3) = 0$

10. $(5x - 4)(x - 10) = 0$

Section 11.2

Warm-Up

The following warm-up exercises involve skills that were covered in earlier sections. You will use these skills in the exercise set for this section.

In Exercises 1–4, expand and simplify the given expression.

1. $(x + 2)^2 - 1$

2. $(x + 5)^2 + 3$

3. $(u - 8)^2 + 10$

4. $(v - 3)^2 - 2$

In Exercises 5–10, solve the given equation.

5. $x^2 = \dfrac{1}{4}$

6. $y^2 = \dfrac{9}{16}$

7. $(y - 5)^2 = 36$

8. $(z + 2)^2 = 10$

9. $(2x - 1)^2 = 5$

10. $(4x + 3)^2 = 7$

Section 11.3

Warm-Up

The following warm-up exercises involve skills that were covered in earlier sections. You will use these skills in the exercise set for this section.

In Exercises 1–4, solve the quadratic equation by factoring.

1. $x^2 - 7x + 10 = 0$ **2.** $x^2 + 3x - 18 = 0$

3. $2x^2 + 17x - 30 = 0$ **4.** $6x^2 - 17x + 5 = 0$

In Exercises 5 and 6, solve the quadratic equation by completing the square.

5. $x^2 + x - 4 = 0$ **6.** $3x^2 + 6x + 1 = 0$

In Exercises 7–10, simplify the radical.

7. $\sqrt{16 - 4(3)(1)}$ **8.** $\sqrt{9 - 4(-2)(5)}$

9. $\sqrt{36 - 4(2)(-4)}$ **10.** $\sqrt{100 - 4(2)(-6)}$

Section 11.4

Warm-Up

The following warm-up exercises involve skills that were covered in earlier sections. You will use these skills in the exercise set for this section.

In Exercises 1–6, sketch the graph of the equation.

1. $y = 2x$ **2.** $y = \frac{1}{3}x$ **3.** $y = 2x - 2$

4. $y = \frac{1}{3}x + 2$ **5.** $y = -2x + 5$ **6.** $y = -\frac{1}{3}x + 2$

In Exercises 7–10, solve the equation.

7. $10(x - 3) = 5$ **8.** $20 - 3x = 50 - 4x$

9. $x(x + 4) = 0$ **10.** $x(x + 4) = 3$

Section 11.5

<table>
<tr><td>**Warm-Up**</td><td>*The following warm-up exercises involve skills that were covered in earlier sections. You will use these skills in the exercise set for this section.*</td></tr>
</table>

In Exercises 1–8, solve the given equation.

1. $3(x - 2) = 0$ **2.** $3x(x - 2) = 0$

3. $2n + (n + 2) = 30$ **4.** $2n(n + 2) = 30$

5. $2(x + 8)^2 = 200$ **6.** $t^2 + 3t - 1 = 0$

7. $t + \dfrac{2}{t} = 3$ **8.** $\sqrt{3s + 4} = s$

In Exercises 9 and 10, sketch the graph of the quadratic function.

9. $y = 2 - 2x - x^2$ **10.** $y = (x - 4)^2$

Warm-Up Exercise Answers

Chapter 1

Section 1.1

> **Warm-Up**
>
> **1.** 0   **2.** 4   **3.** 8   **4.** 6   **5.** 5   **6.** 17   **7.** 2
> **8.** 2   **9.** 7   **10.** 3

Section 1.2

> **Warm-Up**
>
> **1.** −100   **2.** 3   **3.** 12   **4.** −4   **5.** −12   **6.** 9
> **7.** $\frac{4}{9}$ **8.** −4   **9.** 6   **10.** 1

Section 1.3

> **Warm-Up**
>
> **1.** −64   **2.** 75   **3.** −7   **4.** 0   **5.** $\frac{1}{8}$
> **6.** $-\frac{1}{2}$ **7.** $\frac{3}{4}$ **8.** $\frac{14}{3}$ **9.** $-\frac{6}{5}$ **10.** $\frac{14}{5}$

Section 1.4

> **Warm-Up**
>
> **1.** 13   **2.** 28   **3.** 9,300,000   **4.** 0.0066
> **5.** 2   **6.** −4   **7.** 35   **8.** −30   **9.** 10
> **10.** 8

Chapter 2

Section 2.1

> **Warm-Up**
>
> **1.** −10   **2.** $-\frac{5}{3}$ **3.** 20   **4.** 42   **5.** 5
> **6.** 12   **7.** Distributive Property
> **8.** Commutative Property of Multiplication
> **9.** Associative Property of Addition
> **10.** Multiplicative Inverse Property

Section 2.2

> **Warm-Up**
>
> **1.** $(3z)^4$ **2.** $8^3 x^3 y^2$
> **3.** $3 \cdot 3 \cdot x \cdot x \cdot x \cdot x \cdot y \cdot y \cdot y$
> **4.** $5 \cdot uv \cdot uv \cdot uv \cdot uv$ **5.** v^5 **6.** u^6
> **7.** $4x^6$ **8.** $8y^5$ **9.** $5z^7$ **10.** $(a + 3)^7$

Section 2.3

> **Warm-Up**
>
> **1.** $-8x + 20$ **2.** $30t - 40$ **3.** $-2x^2 y + xy^3$
> **4.** $-xz^2 + 2y^2 z$ **5.** $-9 + 6x$ **6.** $-10x - 8z$
> **7.** $11s - 5t$ **8.** $-x^2 + 1$ **9.** $x - 4$
> **10.** $3x^2 y - 5xy - xy^2$

Section 2.4

> **Warm-Up**
>
> **1.** $-9x + 11y$ **2.** $8v - 4$ **3.** $-y^4 + 2y^2$
> **4.** $10t - 4t^2$ **5.** 7   **6.** 0   **7.** 0
> **8.** Division by zero is undefined.   **9.** 4   **10.** 9

Section 2.5

Warm-Up

1. $-8y^6$ 2. $12a^3b$ 3. $x + 10$ 4. $-9x$
5 (a) 0   (b) 4   6. (a) 1   (b) $\frac{1}{5}$
7. $2x + 4$ 8. $2(x - 10)$ 9. $(x - 5)^2$
10. $\dfrac{x + 25}{2}$

Chapter 3

Section 3.1

Warm-Up

1. 3   2. -40 3. $\frac{3}{2}$ 4. $\frac{7}{6}$ 5. $-\frac{1}{16}$
6. $\frac{8}{5}$ 7. Solution   8. Solution
9. Solution   10. Not a solution

Section 3.2

Warm-Up

1. -100 2. 4   3. $-10,000$ 4. 900
5. $-\frac{1}{100}$ 6. $\frac{1}{4}$ 7. 200   8. 40   9. $\frac{1}{5}$
10. $\frac{2}{5}$

Section 3.3

Warm-Up

1. -6 2. 5   3. 35   4. -8 5. 14
6. 0   7. 0   8. $-\frac{1}{3}$ 9. 2   10. $\frac{5}{2}$

Section 3.4

Warm-Up

1. $\frac{3}{5}$ 2. $\frac{6}{7}$ 3. $\frac{11}{41}$ 4. $\frac{2}{3}$ 5. $\frac{4}{3}$ 6. 4
7. $\frac{8}{5}$ 8. $\frac{15}{2}$ 9. 10   10. $\frac{32}{5}$

Section 3.5

Warm-Up

1. $-\frac{1}{2} > -7$ 2. $-\frac{1}{3} < -\frac{1}{6}$ 3. $-2 > -3$
4. $-6 > -\frac{13}{2}$ 5. $-\frac{5}{2}$ 6. 24   7. 36
8. $\frac{1}{2}$ 9. 2   10. 6

Chapter 4

Section 4.1

Warm-Up

1. $-4 < 3$

$$\underset{-4\;-1\;\;0\;\;2\;\;4}{\bullet\quad\quad\quad\bullet\quad}$$

2. $\frac{8}{3} > 2$

$$\underset{0\;\;1\;\;2\;\;3}{\quad\quad\bullet\;\bullet}$$

3. $-2 > -6$

$$\underset{-8\;-6\;-4\;-2\;\;0}{\quad\bullet\quad\bullet\quad}$$

4. $-8 < 0$

$$\underset{-10\;-8\;-6\;-4\;-2\;\;0}{\quad\bullet\quad\quad\quad\quad\bullet}$$

5. $\frac{15}{4} > -\frac{1}{2}$

$$\underset{-1\;\;0\;\;1\;\;2\;\;3\;\;4}{\bullet\quad\quad\quad\quad\quad\bullet}$$

6. $\frac{2}{5} < \frac{15}{16}$

$$\underset{0\quad\;\frac{1}{2}\quad\;1}{\quad\bullet\quad\quad\bullet}$$

7. 1   **8.** 4   **9.** $\frac{9}{2}$ **10.** 5

Section 4.2

Warm-Up

1.

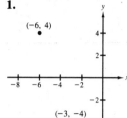

2.

3.

4.

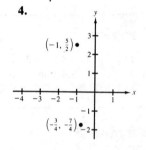

5. $\frac{42}{5}$ **6.** 24   **7.** $y = 14$ **8.** $y = -7$
9. $x = 12$ **10.** $x = 4$

Section 4.3

Warm-Up

1. $y = 4 - 3x$ **2.** $y = x$ **3.** $y = x + 5$
4. $y = x + 4$ **5.** $y = 5 - x$ **6.** $y = x - 5$
7. $y = \frac{2}{3} - \frac{2}{3}x$ **8.** $y = \frac{4}{5}x + \frac{2}{5}$
9. $y = \frac{5}{4} - \frac{3}{4}x$ **10.** $y = 2 - \frac{2}{3}x$

Section 4.4

Warm-Up

1. 14   **2.** 4   **3.** -3 **4.** 4   **5.** 15.5
6. 30   **7.** 200%   **8.** $66\frac{2}{3}\%$ **9.** 455
10. 250

Section 4.5

Warm-Up

1. 4   **2.** -1 **3.** -2 **4.** 1   **5.** 9
6. 10   **7.** $\frac{2}{5}$ **8.** $\frac{10}{3}$ **9.** -9 **10.** $-\frac{11}{5}$

Chapter 5

Section 5.1

Warm-Up

1. $10x - 10$ 2. $12 - 8z$ 3. $-2 + 3x$
4. $-50x + 75$ 5. $10x, -3y, 4$
6. $-2r, 8s, -6$ 7. $\frac{3}{4}$ 8. -8 9. $5x - 2y$
10. $\frac{1}{6}x + 8$

Section 5.2

Warm-Up

1. $(-3)^4$ 2. x^3 3. $\frac{4}{5} \cdot \frac{4}{5} \cdot \frac{4}{5} \cdot \frac{4}{5}$
4. $(4.5)(4.5)(4.5)(4.5)(4.5)$ 5. $x^2 \cdot x^2 \cdot x^2$
6. $y^4 \cdot y^4$ 7. $7x - 8$ 8. $-2y + 14$
9. $-4z + 12$ 10. $-5u - 5$

Section 5.3

Warm-Up

1. $\frac{2}{3}$ 2. $\frac{1}{8}$ 3. $\frac{2}{5}$ 4. $\frac{25}{6}$ 5. $-10x^5$
6. $12y^3 - 4y$ 7. $4z^2 - 1$ 8. $2x^2 + 3x - 20$
9. $x^2 + 14x + 49$ 10. $x^3 + 1$

Section 5.4

Warm-Up

1. 64 2. -32 3. $\frac{4}{9}$ 4. -64 5. 7
6. 144 7. x^5 8. ab 9. u^3v 10. y^2z^5

Chapter 6

Section 6.1

Warm-Up

1. -20 2. -42 3. $24x - 36$
4. $28 - 21x$ 5. $-60 + 42x$ 6. $-2y^2 - 2y$
7. $-3t^2 - 6t$ 8. $-8x^2y^2 - 24xy$ 9. $4 - x^2$
10. $x^2 + 8x + 16$

Section 6.2

Warm-Up

1. $-4x + 24$ 2. $18x + 48$ 3. $y^2 + 2y$
4. $-a^3 + a^2$ 5. $x^2 - 7x + 10$
6. $t^2 + 9t + 18$ 7. $u^2 - 5u - 24$
8. $v^2 - 7v + 6$ 9. $z^2 - 2z - 3$
10. $x^2 + 3x - 28$

Section 6.3

Warm-Up

1. $\frac{1}{3}x + \frac{5}{9} = \frac{1}{9}(3x + 5)$
2. $\frac{5}{8}x - \frac{3}{2} = \frac{1}{8}(5x - 12)$ 3. $6(x + 2)$
4. $4(x - 3)$ 5. $xy(x - y)$
6. $3x(2x^2 - x + 3)$ 7. $(x + 7)(x - 6)$
8. $(x + 7)(x + 6)$ 9. $x(x - 7)(x + 6)$
10. $2y(x + 8)(x - 4)$

Section 6.4

Warm-Up

1. $-12x^2 + 36x$ 2. $-6z^2 + 15z$
3. $x^2 - 100$ 4. $y^2 - a^2$ 5. $x^2 + 10x + 25$
6. $9x^2 - 12x + 4$ 7. $10(x^2 + 7)$
8. $4x(4 - x)$ 9. $(3x - 2)(x - 1)$
10. $(2x + 1)(x - 1)$

Section 6.5

Warm-Up

1. 12   **2.** -3 **3.** $\frac{4}{7}$ **4.** 6   **5.** $x(3x + 7)$
6. $(2x + 5)(2x - 5)$ **7.** $-(x - 15)(x - 7)$
8. $(x + 9)(x - 2)$ **9.** $(2x + 3)(5x - 1)$
10. $(6x - 1)(x - 12)$

Chapter 7

Section 7.1

Warm-Up

1. $2h$ **2.** $3t$ **3.** $2x^2 - 16x + 27$
4. $-3x^2 - 12x - 2$ **5.** $-8x - 10$
6. $-2x^2 + 14x$
7. (a) -4 **8. (a)** 0
   (b) 8   **(b)** 9
   (c) 8   **(c)** 0
   (d) 71   **(d)** 8.75
9. (a) 0   **10. (a)** 2
   (b) Undefined   **(b)** -2
   (c) 8   **(c)** $\frac{5}{2}$
   (d) -6 **(d)** $\frac{101}{10}$

Section 7.2

Warm-Up

1. $\frac{1}{2}$ **2.** $\frac{1}{3}$ **3.** $-\frac{9}{2}$ **4.** $-\frac{13}{3}$ **5.** $\frac{4}{5}$
6. $-\frac{8}{7}$ **7.** $y = \frac{1}{3}(2x - 5)$ **8.** $y = -2x$
9. $y = 3x - 1$ **10.** $y = \frac{2}{3}x + 5$

Section 7.3

Warm-Up

1. $-\frac{1}{4}$ **2.** $\frac{15}{14}$ **3.** Undefined   **4.** $\frac{3}{2}$
5.   **6.**

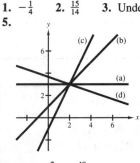

7. $y = -\frac{2}{5}x + \frac{48}{5}$ **8.** $y = \frac{4}{3}x - \frac{29}{3}$
9. $y = -2x + 7$ **10.** $y = x + 3$

Section 7.4

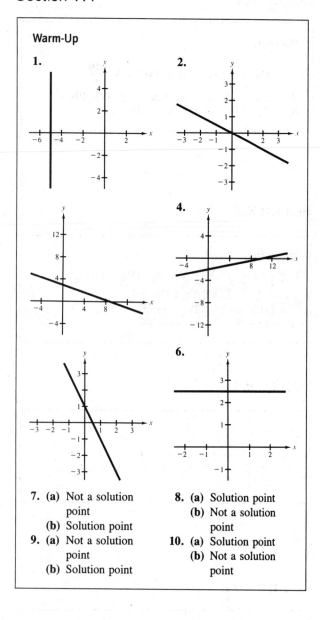

Warm-Up

1.

2.

4.

6.

7. (a) Not a solution
   point
   (b) Solution point
9. (a) Not a solution
   point
   (b) Solution point

8. (a) Solution point
   (b) Not a solution
   point
10. (a) Solution point
    (b) Not a solution
    point

Chapter 8

Section 8.1

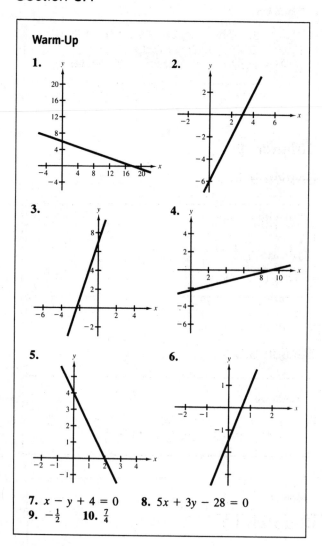

Warm-Up

1.

2.

3.

4.

5.

6.

7. $x - y + 4 = 0$ 8. $5x + 3y - 28 = 0$
9. $-\frac{1}{2}$ 10. $\frac{7}{4}$

Section 8.2

Warm-Up

1. No solution   2. $\frac{1}{3}$ 3. $\frac{5}{11}$ 4. $\frac{14}{11}$
5. $-\frac{1}{6}$ 6. 0   7. $(-2, 0)$ 8. $(0, 3)$
9. $(3, 2)$ 10. $(4, 6)$

Section 8.3

Warm-Up

1. x 2. $-37v$ 3. $2x^2 + 9$ 4. -1
5. 6   6. 6   7. Perpendicular
8. Perpendicular   9. Parallel   10. Neither

Section 8.4

Warm-Up

1. $0.05m + 0.10n$ 2. $0.13l$ 3. $\dfrac{250}{r}$
4. $0.15L$ 5. $3l$ 6. $3.5n$ 7. $(15, 10)$
8. $\left(-2, -\dfrac{8}{3}\right)$ 9. $(28, 4)$ 10. $(4, 3)$

Chapter 9

Section 9.1

Warm-Up

1. -9 2. 48 3. -21 4. -18 5. $\frac{2}{3}$
6. $\frac{3}{7}$ 7. $\frac{6}{7}$ 8. $\frac{7}{11}$ 9. $x(x^2 - 3x + 4)$
10. $x^2(5x + 4)(2x - 1)$

Section 9.2

Warm-Up

1. $\frac{1}{4}$ 2. $\frac{3}{8}$ 3. $\frac{10}{63}$ 4. $-\frac{15}{112}$ 5. $\frac{15}{2}$
6. $-\frac{7}{30}$ 7. $x(2x + 1)(2x - 1)$
8. $(3x + 2)(3x - 2)$ 9. $(5x - 7)(3x + 2)$
10. $(2x - 7)^2$

Section 9.3

Warm-Up

1. 1 2. $\frac{1}{4}$ 3. $\frac{3}{5}$ 4. $-\frac{1}{2}$ 5. $\frac{9}{14}$ 6. $\frac{19}{48}$
7. $-\frac{4}{75}$ 8. $\frac{146}{35}$ 9. $\frac{11}{24}$ 10. $-\frac{7}{18}$

Section 9.4

Warm-Up

1. $\frac{8}{5}$ 2. $-\frac{15}{2}$ 3. 2 4. $\frac{1}{8}$ 5. $-10, 4$
6. $-\frac{3}{2}, 9$ 7. $0, 8$ 8. 4 9. $-8, 7$
10. $0, 5$

Chapter 10

Section 10.1

Warm-Up

1. 225 2. 144 3. 25 4. -25 5. $\frac{4}{9}$
6. $\frac{16}{25}$ 7. $\frac{4}{3}$ 8. $\frac{3}{10}$ 9. $-\frac{1}{81}$ 10. $\frac{1}{625}$

Section 10.2

Warm-Up

1. 100 2. -14 3. $-\frac{13}{5}$ 4. Not possible
5. $32x^5$ 6. $-64x^6y^3$ 7. $\dfrac{16x^2}{y}$ 8. $\dfrac{8y^2}{z^4}$
9. $\dfrac{3x^3}{y^3}$ 10. 1

Section 10.3

Warm-Up

1. $8 + 2x$ **2.** $8x + 5$ **3.** $27 - 5x$
4. $2 - 6x$ **5.** $x^2 + 12x + 36$
6. $4x^2 - 12x + 9$ **7.** $25x^2 - 9$ **8.** $1 - 9x^2$
9. $-2x^2 - 13x$ **10.** $54x - 8x^2$

Section 10.4

Warm-Up

1. 26 **2.** $\frac{18}{7}$ **3.** 3 **4.** 3 **5.** $5, -11$
6. $4, -16$ **7.** $10\sqrt{5}$ **8.** $\sqrt{5}$ **9.** 48
10. $\sqrt{2}$

Chapter 11

Section 11.1

Warm-Up

1. $(3x + 5)(3x - 5)$ **2.** $(2t - 3)^2$
3. $2(x - 5)(x + 1)$ **4.** $(2s + 3)(2s - 3)$
5. $(3x - 5)(x - 2)$ **6.** $4x(x - 4)(x + 1)$
7. $-\frac{4}{5}$ **8.** $\frac{3}{2}$ **9.** $-7, \frac{3}{2}$ **10.** $\frac{4}{5}, 10$

Section 11.2

Warm-Up

1. $x^2 + 4x + 3$ **2.** $x^2 + 10x + 28$
3. $u^2 - 16u + 74$ **4.** $v^2 - 6v + 7$ **5.** $\frac{1}{2}, -\frac{1}{2}$
6. $\frac{3}{4}, -\frac{3}{4}$ **7.** $11, -1$
8. $-2 + \sqrt{10}, -2 - \sqrt{10}$
9. $\frac{1}{2} + \frac{\sqrt{5}}{2}, \frac{1}{2} - \frac{\sqrt{5}}{2}$
10. $-\frac{3}{4} + \frac{\sqrt{7}}{4}, -\frac{3}{4} - \frac{\sqrt{7}}{4}$

Section 11.3

Warm-Up

1. $2, 5$ **2.** $3, -6$ **3.** $\frac{3}{2}, -10$ **4.** $\frac{1}{3}, \frac{5}{2}$
5. $\dfrac{-1 \pm \sqrt{17}}{2}$ **6.** $\dfrac{-3 \pm \sqrt{6}}{3}$ **7.** 2 **8.** 7
9. $2\sqrt{17}$ **10.** $2\sqrt{37}$

Section 11.4

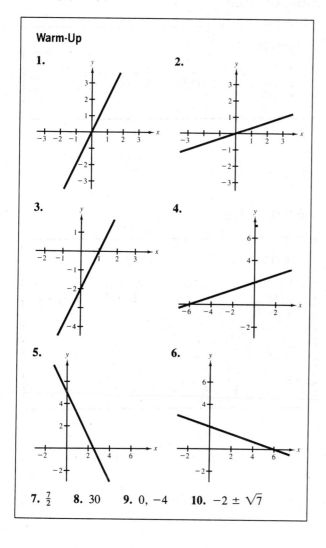

Warm-Up

1.

2.

3.

4.

5.

6.

7. $\frac{7}{2}$ **8.** 30   **9.** 0, −4   **10.** −2 ± $\sqrt{7}$

Section 11.5

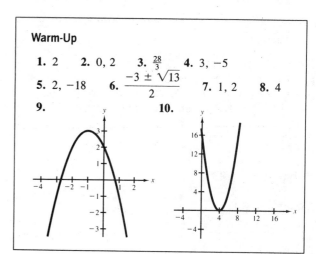

Warm-Up

1. 2   **2.** 0, 2   **3.** $\frac{28}{3}$ **4.** 3, −5

5. 2, −18   **6.** $\dfrac{-3 \pm \sqrt{13}}{2}$ **7.** 1, 2   **8.** 4

9.   **10.**